Martingales and stochastic integrals

Martingales and stochastic integrals

P. E. KOPP

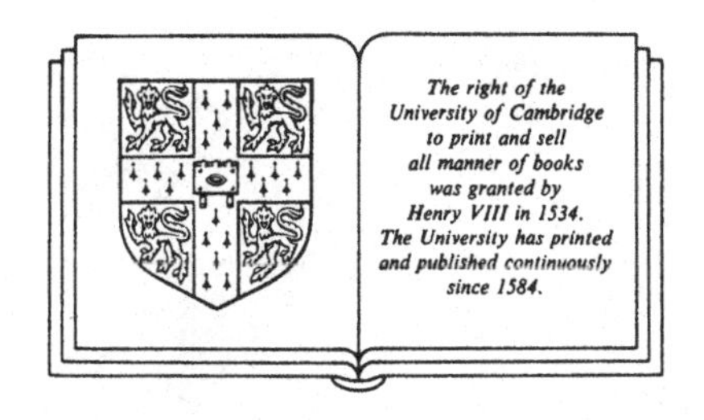

CAMBRIDGE UNIVERSITY PRESS
Cambridge
London New York New Rochelle
Melbourne Sydney

CAMBRIDGE UNIVERSITY PRESS
Cambridge, New York, Melbourne, Madrid, Cape Town, Singapore, São Paulo, Delhi

Cambridge University Press
The Edinburgh Building, Cambridge CB2 8RU, UK

Published in the United States of America by Cambridge University Press, New York

www.cambridge.org
Information on this title: www.cambridge.org/9780521247580

First published 1984
This digitally printed version 2008

A catalogue record for this publication is available from the British Library

Library of Congress Catalogue Card Number: 83-24083

ISBN 978-0-521-24758-0 hardback
ISBN 978-0-521-09033-9 paperback

For Heather, Anna and Emily

cuts. Mathematics in Britain has indeed suffered far more grievous losses than this, but it remains galling for mathematicians to be forced into a defence of their subject in terms of 'market forces', when there are manifestly so many better things for them to do.

My family has been neglected unreasonably through my pre-occupations while preparing this book. Their support and affection has helped me to see it through to the end.

Hull, June 1983

0

Probabilistic background

This chapter summarises some aspects of measure theory and discusses the construction of canonical stochastic processes. We then turn to Brownian motion and Poisson processes to motivate some of the results of Chapters 3 and 4. The development of those chapters is independent of these examples, but since they inspired much of the general theory some knowledge of their properties will greatly aid understanding of that theory.

0.1. Measure and probability

The following concepts should be familiar, but are collected here for ease of reference (further details can be found, for example, in [8], [46]).

0.1.1. ***Definition:*** A *measure space* is a triple $(\Omega, \mathscr{F}, P)$, where Ω is a set, $\mathscr{F}$ a *σ-field* of subsets of Ω (that is, $\Omega \in \mathscr{F}$ and $\mathscr{F}$ is closed under the formation of complements and countable unions) and P is a set function $\mathscr{F} \to [0, \infty]$ which is *countably additive*: if $(A_i)_{i \geq 1}$ is a sequence in $\mathscr{F}$ with $A_i \bigcap A_j = \varnothing$ when $i \neq j$, then $P(\bigcup_{i=1}^{\infty} A_i) = \sum_{i=1}^{\infty} P(A_i)$. We shall deal almost exclusively with *probability spaces*, where P has range in $[0,1]$, and also $P(\Omega) = 1$.. Unless otherwise indicated, we shall also take $(\Omega, \mathscr{F}, P)$ to be *complete*: this means that if $F \in \mathscr{F}$ has $P(F) = 0$ and $G \subseteq F$, then G must necessarily belong to $\mathscr{F}$ (and, of course, $P(G) = 0$). If $(\Omega, \mathscr{F}, P)$ is complete, a sub-σ-field $\mathscr{G}$ of $\mathscr{F}$ (in our framework this implies that $\Omega \in \mathscr{G}$) is said to be *complete* if it contains all $F \in \mathscr{F}$ with $P(F) = 0$. These sets are referred to as *P-null sets* of $\mathscr{G}$.

It is worth recalling that any probability space $(\Omega, \mathscr{F}, P)$ can be 'completed' as follows: the *completion* $(\Omega, \bar{\mathscr{F}}, P)$ of $(\Omega, \mathscr{F}, P)$ is defined by putting $F \in \bar{\mathscr{F}}$ if there exist F_1, F_2 in $\mathscr{F}$ with $F_1 \subseteq F \subseteq F_2$ and $P(F_1) = P(F_2)$, and defining $P(F) = P(F_1) = P(F_2)$. It is clear that $(\Omega, \bar{\mathscr{F}}, P)$ is then a complete probability space.

The completion is closely related to the (Caratheodory) *inner measure* P_* and *outer measure* P^* induced by P on arbitrary subsets of Ω: if $A \subseteq \Omega$, let $P_*(A) := \sup\{P(F): F \subseteq A, F \in \mathcal{F}\}$ and $P^*(A) := \inf\{P(F): F \supseteq A, F \in \mathcal{F}\}$. Then $\bar{\mathcal{F}}$ can be characterised as $\bar{\mathcal{F}} = \{A \subseteq \Omega: P^*(A) = P_*(A)\}$ and the common value of P^* and P_* at $A \in \bar{\mathcal{F}}$ defines the extension of P to A.

0.1.2. Although we shall discuss martingale theory in the context of complete probability spaces, the reader should be aware that this restriction precludes discussion of some of the subtler concepts and extensions of the theory developed in recent years (see [19]), and we thus do not discuss some of the most interesting facets of Brownian motion, in particular, which have given rise to these extensions (see [83] for further discussion of these matters).

Now fix a complete probability space $(\Omega, \mathcal{F}, P)$.

0.1.3. ***Definition:*** A measurable function $f: \Omega \to \mathbf{R}$ or *random variable* satisfies $f^{-1}(B) \in \mathcal{F}$ for all Borel sets $B \subseteq \mathbf{R}$. (The Borel σ-field $\mathcal{B}(\mathbf{R})$ is that generated by the open intervals in $\mathbf{R}$.) Two random variables f and g will normally be identified if the set $\{\omega \in \Omega: f(\omega) \neq g(\omega)\}$ is P-null. We say that $f = g$ a.s. (*almost surely*). By abuse of notation we shall identify the random variable f with the equivalence class $\{g: f = g \text{ a.s.}\}$; this is unlikely to cause any confusion. Thus equations, inequalities, etc., between random variables are assumed to hold a.s. without explicit mention. A sequence (f_n) of random variables *converges a.s.* to f (which is then trivially also a random variable) iff $f_n(\omega) \to f(\omega)$ for *almost all* ω (i.e. except possibly on a P-null set). The vector space of all random variables on $(\Omega, \mathcal{F}, P)$ is denoted by $\mathcal{L}^0 := \mathcal{L}^0(\Omega, \mathcal{F}, P)$.

The space of equivalence classes of functions in $\mathcal{L}^0$, under the equivalence relation '$f \sim g$ iff $f = g$ a.s.' is denoted by $L^0 := L^0(\Omega, \mathcal{F}, P)$. We shall treat $f \in L^0$ as if it were a random variable (which need only be defined a.s. or can take the values $+\infty$ or $-\infty$ on some P-null set). L^0 can be equipped with the metric d of *convergence in probability*: if $f, g \in L^0$, define $d(f,g) = \int_\Omega \min(1, |f(\omega) - g(\omega)|) dP(\omega)$. Then (L^0, d) is a complete metric space and d-convergence of a sequence (f_n) in L^0 to f is equivalent to the statement: for any given $\varepsilon > 0$ we can find N such that $P\{|f_n - f| \geqslant \varepsilon\} < \varepsilon$ for all $n \geqslant N$. (Here $\{|f_n - f| \geqslant \varepsilon\}$ is the set $|\omega \in \Omega: |f_n(\omega) - f(\omega)| \geqslant \varepsilon\}$. Abbreviations of this type will be used freely in the sequel.)

The Banach spaces $L^p := L^p(\Omega, \mathcal{F}, P)$, for $1 \leqslant p \leqslant \infty$, are defined via the norms $\|f\|_p = (\int_\Omega |f|^p dP)^{1/p}$ for $1 \leqslant p < \infty$ and $\|f\|_\infty = \text{ess sup}_{\omega \in \Omega} |f(\omega)|$. Note that $L^p = \{f \in L^0: \|f\|_p < \infty\}$. It is not hard to show that norm-convergence of a sequence $(f_n)\subseteq L^p$ to f (meaning that $\|f_n-f\|_p\to 0$) implies convergence of (f_n) to f in probability.

The integral $\int_\Omega f dP$ of a random variable $f\in L^1$ is called the *expectation* of f and denoted by $\mathbf{E}(f)$. If we allow $\mathbf{E}(f)$ to take the value $+\infty$, generalised expectations can be defined on L^0_+. (For $p=0$ or $1\leqslant p<\infty$, $L^p_+=\{f\in L^p: f\geqslant 0\}$.) Since $P(\Omega)=1$, $\mathbf{E}(f)$ represents the 'average value' or *mean* of f over Ω. For $1\leqslant p<\infty$, $\|f\|_p^p$ represents the pth moment of f. Of particular importance is the second moment $\|f\|_2^2=(\int_\Omega |f|^2\, dP)$.

The *variance* $\sigma^2=\mathbf{E}((f-\mathbf{E}(f))^2)$ measures the dispersion of f about the mean $\mathbf{E}(f)$, distances being taken in the Hilbert space L^2.

Finally, we recall three well-known convergence theorems for sequences in L^1; these will be in constant use throughout this book. For proofs, see [77].

Monotone convergence theorem: If (f_n) is a monotone increasing sequence in L^1 with a.s. limit f and such that $(\mathbf{E}(f_n))$ is bounded above, then f is in L^1 and $\|f_n-f\|_1\to 0$. Hence also $\mathbf{E}(f_n)\uparrow\mathbf{E}(f)$.

This result extends to L^0_+ if we allow $\mathbf{E}(f)$ to take the value $+\infty$. In that case the boundedness condition is superfluous.

Fatou's lemma: If (f_n) is in L^0_+ then $\mathbf{E}(\liminf_{n\to\infty} f_n)\leqslant\liminf_{n\to\infty}\mathbf{E}(f_n)$.

Dominated convergence theorem: If $f_n\to f$ a.s. and there exists g in L^1 such that $|f_n|\leqslant g$ for all n, then $f\in L^1$ and $\mathbf{E}(f_n)\to\mathbf{E}(f)$.

Exercises:

(1) Let (f_n), $n\geqslant 1$, and f be functions in L^0.
  (i) Show that if $f_n\to f$ in L^p-norm, then $f_n\to f$ in probability.
  (ii) Show that if $f_n\to f$ in probability, then there is a subsequence (f_{n_k}) converging to f a.s.

(2) The following basic facts from elementary probability theory will be useful on occasion. Prove them.
  (i) Chebychev's inequality: Let $f\in L^2$ and $t\in\mathbf{R}$ be given. Then

$$P(|f|>t)\leqslant\frac{\mathbf{E}(f^2)}{t^2}.$$

  (ii) Borel–Cantelli lemmas: if $(A_n)\subset\mathscr{F}$ and $\sum_n P(A_n)<\infty$, then $P(\bigcap_k\bigcup_{n\geqslant k}A_n)=0$. If (A_n) are independent events (see 0.1.4) and $\sum_n P(A_n)=+\infty$, then $P(\bigcap_k\bigcup_{n\geqslant k}A_n)=1$.

0.1.4. ***Conditioning:*** In attempting to model 'reality' by means of the probability space $(\Omega,\mathscr{F},P)$ we can think of the sets in $\mathscr{F}$ as possible 'events', and $P(A)$ is then our assignment of the probability that A occurs.

Our further assignment of probabilities may be influenced by the knowledge that A has occurred (think of the effect of election results upon the stock market!). We define the *conditional probability* of $B\in\mathscr{F}$, given that A has occurred, and $P(A)>0$, as

$$P(B|A)=\frac{P(A\cap B)}{P(A)}.$$

For example, given that a family with exactly two children has at least one boy, what are the chances both children are boys? Here event $A=\{$the family has at least one boy$\}$ has probability $\frac{3}{4}$, assuming that the possible combinations of sexes are all equally likely. On the other hand, if $B=\{$both are boys$\}$, then $P(B\cap A)=P(B)=\frac{1}{4}$, so $P(B|A)=\frac{1}{4}/\frac{3}{4}=\frac{1}{3}$. (If this result seems surprising, consider the respective lengths of file indexes of families with at least one boy, and that of families with two boys. See [31] for a further discussion of such examples.)

Taking the conditional probability with respect to A amounts to choosing A as the new sample space (instead of Ω) and normalising to make the probability of A equal to 1. This indicates that all general theorems for probabilities will have counterparts for conditional probabilities. The distinctive nature of probability theory lies in the study of *independent events*, that is, events A and B for which $P(A|B)=P(B)$, or in other words, where $P(A)\cdot P(B)=P(A\cap B)$. Here the restriction of our 'universe' to A does not alter the likelihood that B occurs. (See [31; Ch. V] for detailed discussions.)

Now if $f\in L^1$ we can define the *conditional expectation* of f, given A in $\mathscr{F}$, as the 'average value'

$$\mathbf{E}(f|A)=\frac{1}{P(A)}\int_A f\,\mathrm{d}P$$

of f on A, by analogy with the definitions of $\mathbf{E}(f)$ and $P(B|A)$. Note that

$$\mathbf{E}(1_B|A)=\frac{1}{P(A)}\int_A 1_B\,\mathrm{d}P=\frac{P(A\cap B)}{P(A)}=P(B|A).$$

We can interpret $\mathbf{E}(f|A)$ as our 'best estimate' of the values of f, given only the 'information' contained in A (and hence in its complement, A^c). In a finite sample space Ω, this information amounts to knowing whether a given $\omega\in\Omega$ belongs to A or not. Now the event A generates the σ-field $\{\varnothing,A,A^c,\Omega\}$. More generally, we can regard any sub-σ-field $\mathscr{G}$ of $\mathscr{F}$ as containing some information – whether relevant to f or not. This also enables us to measure the 'amount of information' given: the larger the σ-

field the more information it contains (think of the σ-fields generated by ever finer partitions of Ω). The conditional expectation $\mathbf{E}(f|\mathscr{G})$ will then represent our 'best guess' at the values of f, given only the information in $\mathscr{G}$. The usual construction of $\mathbf{E}(f|\mathscr{G})$ for $f \in L^1$ (or even $f \in L^0_+$) as the unique $\mathscr{G}$-measurable integrable function (write $L^1(\mathscr{G})$ for $L^1 \bigcap L^0(\mathscr{G})$) such that $\int_G f \mathrm{d}P = \int_G \mathbf{E}(f|\mathscr{G}) \mathrm{d}P$ for all $G \in \mathscr{G}$, is via the *Radon–Nikodym theorem.* This states that if μ is a bounded measure on $(\Omega, \mathscr{F})$ which is absolutely continuous with respect to P (i.e. $\mu(A)=0$ whenever $A \in \mathscr{F}$ satisfies $P(A)=0$), then there exists a unique $X \in L^1_+$ with $\int_F X \,\mathrm{d}P = \mu(F)$ for all $F \in \mathscr{F}$. It is easy to extend this result to bounded signed measures (countably additive real-valued set functions), where $X \in L^1$ need no longer be positive. Apply this with P restricted to the sub-σ-field $\mathscr{G}$ of $\mathscr{F}$ and μ on $\mathscr{G}$ defined by $\mu(G) = \int_G f \,\mathrm{d}P$, to obtain $X = \mathbf{E}(f|\mathscr{G}) \in L^1(\Omega, \mathscr{G}, P)$ such that $\mu(G) = \int_G X \,\mathrm{d}P$ for all $G \in \mathscr{G}$.

We shall deduce the Radon–Nikodym theorem as a consequence of the martingale convergence theorem in Chapter 2. For this reason we include in Chapter 2 a definition of $\mathbf{E}(f|\mathscr{G})$ which does not require the Radon–Nikodym theorem, but is based instead upon the characterisation of the operator $\mathbf{E}(\cdot|\mathscr{G})$ in L^2 as the orthogonal projection onto the subspace $L^2(\mathscr{G})$. This will exhibit $\mathbf{E}(f|\mathscr{G})$ as the $\mathscr{G}$-measurable function 'nearest' to f in the least-squares sense. Thus $\mathbf{E}(f|\mathscr{G})$ represents our 'best estimate' of f given only the information contained in $\mathscr{G}$.

0.1.5. ***The Monotone Class Theorem:*** Suppose that we wish to prove that all sets or functions in some class $\mathscr{C}$ have a property $(*)$. One way of doing this is to find a collection $\mathscr{C}_0$ of sets or functions which 'generates' $\mathscr{C}$, so that each element of $\mathscr{C}$ can be constructed from $\mathscr{C}_0$ using certain operations. If each element of $\mathscr{C}_0$ has $(*)$ and the class of all sets or functions which have $(*)$ is closed under these operations, then each element of $\mathscr{C}$ has $(*)$. We shall repeatedly use this procedure for σ-fields of sets and vector spaces of measurable functions using the following two versions of the *Monotone Class Theorem* (there are many versions with this name: see [19; Ch. I]):

Let Ω be a set, $\mathscr{S}$ a collection of subsets of Ω, closed under finite intersections.

(*a*) Let $\mathscr{M}(\mathscr{S})$ be the smallest collection of subsets of Ω which contains $\mathscr{S}$ and satisfies

(i) $\Omega \in \mathscr{M}(\mathscr{S})$,

(ii) if $A, B \in \mathscr{M}(\mathscr{S})$ and $A \subseteq B$, then $B \backslash A \in \mathscr{M}(\mathscr{S})$,

(iii) if (A_n) is an increasing sequence in $\mathscr{M}(\mathscr{S})$, then $\bigcup_n A_n \in \mathscr{M}(\mathscr{S})$.

Under these conditions $\mathscr{M}(\mathscr{S})$ is the smallest σ-field containing $\mathscr{S}$.

(*b*) Let $\mathscr{H}$ be a vector space of functions from Ω to $\mathbf{R}$ satisfying

(i) $1 \in \mathscr{H}$ and $1_A \in \mathscr{H}$ for $A \in \mathscr{S}$,

(ii) if (f_n) is an increasing sequence of non-negative functions in $\mathscr{H}$ with bounded supremum, then $\sup_n f_n \in \mathscr{H}$.

Then $\mathscr{H}$ contains all bounded $\sigma(\mathscr{S})$-measurable real functions on Ω.

Proof: (*a*) If $\sigma(\mathscr{S})$ is the σ-field generated by $\mathscr{S}$, it satisfies (i)–(iii) trivially and contains $\mathscr{S}$, hence $\sigma(\mathscr{S}) \supseteq \mathscr{M}(\mathscr{S})$. To prove the converse inclusion, it will be enough to show that $\mathscr{M}(\mathscr{S})$ is closed under *finite* intersections. For then we can express any countable union of sets (M_i) in $\mathscr{M}(\mathscr{S})$ as follows: set $N_k = \bigcup_{i=1}^k M_i$, which is in $\mathscr{M}(\mathscr{S})$ since $N_k = \Omega \backslash (\Omega \backslash \bigcup_{i=1}^k M_i) = \Omega \backslash \bigcap_{i=1}^k (\Omega \backslash M_i)$. So by (iii), $\bigcup_{i=1}^\infty M_i = \bigcup_{k=1}^\infty N_k \in \mathscr{M}(\mathscr{S})$. Thus $\mathscr{M}(\mathscr{S})$ is a σ-field.

To prove that $\mathscr{M}(\mathscr{S})$ is closed under finite intersections, first set $\mathscr{D}_1 = \{B \in \mathscr{M}(\mathscr{S})\colon B \bigcap A \in \mathscr{M}(\mathscr{S})$ for all $A \in \mathscr{S}\}$. Since $\mathscr{S}$ is closed under finite intersections by hypothesis, $\mathscr{D}_1 \supset \mathscr{S}$. We can now check that $\mathscr{D}_1$ satisfies (i)–(iii) to conclude that $\mathscr{D}_1 = \mathscr{M}(\mathscr{S})$. (Exercise!) Finally, let $\mathscr{D}_2 = \{B \in \mathscr{M}(\mathscr{S})\colon B \bigcap A \in \mathscr{M}(\mathscr{S})$ for all $A \in \mathscr{M}(\mathscr{S})\}$. Again one may check easily that D_2 satisfies (i)–(iii). Moreover, if $A \in \mathscr{S}$, $B \bigcap A \in \mathscr{M}(\mathscr{S})$ for all $B \in \mathscr{D}_1 = \mathscr{M}(\mathscr{S})$, so $\mathscr{S} \subseteq \mathscr{D}_2$. Hence $\mathscr{D}_2 = \mathscr{M}(\mathscr{S})$, and this means that $\mathscr{M}(\mathscr{S})$ is closed under finite intersections.

(*b*) Let $\mathscr{M} = \{A\colon 1_A \in \mathscr{H}\}$. Then $\mathscr{S} \subseteq \mathscr{M}$, $\Omega \in \mathscr{M}$ and $\mathscr{M}$ is closed under relative complements (if $A, B \in \mathscr{M}$, $A \subseteq B$, then $1_{B \backslash A} = 1_B - 1_A \in \mathscr{H}$). Also, if (A_i) is an increasing sequence in $\mathscr{M}$, and $A = \bigcup_{i=1}^\infty A_i$, then $1_A = \sup_{i \geq 1} 1_{A_i} \in \mathscr{H}$. By part (*a*), $\mathscr{M} = \sigma(\mathscr{S})$. Now if $f\colon \Omega \to \mathbf{R}$ is $\sigma(\mathscr{S})$-measurable and bounded, let $f = f^+ - f^-$. Each of f^+ and f^- is the supremum of a sequence of $\mathscr{M}$-simple functions, hence belongs to $\mathscr{H}$ by (iii). So $f \in \mathscr{H}$ as required.

0.1.6. ***Stochastic processes and their distributions:*** A random variable X induces a probability measure P_X on $(\mathbf{R}, \mathscr{B}(\mathbf{R}))$, the *distribution* of X, by $P_X(B) = P(X^{-1}(B))$ for $B \in \mathscr{B}(\mathbf{R})$. This Lebesgue–Stieltjes measure is generated by the increasing right-continuous function F_X, the *distribution function* of X, by $F_X(t) = P\{X \leqslant t\}$. If $X \in L^1(\Omega, \mathscr{F}, P)$, $\mathbf{E}(X) = \int_{\mathbf{R}} x \mathrm{d}P_X(x)$.

Given a finite sequence $X_1, X_2, \ldots, X_n$ of random variables, let $Z(\omega) = (X_1(\omega), X_2(\omega), \ldots, X_n(\omega))$ for all $\omega \in \Omega$. This defines a measurable function $Z\colon \Omega \to \mathbf{R}^n$, where $\mathbf{R}^n$ is given the Borel σ-field $\mathscr{B}(\mathbf{R}^n)$. Hence Z induces a probability measure $P_Z = P_{X_1, \ldots, X_n}$ on $(\mathbf{R}^n, \mathscr{B}(\mathbf{R}^n))$, the *n-dimensional joint distribution* of $X_1, \ldots, X_n$, by $P_Z(B) = P(Z^{-1}(B))$.

We can think of a stochastic process $X = (X_t)_{t \in \mathbf{T}}$ as a family of random variables indexed by some $\mathbf{T} \subseteq \mathbf{R}$. (But see also section 3.1.) Usually we take $\mathbf{T} = \mathbf{N}$ or as an interval in $\mathbf{R}^+$. If $\mathbf{T}$ models the passage of time and X models the time-evolution of some observed system, an immediate practical difficulty is that we can only make finitely many observations. Thus we only observe $X_{t_1}, X_{t_2}, \ldots, X_{t_n}$ for some $t_1, \ldots, t_n$ in $\mathbf{T}$. The question arises to what extent these observations determine X, i.e. how many different models can be built upon the same sets of observations? Writing $T = (t_1, \ldots, t_n)$ we can define the measurable function $X_T = (X_{t_1}, \ldots, X_{t_n})$ as above and determine the joint distribution P_{X_T}. Doing this for all possible choices of n and T then yields the set of all *finite-dimensional distributions* of X. We can now rephrase our question: can a process X be constructed uniquely to have a given set of finite-dimensional distributions?

Kolmogorov's extension theorem provides an explicit canonical construction of X on the product space $\mathbf{R}^{\mathbf{T}}$ when we have a *projective system* of probability measures: for each pair of finite subsets $S \subseteq T$ of $\mathbf{T}$, $P_S = P_T \circ \Pi_{TS}^{-1}$, where Π_{TS}: $\mathbf{R}^T \to \mathbf{R}^S$ is the natural projection map. This allows us to construct a unique probability measure μ on $\Lambda = \mathbf{R}^{\mathbf{T}}$ as the *projective limit* of the system $\{P_S : S \subset \mathbf{T}, \text{ finite}\}$, so that for each finite $S \subseteq \mathbf{T}$, $P_S = \mu \circ \Pi_S^{-1}$, where $\Pi_S : \Lambda \to \mathbf{R}^S$ is the natural projection map.

The construction of such a projective limit measure μ proceeds from the *Caratheodory extension theorem* for measures: if $\mathscr{E}$ is a *field* of subsets of Ω (replacing countable unions by finite unions in the definition of a σ-field yields the definition of a field) and μ is a probability measure on $\mathscr{E}$ (so if $\bigcup_{i=1}^{\infty} E_i \in \mathscr{E}$ for disjoint E_i, then $\mu(\bigcup_{i=1}^{\infty} E_i) = \sum_{i=1}^{\infty} \mu E_i$), then μ extends uniquely to the σ-field $\sigma(\mathscr{E})$ generated by $\mathscr{E}$. (See [37] for a proof.)

To use this result, we define the field $\mathscr{C}$ of *cylinder sets* of $\Lambda = \mathbf{R}^{\mathbf{T}}$, given by the finite-dimensional projection maps: given a finite set $S \subseteq \mathbf{T}$, let $\mathscr{C}_S = \Pi_S^{-1}(\mathscr{B}(\mathbf{R}^S))$, i.e. $C \in \mathscr{C}_S$ iff $C = \{\omega \in \Lambda : \Pi_S(\omega) \in B\}$ for some Borel set $B \subseteq \mathbf{R}^S$. Then each $\mathscr{C}_S$ is a σ-field (Exercise!). We set $\mathscr{C} = \{C \in \mathscr{C}_S : S \subseteq \mathbf{T}, \text{ finite}\}$.

To prove that $\mathscr{C}$ is a field one obviously requires consistency conditions. Thus, given a family $\{P_S : S \subseteq \mathbf{T}, \text{ finite}\}$ of finite-dimensional probability distributions, we require that

(i) if $S_1 = \sigma(S)$ is a permutation of the elements of S, then $P_{S_1}(B) = P_S(f_\sigma^{-1}(B))$ for any Borel set $B \subseteq \mathbf{R}^S$, where $f_\sigma(x_1, \ldots, x_n) = (x_{\sigma(1)}, \ldots, x_{\sigma(n)})$.

(ii) if $S = \{s_1, \ldots, s_n\}$ and $T = \{s_1, \ldots, s_n, t_{n+1}\}$, then $P_S(B) = P_T(B \times \mathbf{R})$ for all Borel sets $B \subseteq \mathbf{R}^S$.

(This is of course just an explicit statement of the requirement that the probability distributions form a projective system.)

The measure μ on $\mathscr{C}$ is now defined by setting $\mu = P_S \circ \Pi_S$ for each finite $S \subseteq \mathbf{T}$. The consistency conditions ensure that μ is well-defined, since any two representations of a cylinder set can be related by projections and permutation of indices. To show that μ is countably additive on $\mathscr{C}$ we need only prove that $\mu(C_n)\to 0$ when $(C_n)\subseteq\mathscr{C}$ is a decreasing sequence with empty intersection. But this follows because for each Borel set $B\subseteq\mathbf{R}^S$ we can find a compact set $K\subset B$ such that $P_S(B\setminus K)$ is arbitrarily small (this expresses the fact that each P_S is *tight* – see [4], [83, p. 25ff]). For if $\mu(C_n)\to\alpha>0$ we can assume that each $B_n=\Pi_{S_n}(C_n)$ is compact and that the index sets S_n defining C_n increase with n. Taking $\omega_n\in C_n$ we can find convergent subsequences $\{\Pi_\alpha(\omega_n)\}$ for each $\alpha\in\mathbf{T}$, and a diagonal argument provides a point $\omega\in\bigcap_{n=1}^{\infty}C_n$. (The details may be found in [52], a more sophisticated proof in [67].)

So μ is a probability measure on $\mathscr{C}$, hence extends to a probability measure on $\sigma(\mathscr{C})$ by Caratheodory's theorem. Thus we have constructed the probability space $(\Lambda,\sigma(\mathscr{C}),\mu)$. Finally we define the stochastic process X on $(\Lambda,\sigma(\mathscr{C}),\mu)$ by setting $X_t(\omega)=\omega(t)$ for $t\in\mathbf{T}$, $\omega\in\Lambda$, where $\omega(t)=\Pi_{\{t\}}(\omega)$. It is then clear that X_t is $\sigma(\mathscr{C})$-measurable and that for S finite, $P_{X_S}=\mu\circ\Pi_S^{-1}=P_S$. We have 'proved' the following result!

0.1.7. ***Theorem (Daniell–Kolmogorov):*** Given a projective system of finite-dimensional probability distributions $\Phi=\{P_S: S\subseteq\mathbf{T}, \text{finite}\}$ there is a stochastic process X having Φ as its system of finite-dimensional distributions. Moreover, the process X can be defined uniquely on the probability space $(\mathbf{R}^{\mathbf{T}},\sigma(\mathscr{C}),\mu)$, by setting $X_t(\omega)=\omega(t)$ for $\omega\in\mathbf{R}^{\mathbf{T}}$, $t\in\mathbf{T}$. Thus if $Y=(Y_t)_{t\in\mathbf{T}}$ is any stochastic process on a probability space $(\Omega',\mathscr{F},P)$ with Φ as its system of finite-dimensional distributions, then Y has a *canonical representation* X on $(\mathbf{R}^{\mathbf{T}},\sigma(\mathscr{C}),\mu)$.

It is clear that Theorem 0.1.7 is fundamental in the construction of stochastic processes. It is now natural to say that two stochastic processes are *equivalent* if they have the same system of finite-dimensional distributions, since this will ensure that they have the same canonical representation on the function space $\mathbf{R}^{\mathbf{T}}$. Of particular interest is the case when the canonical process 'lives' on a particular subset of $\mathbf{R}^{\mathbf{T}}$, i.e. its *paths* $t\to\omega(t)$ μ-almost surely possess a certain property, such as continuity. The verification of such properties requires much more sophisticated techniques and relies heavily on the form of the given system of finite-dimensional distributions, as we shall see below.

The discussion of the *paths* $t\to X_t(\omega)$ of a stochastic process X will in general require rather stronger notions of equivalence of process than the above. We define two such notions in Exercise 0.1.8. They will be discussed further in Chapter 3.

To what extent these finer distinctions accord with 'reality' naturally remains debatable.

0.1.8. ***Exercises:***

(1) Let X and Y be stochastic processes on $(\Omega, \mathscr{F}, P)$, with parameter set $\mathbf{T}=[0, \infty[$. Suppose that $X_t = Y_t$ a.s. (P) for all $t \in \mathbf{T}$. (We say that Y is a *modification* of X; see Chapter 3.) Show that X and Y are equivalent.

(2) Show that if X and Y have a.s. *continuous* paths so that $t \to X(t,\omega)$, $t \to Y(t,\omega)$ are continuous except on some P-null set) and X is a modification of Y, then there is a single P-null set N such that the paths $t \to X(t,\omega)$ and $t \to Y(t,\omega)$ are identical for all $\omega \notin N$. (First consider $t \in \mathbf{Q}^+$.)

This result is extended to right-continuous processes in section 3.1. Quite generally, we say that two processes X and Y are *indistinguishable* if, for almost all ω, the paths $t \to X(t, \omega)$ and $t \to Y(t, \omega)$ are identical.

0.1.9. ***Definition:*** Let $X=(X_t)_{t\in\mathbf{T}}$ be a stochastic process, with finite-dimensional distributions $\Phi=\{P_{X_S}: S \subseteq \mathbf{T}, \text{ finite}\}$. The measure $\mu_X := \mu$ defined by Φ on $(\mathbf{R}^{\mathbf{T}}, \sigma(\mathscr{C}))$ is called the *distribution* of X. If we view X as a *random function*, that is as a map $X:\Omega \to \mathbf{R}^{\mathbf{T}}$ given by $\omega \to X(\cdot,\omega)$, then for $E \in \mathscr{C}$ we have $X^{-1}(E)=X^{-1}(\Pi_S^{-1}(B)=(\Pi_S X)^{-1}(B)$ for some $B \in \mathbf{R}^S$ ($S \subseteq \mathbf{T}$, finite), so that $\mu_X(E)=P((\Pi_S X)^{-1}(B))=P(X^{-1}(E))$. This identity extends to $\sigma(\mathscr{C})$: since the probability measures μ_X and $P \circ X^{-1}$ agree on $\mathscr{C}$, they also agree on $\sigma(\mathscr{C})$.

Although the canonical representation of X on $(\mathbf{R}^{\mathbf{T}}, \sigma(\mathscr{C}), \mu_X)$ has the advantage that $\sigma(\mathscr{C})$ is defined without reference to the probability space $(\Omega, \mathscr{F}, P)$ on which X was originally defined, it also has serious limitations. First note that each set in $\sigma(\mathscr{C})$ is a *σ-cylinder*, i.e. has the form $\{\omega \in \mathbf{R}^{\mathbf{T}}: (\omega(t_1), \omega(t_2), \dots) \in B\}$ where $T=(t_1, t_2, \dots)$ is a sequence in $\mathbf{T}$ and $B \in \mathscr{B}(\mathbf{R})^T$: It is easy to see that $\mathscr{B}(\mathbf{R})^T = \mathscr{B}(\mathbf{R}^T)$ as T is countable, so that $\mathscr{B}(\mathbf{R})^T$ is the σ-field generated by all finite-dimensional rectangles in $\mathbf{R}^T$. It follows (see, e.g., [8; Prop. 12.8]) that the σ-cylinders form a σ-field, which must therefore be $\sigma(\mathscr{C})$.

But this means that the only sets we can guarantee to be measurable in our standard space $(\mathbf{R}^{\mathbf{T}}, \sigma(\mathscr{C}))$ are those which depend on countably many values of the process X, i.e.

$$\{\omega : (\omega(t_1), \dots, \omega(t_n), \dots) \in B\} = \{\omega : (X_{t_1}(\omega), \dots, X_{t_n}(\omega), \dots) \in B\}$$

for some Borel set B in $\mathbf{R}^{\mathbf{T}}$. We can complete $\sigma(\mathscr{C})$ relative to μ_X, but this will still exclude sets such as

$$A = \{\omega : X_t(\omega) = 0 \text{ for some } t \in \mathbf{T}\} = \bigcup_{t \in \mathbf{T}} \{\omega : \omega(t) = 0\}$$

or

$$W=\{\omega : t \to X_t(\omega) \text{ is continuous on } \mathcal{T}\}.$$

Thus if we want to study *path properties* of X, the canonical representation is not very useful. We need to identify subsets of $\mathbf{R}^{\mathrm{T}}$ which 'carry' the measure μ_X when X satisfies path regularity properties. In the case we shall study in the next section, such properties derive in turn from the special nature of the finite-dimensional distributions of the process.

0.2. Brownian motion and the Itô integral

There are many excellent treatises on this most important stochastic process, and we shall not attempt to duplicate them. Conceived as a mathematical model for the highly irregular motions of particles in colloidal suspensions, which experience so many collisions that their motion appears to be quite random, Brownian motion has become the paradigm of a 'random process', and finds applications in fields as disparate as filtering theory in electronics and the fluctuations of the stock market. We shall not attempt to discuss how 'realistic' these models are, but discuss, with only sketched proofs, the construction and basic properties of this process. For fuller treatments we refer to [33], [40], [53], [57], [83].

0.2.1: A discrete model for random behaviour (e.g. coin-tossing) is given by symmetric simple *random walk* on the line: let $\Omega=[0,1]$, $\mathcal{F}=\mathcal{B}([0,1])$ and let P be Lebesgue measure. Each $\omega\in\Omega$ is described by its binary expansion

$$\omega=\sum_{i=1}^{\infty}\frac{\varepsilon_i}{2^i},$$

where $\varepsilon_i=0$ or 1. Define a sequence (R_i) on Ω by

$$R_i(\omega)=\begin{cases} 1 \text{ if } \varepsilon_i=1\\ -1 \text{ if } \varepsilon_i=0\end{cases}$$

and let $S_0=0$, $S_n(\omega)=\sum_{i=1}^{n}R_i(\omega)$. If ω has two binary expansions, set $R_i(\omega)=0$ for all i. This only involves a P-null set of ω's. The functions (R_i) give a model for coin-tossing. Alternatively we can regard (S_n) as giving the position of a particle, starting at the origin, which at each time $t=1,2,3,\ldots$ moves instantaneously to the right or left with probability $\frac{1}{2}$. The (R_i) are obviously independent and identically distributed, and have mean 0 and variance 1. The simplest form of the *Central Limit Theorem* (see e.g. [8; Th. 1.17]) then states that

$$\lim_n P\left(\frac{S_n}{\sqrt{n}}<x\right)=\frac{1}{\sqrt{2\pi}}\int_{-\infty}^{x} e^{-y^2/2}\,dy.$$

To make this a model for the displacements of particles in suspensions, the discrete process (S_n) should be extended to $\mathbf{T}=[0,\infty[$ (the 'time axis'.) This is most easily done by linear interpolation:

$$S_t(\omega):=(t-k)S_{k+1}(\omega)+(k+1-t)S_k(\omega) \text{ if } t\in[k,k+1[, k\geqslant 0.$$

Now, introducing a change in our timescale, we consider the process $Z_t=(\Delta x)S_{nt}$. This represents a particle subject to a displacement $\pm\Delta x$ at times k/n, $k=1,2,\ldots$. For $t=k/n$, $\mathbf{E}(Z_t^2)=(\Delta x)^2(nt)$, so to obtain a finite variance for Z we need Δx to be of order $1/\sqrt{n}$. Hence we define the *normalised* random walk $X^{(n)}$ by $X_t^{(n)}=(1/\sqrt{n})S_{nt}$. The Central Limit Theorem then yields

$$\lim_n P(X_t^{(n)}<x)=\frac{1}{\sqrt{2\pi t}}\int_{-\infty}^{x} e^{-y^2/2t}\,dy.$$

In other words, the distributions $P_{X^{(n)}}$ converge on each cylinder set which arises from open intervals in $\mathbf{R}$ to a normal distribution, $\mathcal{N}(0,t)$. Note that in this construction each $X^{(n)}$ has continuous paths, so that the distribution of $X^{(n)}$ is carried by $\mathscr{W}$, the space of continuous maps $\mathbf{T}\to\mathbf{R}$.

The general (functional) form of the Central Limit Theorem concerns the convergence of the normalised random walks $X^{(n)}$:

Donsker's theorem: The sequence $X^{(n)}$ converges in the following sense: there exists a measure W on $(\mathscr{W},\mathscr{A})$, where $\mathscr{A}$ denotes the σ-field of σ-cylinders in $\mathscr{W}$, such that for each bounded continuous function h on $\mathscr{W}$, $X^{(n)}(h)\to W(h)$ as $n\to\infty$.

The Central Limit Theorem ensures that the finite-dimensional distributions of W will be given by densities of the form

$$\prod_{k=1}^{n}(2\pi(t_k-t_{k-1}))^{-1/2}\exp\left(\frac{-(x_k-x_{k-1})^2}{2(t_k-t_{k-1})}\right),\quad t_0=x_0=0.$$

Hence we can define *Wiener measure* W as the weak*-limit (see section 1.4) of normalised random walks, the convergence taking place in the space of probability measures on $\mathscr{W}$. Many modern texts use the term 'narrow convergence' (see [83; I37] where a proof of Donsker's theorem may also be found.)

0.2.2. ***Remark:*** A rather intuitive approach to the existence of Wiener measure is possible via non-standard analysis (see [1], [14]): instead of approximating when $\Delta t\to 0$, one can set $t=1/H$, where H is a hyperfinite integer, and describe Brownian motion as a random walk with 'infinitesimal steps'. This again leads to a proof of Donsker's theorem. Other constructions of Wiener measure can be found in [16], [40], [57].

0.2.3: The abstract characterisation of Brownian motion on a given probability space $(\Omega, \mathcal{F}, P)$ can be given in terms of a *Gaussian family* of random variables, i.e. a family $\mathcal{L}$ such that each element in the linear span of $\mathcal{L}$ has a normal distribution. For such families the mean $\mathbf{E}(X_t)$ and covariance $\mathbf{E}(X_s X_t)$ for all pairs X_s, X_t in $\mathcal{L}$ determine the finite-dimensional distributions (see [84]).

0.2.4. ***Definition:*** A (standard) *Brownian motion* ($BM_0(\mathbf{R})$) is a Gaussian family $B=(B_t)_{t\in\mathbf{T}}$ of random variables satisfying:

(i) $B_0=0$;

(ii) $\mathbf{E}(B_t)=0$, $\mathbf{E}(B_s B_t)=\min(s,t)$ for s,t in $\mathbf{T}$;

(iii) $P\{t\to B_t \text{ is continuous on } \mathbf{T}\}=1$.

The canonical process $X_t(\omega)=\omega(t)$ defined by Wiener measure on $(\mathcal{W}, \mathcal{A}, W)$ is a representation of $BM_0(\mathbf{R})$, since it clearly satisfies (i), (ii) and (iii). The existence of a process satisfying (i) and (ii) follows already from the Daniell–Kolmogorov theorem. However, that result does not guarantee that the process can be taken to have continuous paths. The construction of a process satisfying (i)–(iii), as we have done for the case of the canonical space $(\mathcal{W}, \mathcal{A}, W)$, is thus a non-trivial problem. We shall indicate the main steps in a more direct construction below.

0.2.5. ***Exercise:*** Show that $BM_0(\mathbf{R})$ has the following symmetries: If c, u in $\mathbf{R}$ are given and B is standard Brownian motion, the following processes are also $BM_0(\mathbf{R})$:

(*a*) $\{-B_t : t\geqslant 0\}$ (reflection),

(*b*) $\{(1/c)B(c^2t) : t\geqslant 0\}$ (scaling),

(*c*) $\{B_{t+u}-B_u : t\geqslant 0\}$ (stationary increments),

(*d*) $\{t\cdot B_{1/t} : t\geqslant 0$, and 0 if $t=0\}$ (inversion).

In all cases it is easiest to check that these are Gaussian families with the required covariance property.

For $s<t\leqslant u<v$ the $\mathbf{R}^2$-valued random vector (B_t-B_s, B_v-B_u) has a bivariate normal distribution with mean $(0,0)$ and covariance 0. It follows that the coordinates B_t-B_s and B_v-B_u are independent (even orthogonal). So (B_t) has independent increments and since each B_t has zero mean (expectation) the variance of (B_t-B_s) is easily shown to be $(t-s)$. So $BM_0(\mathbf{R})$ is characterised as follows:

0.2.6. ***Proposition:*** If B is a continuous process with $B_0=0$ a.s. and, for $0<s<t$, (B_t-B_s) is normally distributed with mean 0 and variance $(t-s)$, and is independent of $\{B_u: u\leqslant s\}$, then B is $BM_0(\mathbf{R})$.

This characterisation allows us to give an outline of the simplest proof of the *existence* of $BM_0(\mathbf{R})$ – it is somewhat less intuitive than the approximation argument leading to Donsker's theorem. (See [15] for detailed proofs.)

We restrict our attention to $[0,1]$. Associated with dyadic partitions of $[0,1]$ one can construct a complete orthonormal basis of $L^2[0,1]$, the *Haar functions*, by setting:

$$f_0\equiv 1,$$

$$f_{k/2^n}=\begin{cases} 2^{(n-1)/2} \text{ if } \dfrac{k-1}{2^n}\leqslant t<\dfrac{k}{2^n},\\ -2^{(n-1)/2} \text{ if } \dfrac{k}{2^n}\leqslant t<\dfrac{k+1}{2^n},\\ 0 \text{ otherwise,}\end{cases}$$

for all *odd* $k\leqslant 2^n-1$. Also let $(\xi_n)_{n\geqslant 0}$ be a sequence of independent $\mathcal{N}(0,1)$-distributed random variables defined on a probability space $(\Omega,\mathcal{F},P)$. (An example of such a space is given via the Kolmogorov theorem.) Denote the Haar functions by $(\phi_n)_{n\geqslant 0}$ for convenience.

For each $t\in[0,1]$ let $B_t^n=\sum_{i=0}^n\xi_i\int_0^t\phi_i(s)\,ds$, and let $I_t=1_{[0,t[}$. In terms of the inner product $\langle,\rangle$ on $L^2[0,1]$ we have $\int_0^t\phi_i(s)\,ds=\langle I_t,\phi_i\rangle$, and as the (ϕ_i) are an orthonormal basis, $I_t=\sum_{i=1}^\infty\langle I_t,\phi_i\rangle\phi_i$ and $t=\|I_t\|^2=\sum_{i=1}^\infty\langle I_t,\phi_i\rangle^2$. Since the (ξ_i) are independent and $\mathbf{E}(\xi_i^2)=1$, it follows that for $n>m$, $\mathbf{E}(B_t^n-B_t^m)^2=\mathbf{E}(\sum_{i=m+1}^n\xi_i\int_0^t\phi_i(s)ds)^2=\sum_{i=m+1}^n\langle I_t,\phi_i\rangle^2$, so that (B_t^n) is a Cauchy sequence in $L^2[0,1]$. Denote the L^2-limit by (B_t), then

$$\langle B_s,B_t\rangle=\sum_{i=1}^\infty\langle I_s,\phi_i\rangle\langle I_t,\phi_i\rangle=\langle I_s,I_t\rangle=s\wedge t.$$

Thus (B_t) has the required covariance function. The independence of the increments, etc., follows easily. The *continuity* of the paths of B is more delicate, since it requires the *uniform* convergence on $[0,1]$ of the sequence $(B_t^n(\omega))_n$ to $B_t(\omega)$, for almost all $\omega\in\Omega$: one sets $S_n=\sup_{t\in[0,1]}|B_t^n(\omega)|$ and shows that for $\theta>1$, the sets (A_n) defined by $A_n=\{S_n>\theta\sqrt{(2^{-n}\log 2^n)}\}$ have $\sum_n P(A_n)<\infty$. Then the Borel–Cantelli lemma implies that $P\{\omega:\omega\in A_n$ infinitely often$\}=0$, and the result follows. (See [57] for a proof of the above estimate.)

0.2.7: Although B has continuous paths, the paths are *nowhere* differentiable. This can be proved directly (see [40]) or deduced from the fact that the variation of the paths is infinite on any finite interval. (This shows up a limitation of the use of B as a model of particle behaviour, since infinite path length over a finite time interval would imply infinite velocity!)

To prove this result, we first consider the *quadratic variation* of B, defined on $[0,1]$ as $\lim_n Q_n$, where $Q_n=\sum_{k=1}^{2^n}(B_{k/2^n}-B_{(k-1)/2^n})^2$. Since $\{(B_{k/2^n}-B_{(k-1)/2^n}):k\geqslant 1\}$ is an independent family, with normal distribution $\mathcal{N}(0,1/2^n)$ for each member, $\mathbf{E}(Q_n)=\sum_{k=1}^{2^n}1/2^n=1$ for all n. Moreover if $X\sim\mathcal{N}(0,t^2)$, $\mathbf{E}(X^4)=(2\pi t)^{-\frac{1}{2}}\int_{-\infty}^{\infty}x^4e^{-x^2/2t}dx=3t^4$ (exercise!). Hence $\operatorname{var}(X^2)=3t^4-(t^2)^2=2t^4$. Thus we have

$$\mathbf{E}((Q_n-1)^2)=\operatorname{var}(Q_n)=\sum_{k=1}^{2^n}\mathbf{E}(B_{k/2^n}-B_{(k-1)/2^n})^4=\sum_{k=1}^{2^n}2\left(\frac{1}{2^{n/2}}\right)^4=\frac{1}{2^{n-1}}\to 0,$$

so that $Q_n\to 1$ in L^2-norm. Replacing $[0,1]$ by $[0,t]$, we can form sums $Q_n(t)$ similarly and conclude that the quadratic variation of B on $[0,t]$ is t. The Chebyshev inequality allows us to estimate $P\{|Q_n(t)-t|>\varepsilon\}$ for any $\varepsilon>0$, and the Borel–Cantelli lemma ensures that $Q_n(t)\to t$ a.s. Heuristically, this result should be clear: $\mathbf{E}((B_{s+t}-B_s)^2)=t$, so $|B_{s+t}-B_s|$ is at least roughly approximated by $\sqrt{t}$. If the partition (t_k) of $[0,t]$ is fine enough, one would 'expect' that $\sum(B_{t_{k+1}}-B_{t_k})^2\simeq\sum(t_{k+1}-t_k)=t$. On the other hand, the (total) variation of B can be defined on $[0,1]$ by $V(B)=\lim_n\sum_{k=1}^{2^n}|B_{k/2^n}-B_{(k-1)/2^n}|$. Since $Q_n\leqslant(\max_{k\leqslant 2^n}|B_{k/2^n}-B_{(k-1)/2^n}|)\sum_{k=1}^{2^n}|B_{k/2^n}-B_{(k-1)/2^n}|$ and by the continuity of B the first term goes to 0 as $n\to\infty$, while $Q_n\to 1$, we conclude that $V(B)\to\infty$.

0.2.8. ***Definition:*** Let $X=(X_t)$ be a stochastic process on $(\Omega,\mathcal{F},P)$. Associate families of σ-fields $\mathcal{F}_t$, $\mathcal{F}(X_t)$, $\mathcal{G}_t$ for $t\in[0,\infty[$ with X by adding the P-null sets to each of $\sigma(X_s:s\leqslant t)$, $\sigma(X_t)$, $\sigma(X_s:s\geqslant t)$. (Think of these as the 'past, present and future' of X!) We say that X is a *Markov process* (or has the weak Markov property) if, for all t, $\mathcal{F}_t$ and $\mathcal{G}_t$ are conditionally independent, given $\mathcal{F}(X_t)$, i.e. $P(F\cap G|\mathcal{F}(X_t))=P(F|\mathcal{F}(X_t))P(G|\mathcal{F}(X_t))$ whenever $F\in\mathcal{F}_t$, $G\in\mathcal{G}_t$. (In other words, what X does *after* time t depends only on where it is *at* time t, not where it was *before* then.)

One of the most useful features of Brownian motion is that it has this property in a very strong form: in fact, we need only check that $P(W_t\in A|\mathcal{F}_s)=P(W_t\in A|\mathcal{F}(X_s))$ for $s<t$ and any cylinder set A in $\mathcal{F}_s$ to see that the Wiener process (W_t) has this property. (The *strong Markov property* extends this to the σ-fields generated by stopping times (see 2.4) and, when applied to Brownian motion B implies, for example, that the process 'started' at the time B 'hits' a given set in $\mathbf{R}$ is again $BM_0(\mathbf{R})$.)

0.2.9: The other crucial property of B which has led to many generalisations and which is central to this book follows from the independence of $(B_t - B_s)$ from $\mathscr{F}_s$ for $s \leqslant t$: we conclude that $\mathbf{E}(B_t|\mathscr{F}_s) = B_s$ a.s. In Chapters 2 and 3 we shall study abstract processes $X = (X_t)$ and σ-fields $(\mathscr{F}_t)$ such that $X_s \in L^1(\mathscr{F}_s)$ and $\mathbf{E}(X_t|\mathscr{F}_s) = X_s$ for all $t > s$. We shall say that $X = (X_t)$ is a *martingale* relative to $(\mathscr{F}_t)$. Hence in the present example, Brownian motion B is a martingale relative to the σ-fields $(\mathscr{F}_t)$ of 0.2.8. Moreover, it is easy to see that for $0 < s < t$, $t - s = \mathbf{E}((B_t - B_s)^2|\mathscr{F}_s) = \mathbf{E}(B_t^2|F_s) - B_s^2$, so that $\mathbf{E}(B_t^2 - t|\mathscr{F}_s) = B_s^2 - s$. Hence $(B_t^2 - t)_{t \geqslant 0}$ is also a martingale, a fact which will serve to explain the nature of the (deterministic!) quadratic variation 'process' $(t, \omega) \longrightarrow t$ of B. The fact that this variation process is finite although B is *not* of bounded variation in the normal sense serves to motivate the following constructions:

0.2.10. ***The Itô integral:*** Let $B = (B_t)_{t \geqslant 0}$ be a standard Brownian motion which is thus a martingale relative to the family of σ-fields $(\mathscr{F}_t)_{t \geqslant 0}$ – recall that $\mathscr{F}_t$ is the σ-field generated by $\{B_s : s \leqslant t\}$ and the P-null sets – and consider the following class of stochastic processes $(f_t)_{t \geqslant 0}$:

(i) $(t, \omega) \longrightarrow f_t(\omega)$ is measurable for the product σ-field $B(\mathbf{R}^+) \times \mathscr{F}$,

(ii) for each t, f_t is $\mathscr{F}_t$-measurable (we say (f_t) is *adapted* to $(\mathscr{F}_t)$),

(iii) $\mathbf{E}(\int_0^\infty f_t^2 \, dt) < \infty$.

The class of processes $f = (f_t)$ satisfying (i)–(iii) is denoted by $\mathscr{L}^2(B)$. In constructing an integral $I(f) = \int_0^\infty f \, dB$ we first consider *step processes* f of the form: $f_t(\omega) = f_{t_j}(\omega)$ for $t_j \leqslant t < t_{j+1}$, where (t_k) is a finite partition of $\mathbf{R}^+$ and f_{t_j} is $\mathscr{F}_{t_j}$-measurable for all j. Then set $I(f) = \sum_{j=1}^n f_{t_j}(B_{t_{j+1}} - B_{t_j})$. It is clear that $I(f)$ has mean 0, as $\mathbf{E}(B_{t_{j+1}} - B_{t_j}) = 0$ for all j. Moreover, since the increments of B are independent and (f_t) is adapted to $(\mathscr{F}_t)$,

$$\mathbf{E}(I(f)^2) = \sum_{j=1}^m \mathbf{E}(f_{t_j}^2(B_{t_{j+1}} - B_{t_j})^2) = \sum_{j=1}^m \mathbf{E}(f_{t_j}^2)(t_{j+1} - t_j) = \mathbf{E}\left(\int_0^\infty f_t^2 \, dt\right).$$

The map $f \longrightarrow I(f)$ is therefore an *isometry* between the space $\mathscr{E}$ of step processes in $\mathscr{L}^2(B)$ and $L^2(P)$. This isometry extends to all of $\mathscr{L}^2(B)$ since $\mathscr{E}$ *is dense in* $\mathscr{L}^2(B)$.

We prove that if $f \in \mathscr{L}^2(B)$ is given, then there is a sequence $f^{(n)}$ in $\mathscr{E}$ such that $\mathbf{E}(\int_0^\infty (f_t - f_t^{(n)})^2 dt) \longrightarrow 0$ as $n \longrightarrow \infty$. For this we can assume that the (f_t) are uniformly bounded and that $f_t = 0$ for t outside a finite interval $[a, b] \subseteq \mathbf{R}^+$. Also set $f_t = 0$ for $t < 0$. Since $\mathbf{E}(\int_0^\infty f_t^2 \, dt) < \infty$, $\int_{\mathbf{R}} f_t^2(\omega) \, dt < \infty$ a.s. (P). Now as $h \longrightarrow 0$ we have $\mathbf{E}(\int_{\mathbf{R}} |f_{t+h} - f_t|^2 dt) \longrightarrow 0$, since the inner integral converges to 0 a.s. and is bounded by $4\int_{\mathbf{R}} f_t^2 \, dt \in L^1(P)$. We deduce that $\mathbf{E}(\int_{\mathbf{R}} |f_{s+k_n(t)} - f_{s+t}|^2 ds) \longrightarrow 0$ as $n \longrightarrow \infty$, for fixed t, where $k_n(t) = i/2^n$ on $[i/2^n, (i+1)/2^n[$, $i \in \mathbf{Z}$. Applying the bounded convergence theorem once more, we see that, as $n\to\infty$, $\int_{a-1}^{b}\mathbf{E}(\int_{\mathbf{R}}|f_{s+k_n(t)}-f_{s+t}|^2\mathrm{d}s)\mathrm{d}t\to 0$. By Fubini's theorem, $\int_{\mathbf{R}}\int_{a-1}^{b}\mathbf{E}(|f_{s+k_n(t)}-f_{s+t}|^2)\mathrm{d}t\,\mathrm{d}s\to 0$, and we can extract a subsequence (l_i) such that $\int_{a-1}^{b}\mathbf{E}(|f_{s+l_i(t)-s}-f_t|^2\mathrm{d}t\to 0$ for almost all s. Fix such an $s\in[0,1]$ and make the change of variable $t\to t-s$, then, as $\int_{a-1}^{b}=\int_{a-1+s}^{b+s}$, we have $\int_{a-1}^{b}\mathbf{E}(|f_{s+l_i(t-s)}-f_{s+t}|^2)\mathrm{d}t\to 0$. So $f_t^{(i)}=f_{s+l_i(t-s)}$ defines a sequence of step processes $f^{(i)}$ in $\mathscr{E}$ satisfying $\lim_i\mathbf{E}(\int_0^\infty(f_t^{(i)}-f_t)^2\mathrm{d}t)=0$.

Having defined $I(f)=\int_0^\infty f_s\mathrm{d}B_s$ for $(f_s)\in\mathscr{L}^2(B)$, we can define a *stochastic process* $(I_t(f))$ by $I_t^f=I(f\cdot 1_{[0,t]})$. Each I_t^f is defined only up to a.s. equivalence. We show that the process $(I_t^f)_{t\geqslant 0}$ is a martingale relative to $(\mathscr{F}_t)$, i.e. that $\mathbf{E}(I_t^f|\mathscr{F}_s)=I_s^f$ for $s\leqslant t$.

First consider the case where $f\in\mathscr{E}$ has the form $f_t(\omega)=g(\omega)\cdot 1_{[u,v]}(t)$ for some $u,v\geqslant 0$ and $\mathscr{F}_u$-measurable function g. Then $I_t=\int_0^t g(\omega)1_{[u,v]}(t)\mathrm{d}B_t(\omega)$ $=g(\omega)(B_{v\wedge t}(\omega)-B_{u\wedge t}(\omega))$, so that $\mathbf{E}(I_t^f|\mathscr{F}_s)=I_s^f$, as may easily be checked by considering the different cases $s<u<t<v$, $u<s<t<v$, etc. Extend to $\mathscr{E}$ by linearity and to $\mathscr{L}^2(B)$ using the fact that $\mathscr{E}$ is dense in $\mathscr{L}^2(B)$ and that $|\mathbf{E}(\int_s^t f_u\mathrm{d}B_u|\mathscr{F}_s)-\mathbf{E}(\int_s^t f_u^{(n)}\mathrm{d}B_u|\mathscr{F}_s)|$ is bounded in expectation by $(\mathbf{E}(\int_0^\infty(f_u-f_u^{(n)})^2\mathrm{d}u))^{1/2}$, for any sequence $f^{(n)}$ in $\mathscr{E}$. (Use Hölder's inequality to establish this bound – *Exercise.*)

The fact that (I_t^f) is a martingale allows us to use Doob's L^2-martingale inequality (see Proposition 3.3.2) to deduce that $t\to I_t^f(\omega)$ can be chosen *continuous* on $\mathbf{R}^+$ for almost all ω: the Doob L^2-inequality, applied to $\int_0^t(f_s-f_s^{(n)})\mathrm{d}B_s$, where $f^{(n)}\in\mathscr{E}$, yields $\mathbf{E}(\sup_t|\int_0^t(f_s-f_s^{(n)})\mathrm{d}B_s|)$ $\leqslant 4\mathbf{E}(\int_0^\infty(f_s-f_s^{(n)})\mathrm{d}B_s)^2$ (the reader is not advised to try to prove this directly, but to take this inequality on trust until Chapter 3!). Hence if $Z_n=\sup_t|\int_0^t(f_s-f_s^{(n)})\mathrm{d}B_s|^2$, the Chebychev inequality shows that $P(Z_n>\varepsilon)\to 0$ for any $\varepsilon>0$. Choosing a subsequence if necessary we have $P(Z_n>1/2^n)<1/2^n$ for all n. So the Borel–Cantelli lemma applied to the sets $\{Z_n>1/2^n\}$ shows that $P\{Z_n>1/2^n$ for infinitely many $n\}=0$. But since $t\to\int_0^t f_s^{(n)}\mathrm{d}B_s$ is clearly continuous for all n, this means that $t\to\int_0^t f_s\mathrm{d}B_s$ can be chosen continuous for almost all $\omega\in\Omega$.

Thus the Itô integral for Brownian motion B has been defined for all $f\in\mathscr{L}^2(B)$ as a continuous square-integrable martingale. Recall that the quadratic variation of the martingale B was the *deterministic* process $(t,\omega)\to t$. We can also characterise the quadratic variation of a continuous square-integrable martingale $M=(M_t)$ as an increasing adapted continuous process $A=(A_t)$, with $A_0=0$, such that M^2-A is again a martingale. This clarifies the role of $(t,\omega)\to t$ in the case $M=B$, since we proved that B_t^2-t is a martingale. Moreover, it is not hard to see that the continuous martingale $I_t^f=\int_0^t f_s\mathrm{d}B_s$ has quadratic variation $A_t=\int_0^t f_s^2\,\mathrm{d}s$: clearly we have $A_t\in L^1$ for all t, since $f\in\mathscr{L}^2(B)$. Moreover, for $s\leqslant t$,

$$\mathbf{E}((I_t^f)^2 - A_t|\mathscr{F}_s) = \mathbf{E}\left(\left(\int_0^t f_u\,\mathrm{d}B_u\right)^2 |\mathscr{F}_s\right) - \mathbf{E}\left(\int_0^t f_u^2\,\mathrm{d}u|\mathscr{F}_s\right)$$

$$= \mathbf{E}\left(\int_0^s f_u\,\mathrm{d}B_u\right)^2 - \int_0^s f_u^2\,\mathrm{d}u + \mathbf{E}\left(\left(\int_s^t f_u\,\mathrm{d}B_u\right)^2 - \int_s^t f_u^2\,\mathrm{d}u|\mathscr{F}_s\right) = (I_s^f)^2 - A_s,$$

where the final term is shown to be zero by approximating f by elements of $\mathscr{E}$.

The Itô integral leads to a calculus which differs markedly from the ordinary calculus, due to the presence of the quadratic variation of B. The main tool is the following 'Fundamental Theorem':

0.2.11. ***Itô formula:*** Let $F:\mathbf{R}\to\mathbf{R}$ be twice continuously differentiable, and let B be standard Brownian motion $BM_0(\mathbf{R})$. Then $F(B_t) - F(B_0) = \int_0^t F'(B_u)\mathrm{d}B_u + \frac{1}{2}\int_0^t F''(B_u)\mathrm{d}u$.

A generalisation of this result will be proved in Chapter 4. For a direct proof, see [57]. As an application, let $F(t) = t^2$. In 'differential' form, the Itô formula states that $\mathrm{d}(B^2) = 2B\cdot\mathrm{d}B + \mathrm{d}t$, so that, as $B_0 = 0$, $\int_0^t B_s\,\mathrm{d}B_s = (B_t^2 - t)/2$.

0.2.12: Finally, we note that $\int_0^t B_s\,\mathrm{d}B_s$ cannot be defined as a Riemann–Stieltjes integral (we know already that B has unbounded variation, but this example indicates that $\int_0^t B_s\,\mathrm{d}B_s$ *cannot* make sense as a limit of 'Riemann sums'). For if we want to define $\int_0^t B_s\,\mathrm{d}B_s$ as a limit of sums $\sum_{k=1}^m B_{s_k} - B_{t_k})$, let D denote the difference between these sums when we choose $s_k = t_{k+1}$ or t_k respectively. Then $D = \sum_{k=1}^m (B_{t_{k+1}} - B_{t_k})^2$ and we have already seen that this tends to t in mean square as the partition lengths tend to 0.

0.3. The Poisson process

Brownian motion B provides an example (in fact the only one!) of a process with continuous paths and independent increments $(B_t - B_s)$ whose distribution depends only on $|t-s|$ (the variance $(t-s)$ determines the distribution, since $(B_t - B_s)$ is $\mathscr{N}(0,(t-s))$. We say that the increments are *stationary*.

0.3.1: The other generic example of a process with stationary independent increments is the *Poisson process* (N_t): here we assume that the increments $(N_t - N_s)$ are integer-valued and that

$$P(N_t - N_s = m) = \frac{e^{-\lambda(t-s)}\lambda^m(t-s)^m}{m!}$$

for $m=0,1,2,\ldots$ and some $\lambda>0$ (the *rate* of the process). It is clear that for any fixed $t_0 \in \mathbf{R}^+$, the process $\{N_t - N_{t_0} : t \geqslant t_0\}$ has the weak Markov property and that $X_t = N_t - N_{t_0} - \lambda t$, $t \geqslant t_0$, is both a Markov process and a martingale relative to the σ-fields $\mathscr{F}_t = \sigma(N_s : s \leqslant t)$.

0.3.2: The expression for $P(N_t - N_s = m)$, $m=0,1,2,\ldots$ shows that

$$P(N_t \geqslant N_s) = \sum_{m=0}^{\infty} \frac{e^{-\lambda(t-s)}\lambda^m(t-s)^m}{m!} = 1,$$

so (N_t) has a.s. non-decreasing paths, which have only countably many jump discontinuities. In fact, each 'jump' has magnitude 1: to see this, we can restrict attention to the interval $[0,\ell]$ for some $\ell>0$ and consider

$$P\left(\max_{i\leqslant j<n}\left(N\left(\frac{j+1}{n}\ell,\omega\right)-N\left(\frac{j-1}{n}\ell,\omega\right)\right)>1\right)\leqslant \sum_{j=1}^{n-1} P\left(N\left(\frac{j+1}{n}\ell,\omega\right)-N\left(\frac{j-1}{n}\ell,\omega\right)>1\right)=\sum_{j=1}^{n-1}\sum_{m=2}^{\infty}\frac{e^{-(2\ell\lambda/n)}\lambda^m(2\ell/n)^m}{m!}$$

$$=(n-1)\left(1-e^{-2\ell\lambda/n}-\frac{2\ell\lambda}{n}e^{-2\ell\lambda/n}\right)\to 0 \text{ as } n\to\infty.$$

If $t \to N_t(\omega)$ had a jump of magnitude greater than 1 in the interval, the maximum would exceed 1 for all n.

0.3.3: The 'time' at which the ith jump occurs is a random variable T_i. We can examine the nature of T_i by computing conditional probabilities: suppose we know that for fixed s,t in $\mathbf{T}$ with $s<t$, $N_t - N_s = m$. Also let $I_1, I_2, \ldots, I_n$ be disjoint intervals in $]s,t[$ of lengths $\ell_1, \ell_2, \ldots, \ell_n$, so that $t-s=\sum_{i=1}^n \ell_i$. Then the probability that on each I_i there will be m_i jumps of magnitude 1 is given, because of independence, by

$$\left(\prod_{i=1}^n \frac{e^{-\lambda\ell_i}(\lambda\ell_i)^{m_i}}{m_i!}\right)\Bigg/\left(\frac{e^{-\lambda(t-s)}(\lambda(t-s))^m}{m!}\right).$$

Now if $m_{i_k}=1$ for $k=1,2,\ldots,m$ and all other $m_i=0$, this reduces to

$$\frac{m!}{(t-s)^m}\prod_1^m \ell_{i_k}.$$

This is just the distribution of m points chosen independently from $]s,t[$, each with constant density $1/(t-s)$. (The probability that the points occur at $t_1, t_2, \ldots, t_m$ is $m!/(t-s)^m$.)

Thus the conditional distribution of the jump times T_i is the same as that of m points chosen independently in $]s,t[$ each with constant density $1/(t-s)$. This indicates what we shall make precise in Chapter 3 and prove in Example 4.1.5: the jump times of the Poisson process are 'totally inaccessible', and cannot be predicted in any way from past knowledge of the process.

0.3.4: Consider the case $\lambda=1$. The martingale $X_t=N_t-t$ has mean 0 and covariance $\min(s,t)$, in common with $BM_0(\mathbf{R})$, but has discontinuous paths. One can show that (up to modifications) they are always right-continuous with left limits. Since $t\to X_t$ is of bounded variation, one can form Stieltjes integrals $\int_0^t f_s(\omega)\,\mathrm{d}X_s(\omega)$ for bounded measurable processes $f=(f_s)$, separately for each $\omega\in\Omega$. The question naturally arises whether the process $Y_t=\int_0^t f_s\,\mathrm{d}X_s$ is again a martingale, as was the case for the Itô integral. We show in Example 4.1.5 that this is false in general. The general theory of Chapter 4 also shows that for bounded 'predictable' processes (f_s) one does indeed obtain a martingale.

Using the form of the conditional distribution of the set $\{T_i : N_s \leqslant i < N_t\}$ of the k jump times in (s,t), one can also prove this directly: observe that $Y_t=\sum_{i=1}^{N_t} f(T_i)-\int_0^t f(u)\mathrm{d}u$ and calculate $\mathbf{E}(Y_t-Y_s|\mathscr{F}_s)=\mathbf{E}(\sum_{i=N_s+1}^{N_t} f(T_i) -\int_s^t f(u)\mathrm{d}u|\mathscr{F}_s)=\mathbf{E}(\sum_{i=N_s+1}^{N_t} f(T_i)|\mathscr{F}_s)-\int_s^t f(u)\mathrm{d}u$. Note first that $\sum_{i=N_s+1}^{N_t} f(T_i)$ is *independent* of $\mathscr{F}_s$: to see this, recall that the increments (N_t-N_s) are independent of $\mathscr{F}_s$ for $t>s$ and that for fixed $t\neq T_i$ $(i\geqslant 1)$ the time Q_t from t to the next jump time is independent of t (and exponentially distributed) so that $f(T_{N_t})$ is independent of $\mathscr{F}_s$ whenever there is exactly one jump in the interval (s,t). The independence of $\sum_{i=N_s+1}^{N_t} f(T_i)$ of $\mathscr{F}_s$ follows by induction. Hence we have

$$\begin{aligned}\mathbf{E}\left(\sum_{i=N_s+1}^{N_t} f(T_i)|\mathbf{F}_s\right)&=\mathbf{E}\left(\sum_{i=N_s+1}^{N_t} f(T_i)\right)\\ &=\sum_{k\geqslant 1} P(N_t-N_s=k)\mathbf{E}\left(\sum_{i=N_s+1}^{N_t} f(T_i)|N_t-N_s=k\right)\\ &=\sum_{k\geqslant 1}\frac{(t-s)^k}{k!}\mathrm{e}^{-(t-s)}\mathbf{E}\left(\sum_{i=1}^{k} f(\xi_i)\right),\end{aligned}$$

where (ξ_i) are k independently chosen points in (s,t) each with uniform density $1/(t-s)$, by the result of 0.3.3. Since

$$\mathbf{E}\left(\sum_{i=1}^{k} f(\xi_i)\right)=k\frac{1}{(t-s)}\int_s^t f(u)\mathrm{d}u,$$

we have finally that

$$\mathbf{E}((Y_t-Y_s)|\mathscr{F}_s)=\left[\sum_{k\geqslant 1}\frac{k}{(t-s)}\frac{(t-s)^k}{k!}\mathrm{e}^{-(t-s)}\left(\int_s^t f(u)\mathrm{d}u\right)\right]-\int_s^t f(u)\mathrm{d}u=0.$$

So Y is a martingale whenever f is a bounded deterministic function. The same proof will apply when there exists $\varepsilon>0$ such that $f(s,\cdot)$ is $\mathscr{F}_{s-\varepsilon}$-measurable for each s when $f=(f_s)$ is a bounded process, since we need only check the martingale property if $t-s<\varepsilon$. Finally we can approximate integrals of left-continuous adapted processes by integrals of functions of the above form. The details are not quite trivial, but need not detain us here, since the result is a special case of the theory developed in Chapter 4. We shall see in Chapter 3 that the left-continuous adapted processes generate the 'predictable' processes mentioned above.

0.4. Martingales and gambling

Although we begin our formal study of martingales only in section 2.3, Brownian motion and Poisson processes have already provided examples of processes (X_t) and σ-fields $(\mathscr{F}_t)$ satisfying $\mathbf{E}(X_t|\mathscr{F}_s)=X_s$ for all $s\leqslant t$. We comment briefly why this condition should be intrinsically interesting by examining how it models a 'fair game' in gambling.

Coin-tossing can be modelled by the simple random walk on the line, or, more directly, in the product space $\Omega=\{-1,1\}^{\mathbf{N}}$, where each $\omega\in\Omega$ is an infinite sequence of tosses of the coin. The probability measure P on the σ-cylinders of Ω is given by the probabilities p, $(1-p)$, which describe the probability of 'heads' or 'tails', respectively at each toss. The tosses are therefore independent random variables (X_i) on Ω, given by $X_i(\omega)=\omega(i)$, with $P(X_i=1)=p$. Betting b_n on 'heads' at the nth toss yields a strategy, i.e. a function $b_n:\{-1,1\}^n\to\mathbf{R}^+$, and the gambler's fortune (S_n) is given by $S_{n+1}=S_n+X_{n+1}b_n(X_i:i\leqslant n)$.

Let $\mathscr{F}_n=\sigma(S_k:k\leqslant n)$. 'Guessing' after the nth toss (i.e. 'knowing' $\mathscr{F}_n$) what S_{n+1} will be means that we must calculate $\mathbf{E}(S_{n+1}|\mathscr{F}_n)$. Since $\mathscr{F}_n$ depends only on X_i $(i\leqslant n)$ we have $\mathbf{E}(S_{n+1}|\mathscr{F}_n)=\mathbf{E}(S_n+X_{n+1}b_n(X_1,\ldots,X_n)|\mathscr{F}_n)=S_n+b_n(X_1,\ldots,X_n)\mathbf{E}(X_{n+1})$ by the independence of the (X_i). If the coin is 'fair', i.e. $p=\frac{1}{2}$, $\mathbf{E}(X_{n+1})=0$, hence $\mathbf{E}(S_{n+1}|\mathscr{F}_n)=S_n$. In other words, 'on average' we expect to gain or lose nothing in 'fair' coin-tossing. Of course if $p\leqslant\frac{1}{2}$, $\mathbf{E}(S_{n+1}|\mathscr{F}_n)\leqslant S_n$ and the likelihood of our eventual ruin increases.

Exactly the same arguments apply to general 'fair' games. But we also want to answer such questions as: what happens if I choose a strategy of 'skipping' certain games? What sort of games lead to eventual ruin almost surely? Does (S_n) converge to a limit in any sense? What if I adopt a 'rule' for quitting the game (depending on ω) after a certain number of games; does this provide any advantage? All these questions relate in some way to *stability properties* of the relation $\mathbf{E}(S_{n+1}|\mathscr{F}_n)=S_n$ under limits and under transformations of the variables (X_n). We shall see that under appropriate measurability conditions (which forbid 'prescience') all these questions lead to the dismal and empirically accurate conclusion that a fair game cannot be made favourable without cheating! Nonetheless, our fascination with gambling has led to much delightful mathematics – see [8], [12], [26] for examples.

1

Weak compactness and uniform integrability

The construction of canonical processes leads us naturally into infinite-dimensional function spaces. On the other hand, Brownian motion and Poisson processes deal with families of random variables with rather special distributions. For more general processes, convergence theorems will require *a priori* boundedness and compactness conditions. We therefore dip briefly into functional analysis to isolate, in particular, a characterisation of relatively weakly compact sets in $L^1(\Omega, \mathscr{F}, P)$. The reader may wish to omit the details here and move straight to Chapter 2, where we begin the study of martingales in discrete time. Even in that context the need for certain weak compactness properties will become apparent.

1.1. Duality and weak compactness in reflexive Banach spaces

The Banach spaces $L^p = L^p(\Omega, \mathscr{F}, P)$ were introduced in Definition 0.1.3. We assume that the reader is familiar with elementary Banach space theory, but we recall the following ideas (details may be found in [27], [77]). We deal only with *real* Banach spaces.

1.1.1. ***Definition:*** Let E be a real Banach space. The vector space E' of all bounded linear functionals on E, with the norm $\|x'\| = \sup\{x'(x): x \in E, \|x\| \leqslant 1\}$, is a Banach space, the *dual* of E. The *weak topology* on E, $\sigma(E,E')$, is defined by the system of neighbourhoods of 0 of the form $V_\varepsilon = \{x \in E: |x_i'(x)| \leqslant \varepsilon, i \leqslant n\}$, where $n \in \mathbf{N}$, $\varepsilon > 0$ and $x_i' \in E'$ for all $i \leqslant n$. Thus a sequence (x_n) in E *converges weakly* to $x \in E$ if $x'(x_n) \longrightarrow x'(x)$ for all $x' \in E'$.

The *second dual* E'' of E is defined as $(E')'$. Note that E is isometrically embedded in E'' by the map $x \longrightarrow \hat{x}$, where $\hat{x}(x') = x'(x)$ for all $x' \in E'$. E is said to be *reflexive* if this embedding is onto E''.

1.1.2. ***Theorem:*** Let $1 \leqslant p < \infty$ and let $1/p + 1/q = 1$ if $p > 1$, and $q = \infty$ if $p = 1$. Then any bounded linear functional ϕ on L^p corresponds uniquely to an element $g \in L^q$ under the map $\phi(f) = \int_\Omega f.g \, \mathrm{d}P$, and $||\phi|| = ||g||_q$. So we can identify the dual of L^p with L^q. (Note that $f.g \in L^1$ by the *Hölder inequality*, which shows that $||f.g||_1 \leqslant ||f||_p ||g||_q$ for all $f \in L^p$, $g \in L^q$, with equality iff $f = \lambda g$ for some $\lambda \in \mathbf{R}$.)

Proof: *Uniqueness:* if $\int_\Omega f.g_1 \, \mathrm{d}P = \int_\Omega f.g_2 \, \mathrm{d}P$ for all $f \in L^p$, then $g_1 = g_2$ a.s. (take $f = 1_A$ for all $A \in \mathscr{F}$). So $g \in L^q$ is unique if it exists, and then Hölder's inequality shows that for $f \in L^p$, $|\phi(f)| = |\int_\Omega fg \, \mathrm{d}P| \leqslant ||f||_p ||g||_q$, so $||\phi|| \leqslant ||g||_q$.

Existence: For $A \in \mathscr{F}$ let $\lambda(A) = \phi(1_A)$. It is clear that λ is an additive set function. If $A = \bigcup_{i=1}^\infty A_i$ is a disjoint union in $\mathscr{F}$, write $B_k = \bigcup_{i=1}^k A_i$ and note that $||1_A - 1_{B_k}||_p = (P(A \setminus B_k))^{1/p} \to 0$ as $k \to \infty$. But ϕ is a continuous map $L^p \to \mathbf{R}$, so $\lambda(B_k) = \phi(1_{B_k}) \to \phi(1_A) = \lambda(A)$. Hence λ is σ-additive. If $P(A) = 0$, then $||1_A||_p = 0$, so $\lambda(A) = \phi(1_A) = 0$, and $\lambda \ll P$. Hence the Radon–Nikodym theorem provides $g \in L^1(P)$ such that $\lambda(A) = \int_A g \, \mathrm{d}P = \int_\Omega 1_A g \, \mathrm{d}P$ for $A \in \mathscr{F}$. By linearity and continuity of ϕ, $\phi(f) = \int_\Omega fg \, \mathrm{d}P$ for all $f \in L^\infty(P)$, since each $f \in L^\infty$ is a uniform limit of simple functions. If we can show that $g \in L^q$ and $||g||_q \leqslant ||\phi||$, the continuous functions $f \to \int_\Omega fg \, \mathrm{d}P$ and ϕ coincide on the dense subspace L^∞ of L^p, hence are equal. Thus it remains to show that $||g||_q \leqslant ||\phi||$, where $1/p + 1/q = 1$.

First take $p = 1$: if $A = \{g > ||\phi||\}$ has positive measure, then $||\phi|| P(A) < \int_A g \, \mathrm{d}P = \lambda(A) = \phi(1_A) \leqslant ||\phi|| P(A)$, a contradiction. Similarly for $\{g < -||\phi||\}$. Hence $||g||_\infty \leqslant ||\phi||$. Finally, let $1 < p < \infty$: let

$$\alpha = \operatorname{sgn}(g) = \begin{cases} 1 \text{ if } g \geqslant 0, \\ -1 \text{ if } g < 0. \end{cases}$$

Then α is measurable and $\alpha g = |g|$. Define $A_n = \{|g| \leqslant n\}$ and $f = 1_{A_n} |g|^{q-1} \alpha$. Then $f \in L^\infty$ and it is easily checked that $\int_{A_n} |g|^q \, \mathrm{d}P = \int_\Omega fg \, \mathrm{d}P = \phi(f) \leqslant ||\phi|| (\int_{A_n} |g|^q \, \mathrm{d}P)^{1/q}$. By monotone convergence we have $||g||_q \leqslant ||\phi||$.

1.1.3. ***Corollary:*** If $1 < p < \infty$, the Banach space L^p is reflexive. When $p = 2$, in particular we have L^2 as a Hilbert space, with the inner product $\langle f, g \rangle = \int_\Omega fg \, \mathrm{d}P$, and $\phi \in (L^2)'$ has the form $f \to \langle f, g \rangle$.

1.1.4. ***Definition:*** Let E be a Banach space, E' its dual. The *weak*-topology* $\sigma(E', E)$ on E' has neighbourhoods of 0 of the form $U_\varepsilon = \{x' \in E' : |x'(x_i)| < \varepsilon, i \leqslant n\}$, where $\varepsilon > 0$, $n \in \mathbf{N}$, $x_i \in E$, $i \leqslant n$. Thus $\sigma(E', E)$ is induced on $E' \subseteq \mathbf{R}^E$ by the product topology on $\mathbf{R}^E$. By considering the weak*-closure of the unit ball $\{x' : ||x'|| < 1\}$ in E' as a closed subset of

$[-1,1]^E$ in its product topology, one can easily deduce the following result from the Tychonoff theorem on compactness in product spaces:

1.1.5. ***Theorem (Alaoglu):*** The unit ball in E' is relatively weak*-compact (i.e. its closure is compact in the $\sigma(E',E)$ tolopogy on E'.)

The following application of this result will be of use in the next section. Recall that E is *separable* if there is a countable subset of E which is dense in E.

1.1.6. ***Proposition:*** If E is a separable Banach space, the weak*-topology is a metric topology for the unit ball B' of E'.

Proof: If (X_n) is a countable dense subset of E, the function

$$\rho(x',y') = \sum_{n=1}^{\infty} \frac{1}{2^n}\left(\frac{|(x'-y')(x_n)|}{1+|(x'-y')(x_n)|}\right)$$

is easily seen to be a metric on E', inducing a topology τ on B' weaker than the weak*-topology. Hence the identity map $i:(B',\mathrm{w}^*)\to(B',\tau)$ is a continuous map from a compact (by Alaoglu's theorem) to a Hausdorff space, hence a homeomorphism. So the topologies coincide on B'.

Compactness of subsets of E in the $\sigma(E,E')$-topology is a little subtler: the analysis of Alaoglu's theorem actually characterises *reflexive* spaces.

1.1.7. ***Theorem:*** A Banach space E is reflexive iff the unit ball in E is relatively weakly compact.

Proof: We shall show below how the Hahn–Banach theorem implies that the natural embedding map $x\to\hat{x}$ of E into E'' has $\sigma(E'',E')$-dense range for *any* Banach space E. Now suppose the norm-closed unit ball B in E is weakly compact. The map $x\to\hat{x}$ is a homeomorphism when B and $\hat{B}$ both have the weak topology given by E', so $\hat{B}$ is $\sigma(E'',E')$-compact, hence is also closed in this topology. But it is dense in $B''=\{x'':\|x''\|\leqslant 1\}$, hence $\hat{B}$ and B'' are equal. This implies that $E=E''$.

Conversely, let E be a reflexive Banach space. The embedding $x\to\hat{x}$ is then an isometry and $\hat{B}=B''$. The map $x\to\hat{x}$ is also a homeomorphism if B and B'' both have the weak topology given by E', hence by Theorem 1.1.5 B is $\sigma(E,E')$-compact.

1.1.8. ***Corollary:*** If $1<p<\infty$, a set $S\subseteq L^p(\Omega,\mathscr{F},P)$ is relatively weakly compact iff S is norm-bounded.

We shall discuss a weak compactness criterion for subsets of L^1 in the next section. To conclude the present section, we turn to the Hahn–Banach theorem and some of its consequences:

1.1.9. ***Theorem (Hahn–Banach):*** If M is a subspace of a normed vector space E and if f is a bounded linear functional on M, then f can be extended uniquely to a bounded linear functional F on E such that $||F|| = ||f||$. (This means that $F(x) = f(x)$ for all $x \in M$ and that the norms on M and E, respectively, given by $\sup\{|f(x)| : x \in M, ||x|| \leqslant 1\}$ and $\sup\{|F(x)| : x \in E, ||x|| \leqslant 1\}$ are equal.) Proofs of this fundamental theorem may be found in most texts on functional analysis, e.g. [78], [82].

1.1.10. ***Corollary:*** If M is a subspace of E and $x \in E$, then x lies in the (norm-) closure $\bar{M}$ of M iff all bounded linear functionals which vanish on M also vanish at x. In particular, if $x \in E$ is given, there exists a bounded linear functional f on E of norm 1 such that $f(x) = ||x||$. Consequently E' separates the points of E.

Proof: Exercise.

The connection with weak topologies can be seen from the following geometric form of Theorem 1.1.9 (recall that a set $C \subseteq E$ is *convex* if $\lambda x + (1-\lambda)y \in C$ whenever $x, y \in C$ and $0 < \lambda < 1$.) Proofs can be found in [78], [82].

1.1.11. ***Separation theorem:*** Let A and B be disjoint non-empty convex sets in a Banach space E.

(i) If A is open there exist $f \in E'$ and $\alpha \in \mathbf{R}$ such that $f(x) < \alpha$ on A and $f(x) \geqslant \alpha$ on B.

(ii) If A is compact and B is closed there exists $f \in E'$ and $\alpha < \beta$ in $\mathbf{R}$ such that $f(x) < \alpha$ on A and $f(x) > \beta$ on B.

(We say that the *hyperplane* $\{x : f(x) = \alpha\}$ *separates* A and B, and *strictly separates* them in case (ii).)

1.1.12. ***Corollary:*** If C is a convex subset of a Banach space E, the closure $\bar{C}$ of C in the norm topology on E coincides with the closure $\bar{C}^w$ of C in the weak topology ($\sigma(E, E')$).

Proof: Since the norm topology is stronger than the weak topology on E, $\bar{C} \subseteq \bar{C}^w$. Now suppose $x \in E \backslash \bar{C}$, and apply 1.1.11 with $A = \{x\}$, $B = \bar{C}$ (*Exercise:* prove that the closure of a convex set is convex!) to find $f \in E'$ and $\alpha \in \mathbf{R}$ such that $f(x) < \alpha < f(y)$ for all $y \in \bar{C}$. Then the weak neighbourhood $\{y : f(y) < \alpha\}$ of x does not meet C, hence $x \notin \bar{C}^w$.

1.1.13. ***Corollary:*** If $x \to \hat{x}$ denotes the natural embedding of a Banach space E into its second dual E'', and B and B'' denote the respective closed unit balls, then $\hat{B}$ is $\sigma(E'', E')$-dense in B''.

Proof: If B_1 is the weak*-closure of $\hat{B}$, we want to show that $B_1 = B''$. Now B_1 is convex and since B'' is weak*-closed (by Alaoglu's theorem) $B_1 \subseteq B''$. If $\omega \in B'' \setminus B_1$ we can separate $\{\omega\}$ and B_1 strictly by a continuous linear functional f on E'', hence by an element $x' \in E'$ (why?). So there exist $\alpha \in \mathbf{R}$ and $\varepsilon > 0$ such that $\omega(x') = f(\omega) > \alpha + \varepsilon$ and $x''(x') = f(x'') < \alpha$ for $x'' \in B_1$. Now $\hat{B} \subseteq B_1$, so $x'(x) = \hat{x}(x') < \alpha$ for $x \in B$. If $x \in B$, $-x \in B$, so $|x'(x)| < \alpha$ for all $x \in B$, hence $\|x'\| \leqslant \alpha$. But then $|\omega(x')| \leqslant \alpha \|\omega\| \leqslant \alpha$, since $\omega \in B''$. This contradicts $\omega(x') > \alpha + \varepsilon$, hence no such ω exists in B''. So $B_1 = B''$. This corollary justifies the assumption made in the proof of Theorem 1.1.7.

1.1.14. ***Exercises:***

1. Let $\ell^2(N)$ be the Hilbert space of real sequences (x_n) with $\sum_n x_n^2 < \infty$ and the norm $\|(x_n)\|_2 = (\sum_n x_n^2)^{1/2}$. Show that the sequence (e_m) defined by

$$e_{m,n} = \begin{Bmatrix} 1 \text{ if } m = n \\ 0 \text{ if } m \neq n \end{Bmatrix}$$

converges to 0 in the weak topology but not in the norm topology on $\ell^2(\mathbf{N})$. Is it possible to find such examples in $\mathbf{R}^k$?

2. (See, for example, [82]). Show that the unit ball in a normed vector space E is norm-compact iff E is finite-dimensional.

1.2. Uniform integrability and weak compactness in L^1

1.2.1: We have found simple criteria for weak compactness of subsets of reflexive Banach spaces, and hence for the spaces L^p, $1 < p < \infty$. The space L^1 is not reflexive, however, as the following example shows: let m be Lebesgue measure on $[0, 1]$. Then $L^\infty([0, 1], m)$ contains all continuous functions on $[0, 1]$, and any bounded linear functional ϕ on the subspace $\mathscr{C}[0, 1]$ of continuous functions can be extended to $L^\infty[0, 1]$ without increase in norm, by the Hahn–Banach theorem. So let $x_0 \in [0, 1]$ and set $\phi(f) = f(x_0)$ for $f \in \mathscr{C}[0, 1]$. Clearly $\|f\| \geqslant |\phi(f)|$ so $\|\phi\| \leqslant 1$. If L^1 were reflexive, the extension Φ of ϕ to $L^\infty[0, 1]$ would have the form $\Phi(f) = \int_0^1 fg\,dm$ for some $g \in L^1$. Hence for all $f \in \mathscr{C}[0,1]$ we would have $f(x_0) = \phi(f) = \int_0^1 fg\,dm$, where $g \in L^1$ is fixed. Now consider the function $1_{\{x_0\}} \in \mathscr{L}^\infty$, which is a.s. zero, and approximate from above by piecewise linear functions (f_n) (Fig. 1). So then $(f_n) \subseteq \mathscr{C}[0, 1]$ and $f_n \downarrow 0$ a.s. (m). Introduce the absolutely continuous set function ν by $\nu(E) = \int_E g\,dm$ for Borel sets $E \subseteq [0, 1]$, then $f_n \downarrow 0$ a.s. (ν). By the dominated convergence theorem,

Fig. 1

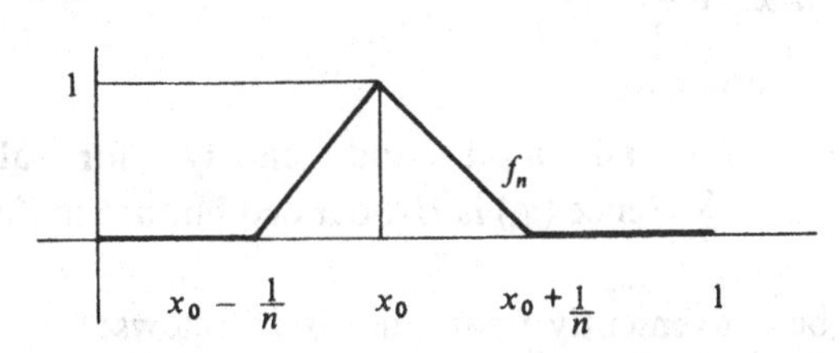

$\int_0^1 f_n \, dv = \int_0^1 f_n g \, dm \to 0$. On the other hand, $\int_0^1 f_n g \, dm = f_n(x_0) = 1$ by construction of f_n. Hence the equation $f(x_0) = \int_0^1 fg \, dm$ for all $f \in \mathscr{C}[0,1]$ cannot be satisfied by any $g \in L^1$. So $\Phi \neq \hat{g}$ for any $g \in L^1$, hence L^1 is not reflexive.

By Theorem 1.1.5, norm-bounded sets in L^1 therefore need not be relatively weakly compact. Our interest in weakly compact subsets of L^1 will become clearer in Chapter 2, where it is shown that this condition suffices for norm-convergence of martingales. The general martingale convergence theorem proved at the end of section 2.6 is really an 'ergodic theorem' in essence; the interested reader can find detailed discussions of mean ergodicity and weak compactness in [29], [27]. Weak compactness is also crucial in our proof of the Doob–Meyer decomposition theorem (Theorem 3.7.1).

The following concept turns out to give an extremely useful weak compactness criterion in L^1, as well as having independent interest. As it is often omitted from a first course in measure theory we give a more detailed discussion here.

1.2.2. ***Definition:*** Let $(\Omega, \mathscr{F}, P)$ be a probability space. A subset $\mathscr{K}$ of $L^1(\Omega, \mathscr{F}, P)$ is *uniformly integrable* if $\sup_{f \in \mathscr{K}} \int_{\{|f| > a\}} |f| \, dP \to 0$ as $a \to \infty$.

1.2.3. ***Examples:***

(1) If $\mathscr{K}$ consists of a single $f \in L^1$, then $\int_{\{|f| > a\}} |f| \, dP \to 0$ as $a \to \infty$ follows at once, since the measure $v(E) = \int_E |f| \, dP$ is P-absolutely continuous. Consequently every *finite* subset of L^1 is uniformly integrable.

(2) An obvious example of a set in L^1 which is *not* uniformly integrable is given by the sequence (f_n), where $f_n(\omega) = n$, on Ω. Note however that (f_n) is not bounded in L^1, i.e. $\sup_n \|f_n\|_1 = +\infty$. But if $\Omega = [0,1]$ and we let

$$g_n(x) = n^2 x \qquad \text{if } 0 \leqslant x \leqslant \frac{1}{n},$$

$g_n(x) = \quad 2n - n^2x \quad \text{if } \frac{1}{n} \leqslant x \leqslant \frac{2}{n},$

$g_n(x) = \quad 0 \qquad \text{otherwise},$

then $\|g_n\|_1 = 2$ for all $n \geqslant 1$ and clearly, for all n, $\int_{\{|g_{2n}|>n\}} |g_{2n}(x)| \, dx = \frac{1}{4}$. Hence (g_n) is L^1-bounded but not uniformly integrable.

The definition can be conveniently reformulated as follows:

1.2.4. ***Proposition:*** $\mathscr{K} \subseteq L^1$ is uniformly integrable iff $\mathscr{K}$ is L^1-bounded and for every $\varepsilon > 0$ we can find $\delta > 0$ with $\sup_{f \in \mathscr{K}} \int_A |f| \, dP < \varepsilon$ whenever $P(A) < \delta$.

Proof: For $f \in L^1_+$, $A \in \mathscr{F}$, $a > 0$ it is clear that

$$\int_A f \, dP = \int_{A \cap \{f \leqslant a\}} f \, dP + \int_{A \cap \{f > a\}} f \, dP \leqslant a \cdot P(A) + \int_{\{f > a\}} f \, dP.$$

Hence

$$\sup_{f \in \mathscr{K}} \int_A |f| \, dP \leqslant a \cdot P(A) + \sup_{f \in \mathscr{K}} \int_{\{|f| > a\}} |f| \, dP \tag{1.1}$$

if the right-hand side is finite. In particular, if $\mathscr{K}$ is uniformly integrable, take $A = \Omega$, then $\sup_{f \in \mathscr{K}} \|f\|_1 \leqslant a + \sup_{f \in \mathscr{K}} \int_{\{|f|>a\}} |f| \, dP < \infty$. Given $\varepsilon > 0$, let $a > 0$ be so large that $\sup_{f \in \mathscr{K}} \int_{\{|f|>a\}} |f| dP < \varepsilon/2$ and let $\delta = \varepsilon/2a$. Hence (1.1) shows that $\sup_{f \in \mathscr{K}} \int_A |f| \, dP < \varepsilon$ if $P(A) < \delta$. Conversely, if $\sup_{\mathscr{K}} \|f\|_1 = M < \infty$ and $\mathscr{K}$ satisfies the above (ε, δ)-condition, let $a = M/\delta$ for given $\varepsilon > 0$. Since $\sup_{\mathscr{K}} P\{|f| > a\} \leqslant M/a = \delta$ we have $\int_{\{|f|>a\}} |f| dP < \varepsilon$ for all $f \in \mathscr{K}$. Hence $\sup_{\mathscr{K}} \int_{\{|f|>a\}} |f| dP \to 0$ as $a \to \infty$.

1.2.5. ***Remark:*** Sometimes the (ε, δ)-condition, which expresses 'uniform absolute continuity', is used as a definition of uniform integrability, without requiring L^1-boundedness. If $(\Omega, \mathscr{F}, P)$ has no atoms, L^1-boundedness follows from this definition, but not in general. Note also that the definition as well as Proposition 1.2.4 remain valid for Bochner-integrable functions, which we shall discuss in section 2.7.

The most useful result for proving that a given set in L^1 is uniformly integrable is the following:

1.2.6. *Proposition:* If $\mathscr{K} \subseteq L^1$ and there is a positive increasing function ϕ defined on $[0,\infty[$ such that $\lim_{t\to\infty}\phi(t)/t = +\infty$ and $\sup_{f\in\mathscr{K}}\int_\Omega(\phi\circ|f|)dP < \infty$ then $\mathscr{K}$ is uniformly integrable.

Proof: Let $M = \sup_{f\in\mathscr{K}}\int_\Omega(\phi\circ|f|)\,dP$, and suppose $\varepsilon > 0$ is given. Put $a = M/\varepsilon$ and then choose t_0 such that $\phi(t)/t \geqslant a$ for $t > t_0$. Hence on the set $\{|f| \geqslant t_0\}$ we have $|f| \leqslant (\phi\circ|f|)/a$ for $f\in\mathscr{K}$. So

$$\int_{\{|f|>t_0\}} |f|\,dP \leqslant \frac{1}{a}\int_{\{|f|>t_0\}} (\phi\circ|f|)\,dP \leqslant \frac{M}{a} = \varepsilon,$$

for all $f\in\mathscr{K}$. We can find t_0 for any given $\varepsilon > 0$, hence $\mathscr{K}$ is uniformly integrable.

1.2.7. *Exercises:*

(1) Show that if $\mathscr{K}$ is L^p-bounded for some $p > 1$, then $\mathscr{K}$ is uniformly integrable. (*Hint:* use Proposition 1.2.6.)

(2) Show that if $\mathscr{K}$ is bounded by some $X\in L^1$ (so that $|f| \leqslant |X|$ for all $f\in\mathscr{K}$) then $\mathscr{K}$ is uniformly integrable. (*Note:* this condition is much stronger than $\sup_{\mathscr{K}}\|f\|_1 < \infty$!)

(3) Verify the claims made in Remark 1.2.5. (For the statement concerning atomless $(\Omega, \mathscr{F}, P)$, use the fact that for all $\varepsilon > 0$ we can find a finite partition of Ω into sets on which P is less than ε.) Deduce that L^1-boundedness in Proposition 1.2.4 can *in general* be replaced by the assumption that the values of all $f\in\mathscr{K}$ are on the atoms of Ω are bounded.

Uniform integrability provides a useful link between convergence in probability and norm-convergence in L^1:

1.2.8. *Theorem:* If $(f_n) \subseteq L^1$ and $f\in L^0$, the following are equivalent:

(i) (f_n) is uniformly integrable and $f_n \to f$ in probability,

(ii) $f\in L^1$ and $f_n \to f$ in L^1-norm.

Proof: (i)⇒(ii): If $f_n \to f$ in probability, some subsequence (f_{n_k}) converges a.s. to f (see Definition 0.1.3). Hence also $|f_{n_k}| \to |f|$ a.s. Now $\mathbf{E}(|f_{n_k}|)$ is bounded since (f_n) is uniformly integrable, hence by Fatou's lemma $\mathbf{E}(|f|) \leqslant \liminf_{k\to\infty}\mathbf{E}(|f_{n_k}|) < \infty$, and so $f\in L^1$. Also for $\varepsilon > 0$, let $A_\varepsilon^n = \{|f_n - f| > \varepsilon\}$, then

$$\|f_n - f\|_1 \leqslant \int_{\Omega\setminus A_\varepsilon^n} |f_n - f|\,dP + \int_{A_\varepsilon^n} |f_n - f|\,dP \leqslant \varepsilon + \int_{A_\varepsilon^n} |f_n|\,dP + \int_{A_\varepsilon^n} |f|\,dP.$$

For all fixed $\varepsilon>0$, $P(A_\varepsilon^n)\to 0$ as $n\to\infty$. Now use Proposition 1.2.4 to deduce that $\int_{A_\varepsilon^n}|f_n|\,dP<\varepsilon$ for n large enough. Finally $\int_{A_\varepsilon^n}|f|\,dP<\varepsilon$ since $f\in L^1$. Hence $f_n\to f$ in L^1-norm.

(ii)⇒(i): If $f_n\to f$ in L^1-norm, then $f_n\to f$ in probability. If $A\in\mathscr{F}$, $\int_A|f_n|dP\leqslant\int_A|f|dP+\|f_n-f\|_1$ for all $n\geqslant 1$, so $\sup_n\int_\Omega|f_n|dP<\infty$ in particular. Given $\varepsilon>0$, take n_0 such that $\|f_n-f\|_1<\varepsilon/2$ for $n\geqslant n_0$. Now consider the *finite* sequence $F=\{f_1,\dots,f_{n_0},f\}$. This is uniformly integrable by Example 1.2.3(1). Hence by Proposition 1.2.4 there is a $\delta>0$ such that $\int_A|g|dP<\varepsilon/2$ whenever $g\in F$ and $P(A)<\delta$. So $\int_A|f_n|dP<\varepsilon$ for all $n\geqslant 1$ if $P(A)<\delta$. Hence, by Proposition 1.2.4 again, (f_n) is uniformly integrable.

Finally, we turn to the main result of this section, the characterisation of weakly compact subsets of L^1. This is by far the 'deepest' theorem proved in this chapter, and it requires several preliminaries:

1.2.9. ***Definition:*** A subset B of a Banach space E is *weakly sequentially compact* if every sequence $(x_n)\subseteq B$ has a subsequence converging weakly to an element of E.

('Relatively' weakly sequentially compact is more accurate, but cumbersome.)

1.2.10. ***Theorem:*** If K is a weakly compact set of a *separable* Banach space E, then K is metrisable in the weak topology and hence weakly sequentially compact.

Proof: By Proposition 1.1.6 the unit ball in E' is weak*-metrisable, and by Alaoglu's theorem (Theorem 1.1.5) it is also weak*-compact. Hence E' is separable for the weak*-topology. So let (x_n') be a weak*-dense sequence in E'. Clearly $x\to|x_n'(x)|$ defines a sequence of seminorms on K, which induce a metrisable topology on K. This topology is weaker than the weak topology on K and, as in Proposition 1.1.6, the topologies coincide. But in a metric space sequential compactness is identical with compactness.

We can now turn to uniformly integrable subsets of L^1. We shall prove:

1.2.11. ***Theorem:*** Let $\mathscr{H}$ be a subset of L^1. The following are equivalent:

(i) $\mathscr{H}$ is uniformly integrable,
(ii) $\mathscr{H}$ is relatively weakly compact in L^1,
(iii) $\mathscr{H}$ is weakly sequentially compact in L^1.

The proof proceeds via the implications (i)⇒(ii)⇒(iii)⇒(i). The equivalence of (ii) and (iii) is part of the Eberlein–Smulian theorem (see [27; Th. V6.1]) and the implication (iii)⇒(i) will require a form of the Vitali–Hahn–Saks theorem ([67], [27]).

Proof that (i)⇒(ii): Embed L^1 in its second dual and let $\hat{\mathscr{K}}$ be the image of $\mathscr{K}$. Consider the weak*-closure $\hat{\mathscr{K}}_w$ of $\hat{\mathscr{K}}$ in $(L^\infty)'$. $\mathscr{K}$ is L^1-bounded and $\mathscr{K}\to\hat{\mathscr{K}}$ is an isometry, so $\hat{\mathscr{K}}_w$ is weak*-bounded and hence (by Alaoglu's theorem) weak*-compact. If $\phi\in\hat{\mathscr{K}}_w$ and $\mu E=\phi(1_E)$ then the uniform integrability of $\mathscr{K}$ ensures (Proposition 1.2.4) that for given $\varepsilon>0$ we can find $\delta>0$ such that $\sup_{f\in\mathscr{K}}\int_E|f|\,dP<\varepsilon/2$ if $P(E)<\delta$. On the other hand, there exists $\psi\in\hat{\mathscr{K}}$ such that for all $h\in L^\infty$, $|\phi(h)-\psi(h)|<\varepsilon/2$, and there is $g\in\mathscr{K}$ such that $\psi(h)=\int_\Omega gh\,dP$. Applying this with $h=1_E$ we see that $P(E)<\delta$ implies $|\mu E|<\varepsilon$. So $\mu\ll P$ and by the Radon–Nikodym theorem there exists $f\in L^1$ such that $\mu E=\int_E f\,dP$ for $E\in\mathscr{F}$. By linearity and continuity we have $\phi(h)=\int_\Omega fh\,dP$ for all $h\in L^\infty$. Hence $\phi=\hat{f}$, and since the weak*-topology on $(L^\infty)'$ induces the topology $\sigma(L^1,L^\infty)$ on L^1, we have shown that $\hat{\mathscr{K}}_w$ is a relatively weakly compact subset of L^1. Hence (i) implies (ii).

Proof that (ii)⇒(iii): First note that any sequence $(f_n)\subseteq L^1(\mathscr{F})$ induces a *countably generated* (e.g. by $\{f_n\leqslant r\}$, $n\geqslant 1$, $r\in\mathbf{Q}$) σ-field $\mathscr{F}_0\subseteq\mathscr{F}$ such that each f_n is $\mathscr{F}_0$-measurable. Since the countable set of $\mathbf{Q}$-valued $\mathscr{F}_0$-simple functions is dense in $L^1(\mathscr{F}_0)$, this is a separable Banach space, and we can apply Theorem 1.2.10 to weakly compact subsets of $L^1(\mathscr{F}_0)$. Now if $\mathscr{K}\subseteq L^1(\mathscr{F})$ is relatively weakly compact and (f_n) is a sequence in $\mathscr{K}$, we can construct $L^1(\mathscr{F}_0)$ as above. The operator $f\to\mathbf{E}(f|\mathscr{F}_0)$ is weakly continuous (see Example 2.2.3(1) for a proof), so the set $\{\mathbf{E}(f|\mathscr{F}_0):f\in\mathscr{K}\}$ is relatively weakly compact in $L^1(\mathscr{F}_0)$. By Theorem 1.2.10 some subsequence (g_{n_k}) of $(\mathbf{E}(f_n|\mathscr{F}_0))_n$ converges weakly to an element g, say, of $L^1(\mathscr{F}_0)$. But since each f_n is $\mathscr{F}_0$-measurable, $g_{n_k}=f_{n_k}$ for all k. Hence for all $h\in L^\infty(\mathscr{F})$, since $\mathbf{E}(\cdot|\mathscr{F}_0)$ is self-adjoint, $\int_\Omega f_{n_k}\cdot h\,dP=\int_\Omega f_{n_k}\mathbf{E}(h|\mathscr{F}_0)dP\to\int_\Omega g\mathbf{E}(h|\mathscr{F}_0)dP=\int_\Omega gh\,dP$ as $k\to\infty$. So (f_{n_k}) converges weakly in $L^1(\mathscr{F})$, as required.

For the final implication (iii)⇒(i) we need a deep result in measure theory, which itself requires an application of the

Baire Category theorem: If X is a complete metric space, then the intersection of every countable collection of dense open subsets of X is dense in X. Consequently a complete metric space cannot be expressed as a countable union of nowhere dense sets. (A set $A\subseteq X$ is *nowhere dense* in X if its closure $\bar{A}$ contains no non-empty open subsets of X.) For a proof of this result, see [77]. We shall have occasion to use it only in the following theorem.

1.2.12. ***Vitali–Hahn–Saks theorem:*** If the sequence (f_n) in L^1 is such that for every $A\in\mathscr{F}$ the sequence $(\int_A f_n\,\mathrm{d}P)$ converges, then (f_n) is uniformly integrable, and converges weakly to some $f\in L^1$.

Proof: The L^1-norm induces a metric d on the set $M\subseteq L^1$ of all indicator functions, $M=\{1_A : A\in\mathscr{F}\}$ (note that this is a set of *equivalence classes* of $\mathscr{F}$-measurable functions) by $d(1_A,1_B)=\|1_A-1_B\|_1=P(A\Delta B)$, where $A\Delta B=(A\setminus B)\bigcup(B\setminus A)$. Since the L^1-limit of a sequence in M is also the a.s. limit of a subsequence, it can take the values 0 and 1 only, and its support is in $\mathscr{F}$. Hence M is closed in L^1, so that (M,d) is a complete metric space. It will sometimes be convenient to identify A and 1_A, making the *measure algebra* $\mathscr{M}$, the quotient of $\mathscr{F}$ by the P-null sets, into a complete metric space with $d(A,B)=P(A\Delta B)$. The map $1_A\to\int_A f\,\mathrm{d}P$ is continuous from M into $\mathbf{R}$: for given $\varepsilon>0$ we can find $\delta>0$ such that $\int_A|f|\,\mathrm{d}P<\varepsilon$ if $P(A)<\delta$. So if $d(A,B)<\delta$ we have $|\int_A f\mathrm{d}P-\int_B f\mathrm{d}P|\leqslant\int_{A\Delta B}|f|\mathrm{d}P<\varepsilon$. Hence each of the sets $G_{m,n}=\{A\in\mathscr{M} : |\int_A f_m\,\mathrm{d}P=\int_A f_n\,\mathrm{d}P|\leqslant\varepsilon/6\}$ is closed in $(\mathscr{M},d)$ and so is $G_p=\bigcap_{m,n\geqslant p}G_{m,n}$ for $p=1,2,\ldots$. But since $(\int_A f_n\,\mathrm{d}P)_n$ converges for all $A\in\mathscr{F}$, the sets G_p make up all of $\mathscr{M}$. By the Baire Category theorem there exists $p\in\mathbf{N}$ such that G_p has non-empty interior. Hence there exists $A\in\mathscr{M}$ and $r>0$ such that $d(A\Delta B)<r$ implies $B\in G_p$.

We can also find $0<\delta<r$ such that $\int_E|f_n|\,\mathrm{d}P<\varepsilon/6$ if $n\leqslant p$ and $P(E)<\delta$. Moreover, $P(E)<\delta$ implies that $d(A,A\bigcup E)\leqslant P(E)<\delta<r$ and $d(A,A\setminus E)<r$, hence $A\bigcup E$ and $A\setminus E$ belong to G_p. But $E=(A\bigcup E)\setminus(A\setminus E)$, so

$$\left|\int_E(f_n-f_p)\mathrm{d}P\right|\leqslant\left|\int_{A\bigcup E}(f_n-f_p)\,\mathrm{d}P\right|+\left|\int_{A\setminus E}(f_n-f_p)\,\mathrm{d}P\right|.$$

Since both sets on the right are in G_p, the right-hand side is less than $\varepsilon/3$ for all $n\geqslant p$. But $\int_E|f_p|\,\mathrm{d}P<\varepsilon/6$, so $|\int_E f_n\,\mathrm{d}P|<\varepsilon/2$ for $n\leqslant p$, whenever $P(E)<\delta$. Apply this with $E\bigcap\{f_n\geqslant 0\}$ and $E\bigcap\{f_n<0\}$ instead of E. Then we have: $\int_E|f_n|\mathrm{d}P<\varepsilon/2$ if $n\geqslant p$ and $P(E)<\delta$. But this also holds for $n\leqslant p$, as we saw above. So given $\varepsilon>0$ there exists $\delta<0$ such that $\sup_n\int_E|f_n|\mathrm{d}P<\varepsilon$ if $P(E)<\delta$. By Proposition 1.2.4 we need only show that (f_n) is bounded in L^1. To do this, construct a finite partition of Ω into atoms $\{A_1,\ldots,A_k\}$ and sets $\{E_1,\ldots,E_j\}$ with $P(E_i)<\delta$ for $i=1,\ldots,j$, where $\delta>0$ is given. (*Exercise:* Show that this can always be done.) As above, we can choose δ so that $\int_E|f_n|\mathrm{d}P\leqslant 1$ for $i\leqslant j$, $n\geqslant 1$. On the other hand, each f_n is constant on each of $A_1,\ldots,A_k$, so $\int_{A_i}|f_n|\mathrm{d}P=|\int_{A_i}f_n\mathrm{d}P|$ for all $n\geqslant 1, i\leqslant k$. But $(\int_{A_j}f_n\mathrm{d}P)_n$ converges for each $i\leqslant k$, hence is bounded. We have therefore shown that $\sup_n\int_\Omega|f_n|\mathrm{d}P<\infty$, so (f_n) is uniformly integrable. Finally, if $E\in\mathscr{F}$ let $\mu(E)=\lim_n\int_E f_n\mathrm{d}P$. This defines an additive set function μ, absolutely continuous relative to P. This is easily seen to imply that μ is σ-additive, hence by the Radon–Nikodym theorem there is $f \in L^1$ with $\mu(E) = \int_E f \, dP = \lim_n \int_E f_n \, dP$. Hence $\int_\Omega f_n h \, dP \to \int_\Omega f h \, dP$ for all $h \in L^\infty$, so $f_n \to f$ weakly.

We return to the proof of Theorem 1.2.11:

Proof that (iii)⇒(i): If $\mathcal{K} \subseteq L^1$ is weakly sequentially compact, $\mathcal{K}$ is norm-bounded. If $\mathcal{K}$ were *not* uniformly integrable, we would (by Proposition 1.2.4) be able to find $\varepsilon > 0$ and a sequence (f_n) in $\mathcal{K}$ and a sequence (E_n) in $\mathcal{F}$ such that $P(E_n) < 1/n$ but $\int_{E_n} |f_n| \, dP \geqslant \varepsilon$ for all $n \geqslant 1$. This means that no subsequence of (f_n) is uniformly integrable.

On the other hand, some subsequence (f_{n_k}) must converge weakly. In particular, for each $A \in \mathcal{F}$, $\int_A f_{n_k} \, dP = \int_\Omega f_{n_k} 1_A \, dP$ defines a convergent sequence in $\mathbf{R}$. But by the Vitali–Hahn–Saks theorem, (f_{n_k}) is then uniformly integrable. This contradiction shows that $\mathcal{K}$ must itself be uniformly integrable.

1.2.13. ***Exercise:*** Prove the following form of the Vitali–Hahn–Saks theorem from 1.2.12: if (P_n) is a sequence of probability measures on $(\Omega, \mathcal{F})$ and for each $F \in \mathcal{F}$, $Q(F) = \lim_n P_n(F)$ exists, then Q is a probability measure on $(\Omega, \mathcal{F})$, and $\sup_n P_n(A)$ decreases to 0 as A decreases to $\varnothing$. (*Hint:* let $P = \sum_n (1/2^n) P_n$ and use the Radon–Nikodym theorem.)

2

Discrete-time martingales

The stability and convergence properties of martingales derive to a large extent from simple combinatorial arguments using properties of the conditional expectation operator. These results are best illustrated in the theory of martingales indexed by the natural numbers. We begin by investigating conditional expectations.

2.1 *The conditional expectation operator*

Fix a probability space $(\Omega,\mathscr{F},P)$, and for $1\leqslant p\leqslant\infty$, denote the usual Lebesgue spaces by $L^p=L^p(\mathscr{F})$. If $\mathscr{G}$ is a sub-σ-field of $\mathscr{F}$, i.e. $\mathscr{G}\subseteq\mathscr{F}$ and $\mathscr{G}$ is a σ-field – note that this means $\Omega\in\mathscr{G}$ – write $L^p(\mathscr{G})$ for $L^0(\mathscr{G})\bigcap L^p$. (Recall that $L^0(\mathscr{G})$ is the set of equivalence classes of $\mathscr{G}$-measurable real functions on Ω.) Then $L^p(\mathscr{G})$ is

(i) A closed subspace of L^p: if $(f_n)\subseteq L^p(\mathscr{G})$ and $\|f_n-f\|_p\to 0$ then some subsequence of (f_n) converges a.s. to f, so that $f\in L^0(\mathscr{G})$.

(ii) A sublattice of L^p which contains 1 and is closed under monotone convergence: this is obvious, for if $f_n\uparrow f$ a.s., where $(f_n)\subseteq L^p(\mathscr{G})$ and $f\in L^p$, then $f\in L^p(\mathscr{G})$ also.

Conversely, these properties characterise subspaces of the form $L^p(\mathscr{G})$: for if $M\subseteq L^p$ has properties (i) and (ii) above, then

$$\mathscr{G}=\{G\in\mathscr{F}:\ 1_G\in M\}$$

is a σ-field and, as in the Monotone Class Theorem 0.1.5, $L^p(\mathscr{G})\subseteq M$. On the other hand, if $f\in M$, $1_{\{f>0\}}=\lim\,(nf^+\wedge 1)\in M$ by (ii), so that $\{f>0\}\in\mathscr{G}$ and, since $1\in M$, also $\{f>\lambda\}=\{f-\lambda 1>0\}\in\mathscr{G}$ for $\lambda\in\mathbf{R}$. Hence $M=L^p(\mathscr{G})$.

The role of these subspaces can be interpreted as follows: think of $f\in L^p$ as an 'observable' of the stochastic system $(\Omega,\mathscr{F},P)$. Stability requirements on our set of observables $M\subset L^p$ would certainly demand that M is a vector space as well as a lattice containing 1 and that M is closed under monotone and L^p-convergence. Thus M would satisfy (i) and (ii) above and so our set of 'observables' would naturally have the form $L^p(\mathscr{G})$, where each f in M is a combination of indicator functions 1_G for G in $\mathscr{G}$. So M consists of mixtures of pure states.

M can also be seen as a subsystem of the 'universe' L^p. The universe can now be studied using knowledge of M and of operators $U: L^p \to M$, which provides a partial analysis of the structure of $(\Omega, \mathscr{F}, P)$, depending on the 'size' of $\mathscr{G}$. These 'prediction operators' U must preserve the vector lattice structure and should not increase the L^p-norm, and also preserve 1. Our analysis of $M = L^p(\mathscr{G})$ then reveals what form such operators U must take.

Assume that U is a positive contraction on L^p (so that $\|U\|_p \leqslant 1$ and $Uf \geqslant 0$ whenever $f \geqslant 0$) and that $U1 = 1$. Then the fixed points of U form a subspace of the above kind: to see this, let $M = \{f \in L^p : Uf = f\}$. M is closed as U is continuous and 1 is in M by assumption. If $f \in M$, $U(f \vee 0) \geqslant (Uf) \vee 0 = f \vee 0$, so that $U(f^+) \geqslant f^+ \geqslant 0$ and therefore $(U(f^+))^p \geqslant (f^+)^p$. On the other hand, $\int_\Omega (U(f^+))^p \mathrm{d}P \leqslant \int_\Omega (f^+)^p \mathrm{d}P$, as U is a contraction. Hence $U(f^+) = f^+$, so that M is a lattice. Finally, if $f_n \uparrow f$ a.s. and $(f_n) \subseteq M$, while $f \in L^p$, then by monotone convergence

$$f_n = Uf_n \xrightarrow{L^p} Uf,$$

hence some subsequence converges to Uf a.s. But $f_n \to f$ a.s., so that $f = Uf$ and $f \in M$. Thus M satisfies (i) and (ii) above, hence has the form $L^p(\mathscr{G})$.

Thus any positive *projection* U on L^p which does not increase the L^p-norm and leaves the function 1 invariant, has *range* of the form $L^p(\mathscr{G})$ for some $\mathscr{G}$. Thus the 'prediction' given by such U is precisely the information gained by 'observing' M (or $\mathscr{G}$).

In particular, given a sub-σ-field $\mathscr{G}$ of $\mathscr{F}$, define the *conditional expectation operator* $\mathbf{E}(\cdot|\mathscr{G})$ first as the orthogonal projection from $L^2(\mathscr{F})$ onto $L^2(\mathscr{G})$, so that

$$\int_\Omega \mathbf{E}(f|\mathscr{G}) g \, \mathrm{d}P = \int_\Omega fg \, \mathrm{d}P$$

for all $g \in L^2(\mathscr{G})$ or, equivalently,

$$\int_G \mathbf{E}(f|\mathscr{G}) \, \mathrm{d}P = \int_G f \mathrm{d}P$$

for all $G \in \mathscr{G}$.

Recall that if H is a Hilbert space and H_0 a closed subspace, the *orthogonal projection* on H_0 is the unique operator T from H onto H_0 satisfying $\langle Tx, y\rangle = \langle x, y\rangle$ for all $y \in H_0$. It is clear that $T^2 = T$ and $\|T\| \leqslant 1$ in this case, and in fact these two properties characterise orthogonal projections – see [38].

The conditional expectation operator $\mathbf{E}(\cdot|\mathcal{G})$ defined above is positive and leaves 1 invariant: for, if $f \geqslant 0$, the function

$$g = 1_{\{\mathbf{E}(f|\mathcal{G})<0\}}$$

belongs to $L^2(\mathcal{G})$, so that

$$0 \leqslant \int fg \, dP = \int_{\{\mathbf{E}(f|\mathcal{G})<0\}} f \, dP = \int_{\{\mathbf{E}(f|\mathcal{G})<0\}} \mathbf{E}(f|\mathcal{G}) \, dP \leqslant 0.$$

Thus $\mathbf{E}(f|\mathcal{G}) \geqslant 0$ a.s., so $\mathbf{E}(\cdot|\mathcal{G})$ is positive. Also, since $1 \in L^2(\mathcal{G})$, $\mathbf{E}(1|\mathcal{G}) = 1$.

Conversely, if T is an orthogonal projection in L^2 satisfying $T \geqslant 0$ and $T1 = 1$, then we have shown above that the range of T has the form $L^2(\mathcal{G})$ for some $\mathcal{G} \subseteq \mathcal{F}$, so that $\mathbf{T} = \mathbf{E}(\cdot|\mathcal{G})$.

The following property of conditional expectations is fundamental. It is intuitively clear if we think of $\mathbf{E}(f|\mathcal{G})$ as the prediction of the values of f given by the information contained in $\mathcal{G}$:

2.1.1. ***Proposition:*** Let $\mathcal{G}$ be a sub-σ-field of $\mathcal{F}$. Then

$$\mathbf{E}(fg|\mathcal{G}) = g . \mathbf{E}(f|\mathcal{G})$$

for all $f \in L^2$, $g \in L^\infty(\mathcal{G})$.

Proof: The function $g.\mathbf{E}(f|\mathcal{G})$ belongs to $L^2(\mathcal{G})$. To show that it is the image of fg under $\mathbf{E}(\cdot|\mathcal{G})$, note that for all $h \in L^2(\mathcal{G})$, $hg \in L^2(\mathcal{G})$ and so by definition

$$\int hg \, \mathbf{E}(f|\mathcal{G}) \, dP = \int hgf \, dP,$$

hence $(g.\mathbf{E}(f|\mathcal{G}) - fg)$ is orthogonal to $L^2(\mathcal{G})$. The result follows.

The notion of conditional expectation is more usually defined via the Radon–Nikodym theorem (see [8], [30], [46]). We shall want to deduce this theorem as a consequence of the convergence theory of martingales. Conditional expectations should be defined for integrable random variables, i.e. L^1-functions. Hence our above definition requires extension from L^2 to L^1. This can be done using the positivity of the operator $\mathbf{E}(\cdot|\mathcal{G})$: first we extend to (equivalence classes of) measurable functions $f: \Omega \to [0, \infty]$. The reason for allowing extended real-valued functions here is that non-negative measurable f need *not* have $\mathbf{E}(f|\mathscr{G})$ finite a.s. We thus *define* $\mathbf{E}(f|\mathscr{G})=\lim_n \mathbf{E}(f\wedge n1|\mathscr{G})$ (pointwise limit, as $(f\wedge n1)_n$ increases in L^2 by positivity of $\mathbf{E}(\cdot|\mathscr{G})$) and this is up to equivalence the only measurable function $\Omega\to[0,+\infty]$ for which the identities

$$\int \mathbf{E}(f|\mathscr{G})g\,\mathrm{d}P=\int fg\,\mathrm{d}P$$

and

$$\mathbf{E}(fh|\mathscr{G})=h\,\mathbf{E}(f|\mathscr{G})$$

are valid whenever g and h are measurable functions $\Omega\to[0,\infty]$ and h is also $\mathscr{G}$-measurable. (Here the value $+\infty$ is permitted for the integrals.) The positive-linearity of $f\to\mathbf{E}(f|\mathscr{G})$ on L^1_+ follows from the uniqueness, monotone convergence and Fatou theorems for $\mathbf{E}(\cdot|\mathscr{G})$ follow easily, and the extension to L^1 follows by letting

$$\mathbf{E}(f|\mathscr{G})=\mathbf{E}(f^+|\mathscr{G})-\mathbf{E}(f^-|\mathscr{G}).$$

The details are left to the reader ([68] has a full account). In summary:

2.1.2. ***Theorem:*** Let $\mathscr{G}$ be a sub-σ-field of $\mathscr{F}$. For all $f\in L^1$ there is a unique element $\mathbf{E}(f|\mathscr{G})$ of $L^1(\mathscr{G})$ such that

$$\int_G \mathbf{E}(f|\mathscr{G})\,\mathrm{d}P=\int_G f\,\mathrm{d}P$$

for all $G\in\mathscr{G}$. The mapping $\mathbf{E}(\cdot|\mathscr{G}):L^1\to L^1(\mathscr{G})$ is called the *conditional expectation operator* and has the following properties:

(i) $\mathbf{E}(\cdot|\mathscr{G})$ is a linear, positive, idempotent contraction and $\mathbf{E}(1|\mathscr{G})=1$.

(ii) (Monotone convergence) If $f_n\uparrow f$ in L^1, then $\mathbf{E}(f_n|\mathscr{G})\uparrow\mathbf{E}(f|\mathscr{G})$ in $L^1(\mathscr{G})$.

(iii) (Fatou) If (f_n), g, h belong to L^1 and if $g\leqslant f_n$ for all n, then $\mathbf{E}(\liminf_n f_n|\mathscr{G})\leqslant\liminf_n \mathbf{E}(f_n|\mathscr{G})$; if $f_n\leqslant h$ for all n, then $\mathbf{E}(\limsup_n f_n|\mathscr{G})\geqslant\limsup_n \mathbf{E}(f_n|\mathscr{G})$.

(iv) (Mean value) If $h\in L^\infty(\mathscr{G})$, then $\mathbf{E}(fh|\mathscr{G})=h\mathbf{E}(f|\mathscr{G})$.

(v) (Tower property) If $\mathscr{G}_1\subset\mathscr{G}$ is a further sub-σ-field of $\mathscr{F}$, then $\mathbf{E}(\mathbf{E}(f|\mathscr{G})|\mathscr{G}_1)=\mathbf{E}(f|\mathscr{G}_1)=\mathbf{E}(\mathbf{E}(f|\mathscr{G}_1)|\mathscr{G})$ a.s.

(i) and (v) follow at once from the operator characterisation above, (iv) is clear by Proposition 2.1.1, and (ii) and (iii) are elementary consequences of the definition and integration theory.

The tower property (v) says simply that the predictions are *consistent*; if we 'know f up to $\mathscr{G}_1$' and 'up to $\mathscr{G}$', then 'ignoring' the additional knowledge given by $\mathscr{G}\backslash\mathscr{G}_1$ we recover our original prediction $\mathbf{E}(f|\mathscr{G}_1)$ out of $\mathbf{E}(f|\mathscr{G})$.

2.1.3. *Exercises:*

(1) Fill in the details of the extension of $\mathbf{E}(\cdot|\mathscr{G})$ from L^2 to L^1.

(2) Show that the definition of conditional expectation reduces to the elementary one if $\mathscr{G}$ is generated by an *atom* G: $\mathbf{E}(f|\mathscr{G})=(1/P(G))\int_G f\,dP$ on G. Deduce that if $\mathscr{G}$ is generated by disjoint atoms $G_1,G_2,\ldots,G_n$ with $\Omega=\bigcup_{i=1}^n G_i$, then $\mathbf{E}(f|\mathscr{G})=(1/P(G_i))\int_{G_i} f\,dP$ on G_i.

(3) Verify that if $f\in L^1(\mathscr{G})$, then $\mathbf{E}(f|\mathscr{G})=f$ a.s.

(4) For f in L^2 define the *variance* of f as $\mathrm{Var}(f)=\mathbf{E}(f^2)-(\mathbf{E}f)^2$. For a sub-$\sigma$-field $\mathscr{G}$ of $\mathscr{F}$ define the *conditional variance* $\mathrm{Var}(f|\mathscr{G})$ by $\mathrm{Var}(f|\mathscr{G})=\mathbf{E}(f^2|\mathscr{G})-(\mathbf{E}(f|\mathscr{G}))^2$. Verify that $\mathrm{Var}(f)=\mathbf{E}(\mathrm{Var}(f|\mathscr{G}))+\mathrm{Var}(\mathbf{E}(f|\mathscr{G}))$. Deduce that the operator $\mathbf{E}(\cdot|\mathscr{G})$ does not increase the variance of f in L^2. Interpret this result heuristically in terms of our 'prediction' model.

(5) (Harder: see [25], [68].) Show that any linear, idempotent, positive contraction on L^1 which preserves 1 is a conditional expectation operator.

2.2. *Jensen's inequality*

One of the most useful inequalities in integration theory relates to the behaviour of integrals under transformation by convex functions:

2.2.1. ***Definition:*** A real function ϕ is *convex* if for all x, y in $\mathbf{R}$ and $0<\lambda<1$,

$$\phi(\lambda x+(1-\lambda)y)\leqslant\lambda\phi(x)+(1-\lambda)\phi(y). \tag{2.1}$$

(Note that this inequality also makes sense if $\phi: C\to\mathbf{R}$, where C is a convex subset of a Banach space and x, y lie in C.)

We can rephrase (2.1) as follows: suppose the random variable X takes values x in $\mathbf{R}$ with probability λ and y with probability $(1-\lambda)$. Then $\mathbf{E}X=\lambda x+(1-\lambda)y$, and (2.1) says that for any convex function ϕ, $\phi(\mathbf{E}X)\leqslant\lambda\phi(x)+(1-\lambda)\phi(y)=\mathbf{E}(\phi(X))$. Jensen's inequality extends this to all X to L^1 and provides a 'conditional' form for this result, where the expectation is replaced by the 'prediction' $\mathbf{E}(X|\mathscr{G})$.

To prove the inequality in general, note first that any convex real function ϕ is continuous and has one-sided derivatives everywhere. Moreover, (2.1) implies that if $x<u<y$ then (see Fig. 2)

$$\frac{\phi(u)-\phi(x)}{u-x}\leqslant\frac{\phi(y)-\phi(x)}{y-x}\leqslant\frac{\phi(y)-\phi(u)}{y-u}.$$

Fig. 2

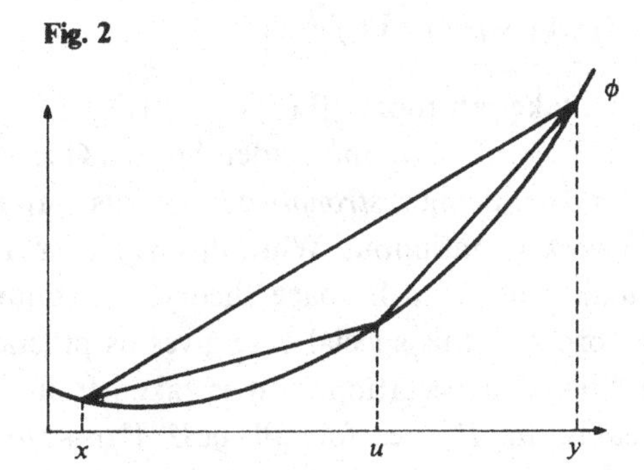

It follows that if $\beta(x)$ denotes the right-hand derivative of ϕ at x, $\phi(y)-\phi(x)\geqslant\beta(x)(y-x)$ for all x and y in $\mathbf{R}$: if $y\geqslant x$, choose u in $]x,y[$, then

$$\frac{\phi(y)-\phi(x)}{y-x}\geqslant\frac{\phi(u)-\phi(x)}{u-x}$$

and as $u\downarrow x$, the right-hand side converges to $\beta(x)$. If $x\geqslant y$, choose $u>x$ and consider

$$\frac{\phi(y)-\phi(x)}{y-x}=\frac{\phi(x)-\phi(y)}{x-y}\leqslant\frac{\phi(u)-\phi(x)}{u-x}\to\beta(x)$$

so that as $(y-x)\leqslant 0$, $\phi(y)-\phi(x)\geqslant(y-x)\beta(x)$ again.

Now apply this inequality to $y=f(\omega)$ and $x=\mathbf{E}(f|\mathscr{G})(\omega):=g(\omega)$, where f in L^1 and ω in Ω are arbitrary. We obtain

$$\phi(f(\omega))\geqslant\phi(g(\omega))+(f-g)(\omega)\beta(g(\omega)) \text{ a.s.}$$

Hence $\phi\circ f\geqslant\phi\circ g+(f-g)\beta\circ g$. Now if $\phi\circ f\in L^1$ we can take conditional expectations and note that $\phi\circ g$ and $\beta\circ g$ are $\mathscr{G}$-measurable since both ϕ and β are Borel-measurable. Therefore $\mathbf{E}(\phi\circ f|)\geqslant\phi\circ g+(\beta\circ g)(\mathbf{E}(f|\mathscr{G})-g)=\phi\circ g$. We have proved the following.

2.2.2. ***Theorem*** **(Jensen's inequality):** If $f\in L^1$ and ϕ is a convex real function such that $\phi\circ f\in L^1$, then for any sub-σ-field $\mathscr{G}$ of $\mathscr{F}$,

$$\phi\circ\mathbf{E}(f|\mathscr{G})\leqslant\mathbf{E}(\phi\circ f|\mathscr{G}) \tag{2.2}$$

Note that if $\mathscr{G}=\{\varnothing,\Omega\}$, (2.2) reduces to: $\phi(\mathbf{E}(f))\leqslant\mathbf{E}(\phi(f))$.

2.2.3. ***Examples:***

(1) $\phi(x)=|x|^p$ is convex for $p\geqslant 1$. Now if $f\in L^p$, then by (2.2)

$$|\mathbf{E}(f|\mathscr{G})|^p = \phi \circ \mathbf{E}(f|\mathscr{G}) \leqslant \mathbf{E}(\phi \circ f|\mathscr{G}) = \mathbf{E}(|f|^p|\mathscr{G}).$$

Integrate over Ω and take pth roots: $||\mathbf{E}(f|\mathscr{G})||_p \leqslant ||f||_p$. For $f \in L^\infty$ the same result may be verified by the reader. So $\mathbf{E}(\cdot|\mathscr{G})$ is an L^p-contraction for $1 \leqslant p \leqslant \infty$. Being a *strongly* continuous map from L^p to itself, it is also *weakly* continuous. While this is an elementary consequence of duality in Banach space theory, the following formulation of the proof in this special case gives us practice in manipulating conditional expectations – it is taken from [83]: suppose $f_\alpha \to f$ weakly in L^p, i.e. for all $g \in L^q$ $(1/p + 1/q = 1)$, $\mathbf{E}(f_\alpha g) \to \mathbf{E}(fg)$. Then

$$\mathbf{E}(g\mathbf{E}(f_\alpha|\mathscr{G})) = \mathbf{E}(\mathbf{E}(g|\mathscr{G}).\mathbf{E}(f_\alpha|\mathscr{G})) =$$
$$\mathbf{E}(f_\alpha.\mathbf{E}(g|\mathscr{G})) \to \mathbf{E}(f.\mathbf{E}(g|\mathscr{G})) = \mathbf{E}(g.\mathbf{E}(f|\mathscr{G}))$$

for all $g \in L^q$, so $\mathbf{E}(f_\alpha|\mathscr{G}) \to \mathbf{E}(f|\mathscr{G})$ weakly, as required. Note that property (iv) of Theorem 2.1.2, and the fact that $\mathbf{E}(\cdot|\mathscr{G})$ is integral-preserving already imply the self-adjointness of $\mathbf{E}(\cdot|\mathscr{G})$.

(2) Convex functions for which Jensen's inequality is often used include: $|x|\log^+|x|$, $(x-\alpha)^+$, $\exp(\alpha x)$, where $\alpha \in \mathbf{R}$. The following application of Jensen's inequality will be important in the sequel:

2.2.4. ***Proposition:*** If $f \in L^1$, the family $\{\mathbf{E}(f|\mathscr{G})\}$, where $\mathscr{G}$ ranges over the sub-σ-fields of $\mathscr{F}$, is uniformly integrable.

Proof: Fix $\lambda > 0$. The set $A_\lambda := \{|\mathbf{E}(f|\mathscr{G})| > \lambda\}$ belongs to $\mathscr{G}$. By (2), $|\mathbf{E}(f|\mathscr{G})| \leqslant \mathbf{E}(|f||\mathscr{G})$, so $\int_{A_\lambda} |\mathbf{E}(f|\mathscr{G})|\,\mathrm{d}P \leqslant \int_{A_\lambda} \mathbf{E}(|f||\mathscr{G})\,\mathrm{d}P = \int_{A_\lambda} |f|\,\mathrm{d}P$. But now $P(A_\lambda) \leqslant (1/\lambda)\int_\Omega |\mathbf{E}(f|\mathscr{G})|\,\mathrm{d}P \leqslant (1/\lambda)||f||_1$, so that $P(A_\lambda)$, and hence $\int_{A_\lambda} |f|\,\mathrm{d}P$, tend to zero *independently* of $\mathscr{G} \subseteq \mathscr{F}$ as $\lambda \to \infty$. Hence the same holds for $\int_{A_\lambda} |\mathbf{E}(f|\mathscr{G})|\,\mathrm{d}P$.

2.3. Martingales: definitions and stability properties

2.3.1. ***Definition:*** We fix a *stochastic base* $(\Omega, \mathscr{F}, P, (\mathscr{F}_n), \mathbf{N})$, where $(\Omega, \mathscr{F}, P)$ is a probability space, $(\mathscr{F}_n)$ is an increasing sequence of sub-σ-fields of $\mathscr{F}$ (indexed by $\mathbf{N}$), where we always assume that $\mathscr{F}_0$ contains all P-null sets and $\mathscr{F}$ is the σ-field generated by $\bigcup_n \mathscr{F}_n$. (The sequence $(\mathscr{F}_n)$ thus represents successive refinements of our predictions, based on knowledge of ever larger sections of our 'universe', reaching certainty, or complete knowledge of the universe, in the limit.) Recall further that we assume all our σ-fields to be complete.

(i) A sequence $X = (X_n)$ of random variables ($\mathscr{F}$-measurable real functions) is an *adapted* stochastic process for this base if, for each

$n \geqslant 0$, X_n is $\mathscr{F}_n$-measurable. X is *predictable* if X_n is $\mathscr{F}_{n-1}$-measurable for $n \geqslant 1$ and X_0 is $\mathscr{F}_0$-measurable.

(ii) $X = (X_n)$ is a *martingale* if $X_n \in L^1$ for each n and $\mathbf{E}(X_m | \mathscr{F}_n) = X_n$ a.s. if $0 \leqslant n \leqslant m$. (Note that this already ensures that X is adapted.)

(iii) An adapted process X is a *supermartingale* (submartingale) if $X_n \in L^1$ for each n and $\mathbf{E}(X_m | \mathscr{F}_n) \leqslant X_n$ a.s. ($\mathbf{E}(X_m | \mathscr{F}_n) \geqslant X_n$ a.s.) if $0 \leqslant n \leqslant m$.

2.3.2. ***Remarks:*** (1) The 'tower property' of the increasing sequence of positive projections $(\mathbf{E}(\cdot | \mathscr{F}_n))$ allows us to simplify the defining relations to $X_n = \mathbf{E}(X_{n+1} | \mathscr{F}_n)$ for $n = 0,1,2,\ldots$ and similarly for sub- and supermartingales.

(2) $(-X_n)$ is a submartingale iff (X_n) is a supermartingale. Consequently most theorems only need proof for one of these classes.

(3) Predictable martingales are constant: since X_{n+1} is $\mathscr{F}_n$-measurable, $X_{n+1} = X_{n+1} \mathbf{E}(1 | \mathscr{F}_n) = \mathbf{E}(X_{n+1} | \mathscr{F}_n) = X_n$ a.s. for all n.

2.3.3. ***Examples:*** (1) Fix $f \in L^1$, and define $X_n = \mathbf{E}(f | \mathscr{F}_n)$ for $n \geqslant 0$. Since $\mathbf{E}(X_{n+1} | \mathscr{F}_n) = \mathbf{E}(\mathbf{E}(f | \mathscr{F}_{n+1}) | \mathscr{F}_n) = \mathbf{E}(f | \mathscr{F}_n) = X_n$ a.s., $X = (X_n)$ is a martingale. We shall see that all uniformly integrable martingales are of this form. The sequence (X_n) represents a *self-consistent* sequence of predictions of the values of f. An obvious example is given by taking successive (e.g. dyadic) partitions of $\Omega = [0,1]$ and approximating f 'symmetrically' by step functions.

(2) Suppose that Q is a further probability measure on $\mathscr{F}$ and that for each n the restriction $Q_n = Q|_{\mathscr{F}_n}$ is absolutely continuous relative to the restriction $P_n := P|_{\mathscr{F}_n}$. Let $X_n = \mathrm{d}Q_n/\mathrm{d}P_n$ for each $n \geqslant 0$. Then if $F \in \mathscr{F}_n$,

$$\int_F X_n \,\mathrm{d}P = Q_n(F) = Q(F) = Q_{n+1}(F) = \int_F = X_{n+1} \,\mathrm{d}P.$$

Thus $\mathbf{E}(X_{n+1} | \mathscr{F}_n) = X_n$, so (X_n) is a martingale.

(3) Suppose $\mathscr{G} \subseteq \mathscr{F}$ is a *separable* sub-σ-field. Then $\mathscr{G}$ is generated by a sequence (G_n) of sets in $\mathscr{F}$. Let $\mathscr{G}_n = \sigma(G_1, \ldots, G_n)$ and let $\mathscr{P}_n$ denote the smallest finite partition of Ω which generates $\mathscr{G}_n$. It is clear that the partitions $(\mathscr{P}_n)$ become finer with increasing n. Now suppose Q is another probability measure on $\mathscr{F}$ and define

$$X_n(\omega) = \sum_{A \in \mathscr{P}_n} \frac{Q(A)}{P(A)} 1_A(\omega),$$

where ratios of the form 0/0 are interpreted as 0. Then $X:=(X_n)_n$ is a positive supermartingale for $(\mathcal{G}_n)_n$: we need only check that

$$\int_A X_{n+1}\,\mathrm{d}P \leqslant \int_A X_n\,\mathrm{d}P$$

for all elements A of the partition $\mathcal{P}_n$. But by definition

$$\int_A X_{n+1}\,\mathrm{d}P = \sum_{\substack{B\in\mathcal{P}_{n+1}\\ B\subseteq A}} \frac{Q(B)}{P(B)}\cdot P(B) \leqslant Q(A) = \int_A X_n\,\mathrm{d}P.$$

Note that the inequality may be strict: the sum does not count $Q(B)$ if $P(B)=0$. Consequently we obtain a *martingale* if each P-null partition set in $\mathcal{P}_n$ is Q-null. This is true in particular if $Q \ll P$ – compare with Example 2.

(4) A more general method of generating σ-fields is given by considering sequences of random variables. If $(Y_n)_n$ is given, defined $\mathcal{G}_n=\sigma(Y_1,\ldots,Y_n)$. In particular, suppose $Y_1,\ldots,Y_n\ldots$ are independent and $\mathbf{E}(Y_i)=0$ for all $i\geqslant 1$. Then the partial sums

$$X_n=\sum_{i=0}^{n} Y_i$$

form a martingale for $(\mathcal{G}_n)$: we have

$$\mathbf{E}(X_{n+1}|\mathcal{G}_n)=\mathbf{E}(Y_1+Y_2+\ldots+Y_n|\mathcal{G}_n)+\mathbf{E}(Y_{n+1}|\mathcal{G}_n)=X_n,$$

since $Y_1+\ldots+Y_n$ is $\mathcal{G}_n$-measurable and Y_{n+1} is independent of $\mathcal{G}_n$.

2.3.4. *Properties:*

(i) A supermartingale $X=(X_n)$ is a martingale iff $\mathbf{E}(X_n)=\mathbf{E}(X_0)$ for all n.

Proof: $\mathbf{E}(X_m|\mathcal{F}_n)\leqslant X_n$ if $m\geqslant n$. We obtain equality iff both sides have the same integral, i.e. $\mathbf{E}(X_m)=\mathbf{E}(X_n)=\ldots=\mathbf{E}(X_0)$.

(ii) If (X_n) and (Y_n) are martingales and $a,b\in\mathbf{R}$ then (aX_n+bY_n) is a martingale. If (X_n) and (Y_n) are supermartingales and a and b are non-negative reals, (aX_n+bY_n) is a supermartingale. Moreover, $(\min(X_n,Y_n))_n$ is a supermartingale. Also $(\max(X_n,Y_n))_n$ is a submartingale, if X, Y are submartingales. The proofs are left to the reader.

(iii) If $X=(X_n)$ is a (sub-)martingale and the real mapping ϕ is (increasing and) convex, and if $\phi\circ X_n$ is integrable for all $n\geqslant 0$, then $\phi\circ X:=(\phi\circ X_n)$ is a submartingale.

Proof: We use Jensen's inequality (Theorem 2.2.2): fix $m \geqslant n \geqslant 0$, then

$$\phi \circ X_n \leqslant \phi \circ \mathbf{E}(X_m | \mathscr{F}_n) \leqslant \mathbf{E}(\phi \circ X_m | \mathscr{F}_n).$$

The first inequality holds in the submartingale case as ϕ is increasing; when X is a martingale this is an equality for any map ϕ. Moreover, $\phi \circ X$ is adapted, since X is. This means in particular that if $X = (X_n)_n$ is a martingale then $(X_n^p)_n$, $(|X_n|)_n$ and $(X_n - a)_n^+$ are submartingales for $a \in \mathbf{R}$, $1 < p < \infty$. (Note however that the result for $(X_n - a)_n^+$ is already implied by (ii) above and the obvious fact that the sequence $(a, a, \ldots a, \ldots)$ is a martingale.)

2.4. Stopping and transforms

The interpretation of a martingale as a 'fair' gambling game leads one to consider how to describe the gambler's strategy for stopping – usually with the aim of maximising his winnings! The strategy should not involve prescience, and – in principle – the gambler could decide not to stop at all. This leads to

2.4.1. ***Definition:*** A *stopping time* relative to $(\mathscr{F}_n)$ is a random variable $T: \Omega \to \mathbf{N} \bigcup \{+\infty\}$ such that $\{T = n\} \in \mathscr{F}_n$ for each n.

The requirement that $\{T = n\}$ belongs to $\mathscr{F}_n$ says that the gambler's decision to stop after the nth game depends only on the history of the game and not on any knowledge of the 'future'. It is trivially equivalent to the condition $\{T \leqslant n\} \in \mathscr{F}_n$, since $\{T \leqslant n\} = \bigcup_{k=0}^n \{T = k\}$. Note, however, that this depends on the countability of the index set and has no equivalent in general for martingales indexed by $\mathbf{R}_+$.

Clearly, any natural number n_0 defines a (constant) stopping time: $T(\omega) = n_0$ a.s. yields that $\{T \neq n_0\} \in \mathscr{F}_0$ (being P-null) and $\{T = n_0\} \in \mathscr{F}_0 \subseteq \mathscr{F}_{n_0}$. If S is any stopping time, so is $S + n_0$. If $X = (X_n)$ is any adapted process, and T is an a.s. *finite-valued* stopping time, the random variable X_T is defined by $X_T(\omega) := X_{T(\omega)}(\omega)$. We can rewrite this in closed form as

$$X_T = \sum_{n \geqslant 0} X_n 1_{\{T = n\}}.$$

It is clear that X_T is measurable, as for any Borel set B,

$$\{X_T \in B\} = \bigcup_{n \geqslant 0} \{X_n \in B\} \bigcap \{T = n\} \text{ belongs to } \mathscr{F}.$$

A stopping time T defines a σ-field $\mathscr{F}_T$ – the events known at time T – by $\mathscr{F}_T := \{A \in \mathscr{F} : A \bigcap \{T = n\} \in \mathscr{F}_n \text{ for all } n\}$. If $T = n_0$ a.s. $\mathscr{F}_T = \mathscr{F}_{n_0}$, so this concept extends the role of the $\mathscr{F}_n$ to $\mathscr{F}_T$. It is any easy exercise to verify that T and X_T (for *adapted* X) are $\mathscr{F}_T$-measurable.

2.4.2. ***Exercise:*** Suppose S and T are stopping times. Verify the following:

(i) $S \vee T$ and $S \wedge T$ are stopping times. (Generalise to sequences.)

(ii) If $A \in \mathscr{F}_S$ then $A \bigcap \{S \leq T\} \in \mathscr{F}_T$. (*Hint:* $A \bigcap \{S \leq T\} \bigcap \{T \leq n\} = \bigcup_{m \leq n} (A \bigcap \{S \leq m\} \bigcup \{T = m\})$.)

(iii) If $S \leq T$ then $\mathscr{F}_S \subseteq \mathscr{F}_T$.

(iv) $\{S < T\}$, $\{S = T\}$, $\{S > T\}$ all belong to $\mathscr{F}_S$ and to $\mathscr{F}_T$.

If X is a martingale, it is natural to ask whether $(X_T, \mathscr{F}_T)$, for varying T, obeys similar relations to $(X_n, \mathscr{F}_n)$. In particular, $(\mathbf{E}X_n)_n$ is constant; what happens to $(\mathbf{E}X_T)$ as T varies?

These questions can be answered very simply by investigating the stability of martingales under *transforms*. These will turn out to be the 'discrete version' of stochastic integrals relative to a martingale and thus have considerable intrinsic importance for us.

2.4.3. ***Definition:*** Suppose $V = (V_n)$ is a predictable process and $X = (X_n)$ is adapted. The *transform of X by V* is the process $Z := V \cdot X$ defined by

$$Z_n = V_0 X_0 + V_1(X_1 - X_0) + \ldots + V_n(X_n - X_{n-1}).$$

2.4.4. ***Theorem:*** If X is a (super-)martingale and V is a (non-negative) predictable process such that $(V \cdot X)_n \in L^1$ for all $n \geq 0$, then the transform $V \cdot X$ is a (super-)martingale.

Proof:

$$\mathbf{E}((V \cdot X)_{n+1} - (V \cdot X)_n | \mathscr{F}_n) = \mathbf{E}(V_{n+1}(X_{n+1} - X_n) | \mathscr{F}_n) = V_{n+1}\mathbf{E}(X_{n+1} - X_n | \mathscr{F}_n)$$

since V is predictable. So this quantity is 0 if X is a martingale, non-positive if X is a supermartingale and V non-negative.

2.4.5. ***Examples:*** (1) Let $Y = (Y_n)$ be as in Example 2.3.3(iv). The martingale $X_n = \sum_{i=0}^n Y_i$ defined there can also be viewed as the transform of the adapted process Y by the predictable process $V = (V_n)$, where $V_n \equiv 1$ for all n.

(2) Consider the gambling model: if the gambler bets 1 unit on each game at times $0, 1, \ldots, n$, with returns $y_0, \ldots, y_n$, his fortune at time n is $Y_n = y_0 + y_1 + \ldots + y_n$. Thus if he bets $V_0, V_1, \ldots, V_n$, instead (these amounts are known *before* playing each game, hence predictable) the return will be $Z_n = V_0 y_0 + V_1 y_1 + \ldots + V_n Y_n = V_0 Y_0 + V_1(Y_1 - Y_0) + \ldots + V_n(Y_n - Y_{n-1}) = (V \cdot Y)_n$. If Y results from a fair game, we would expect the same of Z.

2.4.6. ***An application:*** To illustrate the power of Theorem 2.4.4 we consider a recent result of Pliska [72] which has ramifications in stochastic control theory and in economics. It also serves to introduce the discrete analogue of optional processes and measures acting on them (cf. section 3.4).

Fix a *finite* sequence $(\mathscr{F}_n)_{n \leqslant N}$ with $\mathscr{F}_0$ generated by the P-null sets, and $\mathscr{F}_N = \mathscr{F}$. The optional σ-field Σ_0 is that generated on $\mathscr{U} = \mathbf{T} \times \Omega$ by all adapted processes, where $\mathbf{T} = \{0,1,\ldots,N\}$. Define a measure m on Σ_0 by $m(A) = \mathbf{E}(\sum_{n=0}^N 1_A(n,\omega))$. Let $L^p(\mathscr{U},\Sigma_0,m)$ denote the usual Lebesgue space and let $1/p + 1/q = 1$. Any bounded linear functional f on L^p is then of the form $f(X) = \mathbf{E}(\sum_{n=0}^N X_n Y_n)$ for some $Y \in L^q$.

Let $\mathscr{D}_p$ denote the set of predictable processes in L^p and define $\mathscr{D}_p^\perp = \{Y \in L^q : \mathbf{E}(\sum_{n=0}^N X_n Y_n) = 0$ for all $X \in \mathscr{D}_p\}$ as its orthogonal complement. We show that $\mathscr{D}_p^\perp$ consists of L^p-martingale differences.

For any process $X = (X_n)_{n \leqslant N}$ define $\Delta X_0 = 0$, $\Delta X_n = X_n - X_{n-1}$ for $n = 1,2,\ldots,N$. We want to show that $Y \in \mathscr{D}_p^\perp$ if and only if there exists a martingale $M \in L^p$ with $Y = \Delta M$.

Now if $M \in L^q$ is given, set $Y = \Delta M$. Hence Y gives rise to a linear functional on L^p by $f(X) = \mathbf{E}(\sum_{n=1}^N X_n \Delta M_n)$, since $\Delta M_0 = 0$. Now for $X \in \mathscr{D}_p$, $(X \cdot M)_k = \sum_{n=1}^k X_n \Delta M_n$ is a martingale transform, hence a martingale by Theorem 2.4.4, and $(X \cdot M)_0 = 0$. Hence $f(X) = \mathbf{E}((X \cdot M)_N) = \mathbf{E}((X \cdot M)_0) = 0$. Thus $Y = \Delta M \in \mathscr{D}_p^\perp$.

Conversely if $Y \in \mathscr{D}_p^\perp$ we must have $Y_0 = 0$ (since Y_0 is $\mathscr{F}_0$-measurable, hence constant and $\mathbf{E}(\sum_{n=0}^N X_n Y_n) = 0$ for *all* $X \in \mathscr{D}_p$). The martingale M is then defined by $M_0 = 0$, $M_n = Y_n + M_{n-1}$ for $n = 1,2,\ldots,N$. In fact we need only show that $\mathbf{E}(\Delta M_n | \mathscr{F}_{n-1}) = \mathbf{E}(Y_n | \mathscr{F}_{n-1}) = 0$ for all $n \geqslant 1$. But if $n \geqslant 1$ and $B \in \mathscr{F}_{n-1}$ are given, set $X_n = 1_B$, $X_m = 0$ for $m \neq n$. Then $X \in \mathscr{D}_p$, so $0 = \mathbf{E}(\sum_{k=0}^N X_k Y_k) = \mathbf{E}(1_B Y_n) = \int_B Y_n \, dP$. Since $B \in \mathscr{F}_{n-1}$ was arbitrary, the result follows.

Pliska uses this simple characterisation to discuss the dual variables in optimal control problems, which can be applied, *inter alia*, to consumption–investment problems in economics. A continuous-time analogue of these results would also yield a rich source of applications of 'transforms', i.e. stochastic integrals.

2.4.7. ***Theorem*** (Optional sampling for bounded stopping times): If X is a (super-)martingale and $S \leqslant T$ are bounded stopping times, then $\mathbf{E}(X_T | \mathscr{F}_S) = X_S$, $(\mathbf{E}(X_T | \mathscr{F}_S) \leqslant X_S)$.

Proof: Let $V_n = 1_{\{S < n \leqslant T\}}$. Then $V = (V_n)$ is predictable as

$\{S<n\}\bigcap\{T<n\}^c\in\mathscr{F}_{n-1}$. Also $V_n\geqslant 0$ and $|(V\cdot X)_n|\leqslant|X_0|+|X_1|+\ldots+|X_k|$, where $k\geqslant\|T\|_\infty$. So Theorem 2.4.4 applies to $Z=V\cdot X$, which is a (super-)martingale with $Z_0=0$ and $Z_k=X_T-X_S$. Hence

$$0\overset{(\geqslant)}{=}\mathbf{E}(Z_k)=\mathbf{E}(X_T-X_S).$$

If $A\in\mathscr{F}_S$ we can apply this inequality to the stopping times $S'=1_A S+1_{A^c}k$ and $T'=1_A T+1_{A^c}k$ instead, as $S'\leqslant T'\leqslant k$. Hence

$$\int_A X_T\,dP\overset{(\leqslant)}{=}\int_A X_S\,dP$$

for all $A\in\mathscr{F}_S$, as required.

2.4.8. ***Definition:*** If S is a stopping time and $X=(X_n)$ a process, the *stopped process* X^S is defined by letting $X_n^S=X_{S\wedge n}$ for all $n\geqslant 0$. This is simply the transform $V\cdot X$ where $V_n(\omega)=1$ if $n\leqslant S(\omega)$ and $V_n(\omega)=0$ if $n>S(\omega)$. In other words, $V_n=1_{\{S\geqslant n\}}$. This makes $V=(V_n)$ predictable, as $\{S\leqslant n\}^c=\{S<n\}=\bigcup_{k=0}^{n-1}\{S=k\}\in\mathscr{F}_{n-1}$. Also $X_n^S=(V\cdot X)_n$ is integrable for all n. Finally the following result is immediate from Theorem 2.4.4:

2.4.9. ***Corollary:*** If X is a (super-)martingale and S is a stopping time, then X^S is also a (super-)martingale.

2.4.10. ***Local martingales:*** In fact, all martingale transforms can be generated from martingales by means of stopping times, although this no longer holds in general for continuous-time processes. Consider an adapted process $X=(X_n)$, where we assume $X_0\in L^1$ for simplicity. Then X is a *local martingale* if there is a sequence (T_n) of stopping times, increasing a.s. to $+\infty$, such that each stopped process X^{T_n} is a martingale. We then have the following

2.4.11. ***Proposition:*** Let X be adapted, with $X_0\in L^1$. Then the following are equivalent:

(i) X is a local martingale.

(ii) $\mathbf{E}(|X_{n+1}||\mathscr{F}_n)$ is finite a.s. and $\mathbf{E}(X_{n+1}|\mathscr{F}_n)=X_n$ for all n (where $\mathbf{E}(\cdot|\mathscr{F}_n)$ is extended to L_+^0 as in Proposition 2.1.1),

(iii) X is a martingale transform.

Hence if X is a local martingale with $X_n\in L^1$ for all $n\geqslant 0$, then X is a martingale.

Proof: (i)⇒(ii): If X^{T_n} is a martingale for each n and $T_n \uparrow \infty$ a.s., then for each $A \in \mathscr{F}_m$, $m \geqslant 0$, $A \cap \{T_n > m\} \in \mathscr{F}_m$, while $X_{m+1} 1_{\{T_n > m\}}$ and $X_m 1_{\{T_n > m\}}$ are integrable and, for fixed n, $\int_A X^{T_n}_{m+1} \, dP = \int_A X^{T_n}_m \, dP$, hence

$$\int_{A \cap \{T_n > m\}} X_{m+1} \, dP = \int_{A \cap \{T_n > m\}} X_m \, dP.$$

But as $n \to \infty$, $A \cap \{T_n > m\} \to A$, and (ii) follows, as does the final statement.

(ii)⇒(iii): Write $\Delta X_0 = X_0$, $\Delta X_n = X_{n+1} - X_n$ if $n \geqslant 0$, and set $W_{n+1} = \mathbf{E}(|\Delta X_n| \,|\, \mathscr{F}_n)$, $W_0 = |X_0|$ and finally

$$V_n = \begin{cases} 1/W_n & \text{if } W_n > 0, \\ 0 & \text{if } W_n = 0. \end{cases}$$

Then $V = (V_n)$ is predictable and $Y = V \cdot X$ is well-defined. We now 'invert' this transform using $y_0 = V_0 X_0$, $y_{n+1} = \Delta Y_n = Y_{n+1} - Y_n = V_{n+1} \Delta X_n$: in fact, $\mathbf{E}(y_{n+1} | \mathscr{F}_n) = V_{n+1} \mathbf{E}(\Delta X_n | \mathscr{F}_n) = 0$ by (ii) since also $\mathbf{E}(|y_{n+1}| \,|\, \mathscr{F}_n) \leqslant 1$ a.s. by construction, so that the equation has a meaning. So Y is a martingale and $X = W \cdot Y$ is a transform.

(iii)⇒(i): If $Y = V \cdot X$ is a martingale transform, define (T_n) by $T_n = \inf\{k : |V_{k+1}| > n\}$; note that this is a stopping time as V is predictable. V^{T_n} is then bounded above by n on $\{T_n > 0\}$. Moreover V_0 is constant and $X_0 \in L^1$, so Theorem 2.4.4 applies and $Y^{T_n} = V^{T_n} \cdot X$ is a martingale (see Exercises 2.4.12).

Thus local martingales have little intrinsic interest in the discrete-time framework. In continuous time this concept is the fundamental tool in the extension of stochastic integrals. (See section 4.4.)

2.4.12. ***Exercises:***

(1) Let V be predictable, Y adapted, S a stopping time. Prove that $(V \cdot Y)^S = V \cdot Y^S = V^S \cdot Y$.

(2) Dellacherie and Meyer [19; Ch. VI, p. 31] have an example showing that if $(X_n, \mathscr{F}_n)_{n \geqslant 1}$ is a local martingale, $(X_{n_k}, \mathscr{F}_{n_k})_{k \geqslant 1}$ *need not* be a local martingale. (Similarly, 'discrete skeletons' of continuous-time local martingales are not useful!)

2.5. Skipping and upcrossings

A further application on Theorem 2.4.4 models the situation where the gambler devises a rule for 'skipping' certain games. The transform $V = (V_n)$ thus has $V_0 \equiv 1$ and, for each $n \geqslant 1$, V_n takes only the values 0 and 1.

So $(V\cdot X)_n \in L^1$ for all n and since $\{V_n=1\}\in\mathscr{F}_{n-1}$ we have

$$\begin{aligned}\mathbf{E}(V\cdot X)_n &= \mathbf{E}X_0 + \sum_{k=1}^{n} \int_{\{V_k=1\}} (X_k - X_{k-1})\,\mathrm{d}P \\ &= \mathbf{E}X_0 + \sum_{k=1}^{n} \int_{\{V_k=1\}} \mathbf{E}(X_k - X_{k-1}|\mathscr{F}_{k-1})\,\mathrm{d}P.\end{aligned} \tag{2.3}$$

In particular, if X is a martingale, $\mathbf{E}(V\cdot X)_n = \mathbf{E}X_0$, so that, as we would expect, no skipping rule yields any advantage. Moreover, if X is a submartingale, so that the game is on average favourable to the gambler, the best rule is not to skip at all: since $\mathbf{E}(X_k - X_{k-1}|\mathscr{F}_{k-1}) \geqslant 0$ for all k, (2.3) leads to

$$\mathbf{E}(V\cdot X)_n \leqslant \mathbf{E}X_0 + \sum_{k=1}^{n} \mathbf{E}(X_k - X_{k-1}) = \mathbf{E}X_n. \tag{2.4}$$

We can use (2.4) to derive Doob's estimate of the number of times a submartingale 'upcrosses' an interval:

2.5.1. ***Definition:*** Let $]a,b[$ be any non-empty real interval and suppose $x_1,x_2,\ldots,x_p$ are real numbers. To count the number of times the x_k's move from below a to above b we let $t_0=0$, $t_1=\min(k: k>t_0, x_k\leqslant a)$, $t_2=\min(k: k>t_1, x_k\geqslant b)$, $t_2=\min(k: k>t_2, x_k\leqslant a)$, etc. If the set whose minimum element is sought in defining t_m is empty, we take $t_m=p$. The finite sequence $t_1,t_2,\ldots$ then counts the number U_a^b of *upcrossings* of the open interval $]a,b[$ by $x_1,\ldots,x_p$. (See Fig. 3.)

Now let $Y=(Y_n)$ be an adapted process, S a bounded stopping time. Considering each 'path' $\omega\to Y_n^S(\omega)$ separately as a finite sequence of reals, the above procedure defines a sequence of *stopping times* (T_i). We apply this only in the case when Y is a *non-negative submartingale* and $a=0$. Then

$$T_0\equiv 0,\ T_{2m-1}(\omega)=S(\omega)\wedge\min(k: k>T_{2m-2}(\omega), Y_k(\omega)=0)$$

and

$$T_{2m}(\omega)=S(\omega)\wedge\min(k: k>T_{2m-1}(\omega), Y_k(\omega)\geqslant b).$$

Now $U_0^b(S):\Omega\to\mathbf{N}$ is the random variable counting the number of completed upcrossings of $]0,b[$ by $n\to Y_n^S(\omega)$ for each $\omega\in\Omega$.

In estimating $\mathbf{E}(U_0^b(S))$ we now 'skip' certain $(Y_n - Y_{n-1})$: let

$$V_0\equiv 1,\ V_k(\omega)=\begin{cases}1 \text{ if } T_{2m-1}(\omega)<k\leqslant T_{2m}(\omega) \text{ for some } m\in\mathbf{N},\\ 0 \text{ if } T_{2m}(\omega)<k\leqslant T_{2m+1}(\omega) \text{ for some } m\in\mathbf{N}.\end{cases}$$

Fig. 3

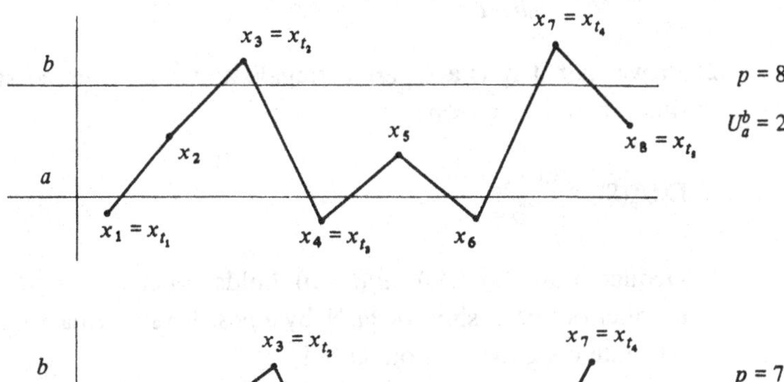

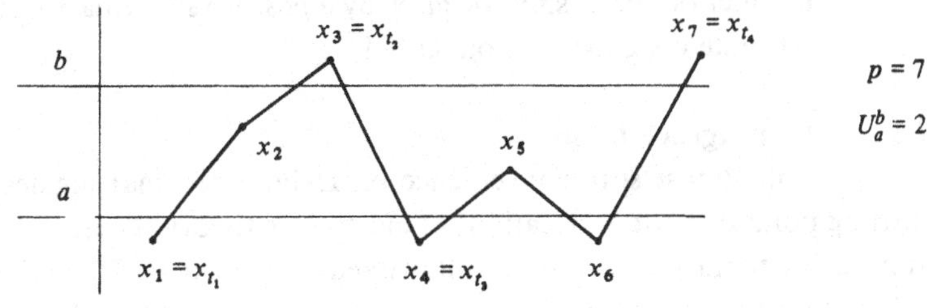

The transform $Z = V \cdot Y$ therefore counts only differences of the form $(Y_{T_{2m}} - Y_{T_{2m-1}})$, that is, the length of each upcrossing of $]0,b[$. Hence $b \cdot U_0^b(S) \leq Z_S$ a.s.

However, applying (2.4) to Y^S instead of X and noting that $(V \cdot Y^S)_n = (V \cdot Y)_n^S$ for all n, we obtain finally

$$b\mathbf{E}(U_0^b(S)) \leq \mathbf{E}Z_S \leq \mathbf{E}Y_S.$$

For we have $Z_S = (V \cdot Y)_S = (V \cdot Y^S)_n$ if $n \geq S(\omega)$, and since S is bounded, $\mathbf{E}Z_S = \mathbf{E}(V \cdot Y^S)_n$ for large enough n. But, since the upcrossings of an arbitrary interval $]a,b[$ by the submartingale X^S correspond to upcrossings of $]0,b-a[$ by the non-negative submartingale Y with $Y_n = (X_n^S - a)^+$, we have proved the famous

2.5.2. ***Doob upcrossing inequality:*** Let X be a submartingale, S a bounded stopping time. For any real interval $]a,b[$

$$\mathbf{E}(U_a^b(S)) \leq \frac{\mathbf{E}(X_S - a)^+}{(b-a)} \leq \frac{\|X_S\|_1 + |a|}{(b-a)}.$$

2.5.3. ***Exercises:***

(1) Prove that the number of *downcrossings* $D_a^b(S)$ of $]a,b[$ by X^S can be estimated as:

$$\mathbf{E}(D_a^b(S)) \leqslant \frac{\mathbf{E}(X_S - b)^+}{b-a}.$$

(2) Prove that if X is a *super*martingale and S a bounded stopping time, then, if $0<a<b$,

$$\mathbf{E}(U_a^b(S)) \leqslant \frac{\mathbf{E}(X_S - a)^-}{b-a}.$$

Deduce that $\mathbf{E}(U_a^b(S)) \leqslant a/(b-a)$ holds whenever $U_a^b(S)$ is the number of upcrossings of $]a,b[$ by a *positive* supermartingale: the estimate depends *only* on $]a,b[$!

2.6. Convergence theorems

The almost sure convergence on L^1-bounded martingales is the starting point of most applications of martingale theory. There are several approaches to the proof of this result; we shall discuss two of these. The first is based on the upcrossing inequality 2.5.2; the second, based on Doob's 'maximal' inequality (Corollary 2.6.9), allows us to prove convergence theorems for sequences of random variables which take their values in a Banach space.

2.6.1. ***Martingale convergence theorem:*** If X is a submartingale and $\sup_n \mathbf{E}(X_n^+) < \infty$, then (X_n) converges a.s. to an integrable limit.

Proof: Fix $\omega \in \Omega$. If the sequence $(X_n(\omega))$ does not converge, there are $a,b \in \mathbf{Q}$ such that

$$\omega \in D_{a,b} := \left\{ \omega : \liminf_{n\to\infty} X_n(\omega) < a < b < \limsup_{n\to\infty} X_n(\omega) \right\}.$$

Applying 2.5.2 to the stopped submartingales X^N, for $N=1,2,\ldots$ and denoting the number of upcrossings of $]a,b[$ by $U_a^b(N)$, we have

$$\mathbf{E}(U_b(N)) \leqslant \frac{\mathbf{E}(X_N - a)^+}{b-a} \leqslant \frac{\sup_n \mathbf{E}(X_n - a)^+}{b-a}.$$

Now the number of upcrossings, U_a^b, of $]a,b[$ by X is the pointwise limit of the increasing sequence $U_a^b(N)$. Hence

$$\mathbf{E}(U_a^b) \leqslant \frac{\sup_n \mathbf{E}(X_n - a)^+}{b-a}$$

by monotone convergence, and this is finite because $(X_n - a)^+ \leqslant X_n^+ + |a|$.

So X upcrosses $]a,b[$ infinitely often only on a P-null set. But then

$$\left\{\omega : \liminf_{n\to\infty} X_n(\omega) < \limsup_{n\to\infty} X_n(\omega)\right\} = \bigcup_{a,b\in\mathbf{Q}} D_{a,b}$$

is also P-null, so (X_n) converges a.s. to some random variable Y.

To see that $Y\in L^1$, note first that (X_n) is in fact *bounded* in L^1: for all n,

$$\begin{aligned}\mathbf{E}(X_n^+) &\leqslant \|X_n\|_1 = \mathbf{E}(X_n^+ + X_n^-)\\ &= 2\mathbf{E}(X_n^+) - \mathbf{E}(X_n) \leqslant 2\mathbf{E}(X_n^+) - \mathbf{E}(X_0)\end{aligned}$$

as X is a submartingale. Hence $\sup_n \|X_n\|_1 < \infty$ if $\sup_n \mathbf{E}(X_n^+) < \infty$. But in that case $|X_n| \to |Y|$ a.s. and, by Fatou's lemma,

$$\|Y\|_1 = \mathbf{E}\left(\lim_n |X_n|\right) \leqslant \liminf_{n\to\infty} \mathbf{E}(|X_n|) \leqslant \sup_n \|X_n\|_1$$

as required.

The above theorem was stated and proved for *sub*martingales because our approach to the upcrossing inequality is best suited to submartingales. We now revert to our usual practice of stating results primarily for *super*martingales. Using $-X$ instead of X above gives the following immediate

2.6.2. ***Corollary:*** Let X be a supermartingale with $\sup_n \mathbf{E}(X_n^-) < \infty$. Then (X_n) converges a.s. to an integrable limit. In particular, each positive supermartingale converges a.s.

2.6.3. ***Corollary:*** Each L^1-bounded martingale converges a.s. to an integrable limit.

We shall see that in many cases the a.s. limit of a (super-)martingale $X = (X_n)$ *closes* the (super-)martingale: in general, if there exists $Y\in L^1$ such that

$$\mathbf{E}(Y|\mathscr{F}_n) \overset{(\leqslant)}{=} X_n$$

for all n, we say that (X_n) is *closed by* Y. Note, however that the choice of such an L^1-function Y may, in general, be non-unique (see [19; Ch. II, p. 6]. The restriction $\mathscr{F} = \sigma(\bigcup_n \mathscr{F}_n)$ we have imposed will resolve this ambiguity in our set-up.

Nonetheless, care is needed: if $X = (X_n)$ is a positive martingale, for example, X is closed by 0 *as a supermartingale*, since $X_n \geqslant \mathbf{E}(0|\mathscr{F}_n)=0$ for all n, but *not*, in general, as a martingale.

We shall now usually denote the a.s. limit of (X_n) by X_∞ and (sometimes) $\mathscr{F}$ by $\mathscr{F}_\infty$. Thus if X_∞ closes (X_n) as a (super)-martingale, $(X_n, \mathscr{F}_n)_{n\in\bar{\mathbf{N}}}$ is a (super-)martingale. (Recall that $\bar{\mathbf{N}}=\mathbf{N}\bigcup\{\infty\}$.)

2.6.4. ***Theorem:*** Suppose X is a uniformly integrable supermartingale. Then $X_\infty = \text{a.s.}\lim_n X_n$ exists and closes (X_n). Moreover, $X_n \to X_\infty$ in L^1-norm.

Proof: X is L^1-bounded as it is uniformly integrable (see Proposition 1.2.4). So X_∞ exists by Corollary 2.6.2. Now (X_n) is uniformly integrable and $X_n \to X_\infty$ a.s., so by Theorem 1.2.8, $X_n \to X_\infty$ in L^1-norm. If $m>n$ and $A\in\mathscr{F}_n$ we have

$$\int_A X_m \, \mathrm{d}P \leqslant \int_A X_n \, \mathrm{d}P.$$

Letting $m\to\infty$ we obtain

$$\int_A X_\infty \, \mathrm{d}P = \lim_m \int_A X_m \, \mathrm{d}P \leqslant \int_A X_n \, \mathrm{d}P$$

as required.

2.6.5. ***Remark:*** Since the operators $\mathbf{E}(\cdot|\mathscr{F}_n)$ extend to the space of positive measurable functions on Ω and Fatou's lemma remains valid, we can also show that X_∞ closes (X_n) when (X_n) is a *positive* supermartingale; for $n \geqslant p$ we have

$$\mathbf{E}(\inf_{m\geqslant n} X_m|\mathscr{F}_p) \leqslant \mathbf{E}(X_n|\mathscr{F}_p) \leqslant X_p,$$

so by Fatou's lemma $\mathbf{E}(X_\infty|\mathscr{F}_p) \leqslant X_p$.

We can now summarise our results for martingales:

2.6.6. ***Theorem:*** Let $X=(X_n)$ be a martingale. The following statements are equivalent:

(i) X is uniformly integrable.
(ii) (X_n) converges a.s. and in L^1-norm.
(iii) (X_n) is closed as a martingale by some $Y\in L^1$.

Proof: (i)⇒(ii): this follows from Theorem 2.6.4 applied to (X_n).

(ii)⇒(iii): as in the proof of Theorem 2.6.4, $X_n \to X_\infty$ in L^1-norm, so

$$\int_A X_\infty \, dP = \int_A X_n \, dP$$

for all $A \in \mathscr{F}_n$, $n \geqslant 0$. Hence (X_n) is closed by X_∞.

(iii)⇒(i): If $X_n = \mathbf{E}(Y|\mathscr{F}_n)$ for all $n \geqslant 0$, $X_n = (X_n)$ is uniformly integrable by Proposition 2.2.3.

2.6.7. ***Remark:*** If $Y \in L^1$ is given, the uniformly integrable martingale (X_n) defined by $X_n = \mathbf{E}(Y|\mathscr{F}_n)$ will converge a.s. to Y: we know that $X_\infty = \text{a.s.}\lim_n X_n$ exists. Let $\mathscr{M}$ be the collection of sets $A \in \mathscr{F}$ such that

$$\int_A Y dP = \int_A X_\infty \, dP.$$

Since $\mathbf{E}(Y|\mathscr{F}_n) = X_n = \mathbf{E}(X_\infty|\mathscr{F}_n)$ for all n, $\mathscr{M}$ includes all the $\mathscr{F}_n$. But $\mathscr{M}$ is clearly a monotone class, so by 0.1.5, $\mathscr{M} = \mathscr{F}$. So $Y = X_\infty$.

Our second proof of the convergence theorem relies on the following simple inequality, which is a form of Doob's 'maximal' inequality and has many varied applications. The name reflects analogies with the maximal ergodic theorem and the Hardy–Littlewood maximal inequality of potential theory – see [27], [35]. Our application of the inequality to convergence theorems is very close to the methods employed in the proof of the ergodic theorems – see [43], [47], [48], [69] for further analogies with ergodic theory.

2.6.8. ***Theorem:*** Let X be a positive submartingale, $\lambda > 0$. For $N \in \mathbf{N}$ write $X_N^* := \max_{n \leqslant N} X_n$, and let $X^* := \sup_n X_n$. Then we have

$$P(X_N^* \geqslant \lambda) \leqslant \frac{1}{\lambda} \int_{\{X_N^* \geqslant \lambda\}} X_N \, dP \text{ for each } N \in \mathbf{N} \tag{2.5}$$

and hence

$$\lambda P(X^* \geqslant \lambda) \leqslant \frac{1}{\lambda} \sup_n \mathbf{E}(X_n). \tag{2.6}$$

Proof: Define the bounded stopping time T_N by setting

$$T_N(\omega) = N \wedge \min(n \leqslant N : X_n(\omega) \geqslant \lambda).$$

By optional sampling (Theorem 2.4.7.)

$$\mathbf{E}(X_N) \geqslant \mathbf{E}(X_{T_N}) = \int_{(X_N^* \geqslant \lambda)} X_{T_N} \, \mathrm{d}P + \int_{(X_N^* < \lambda)} X_{T_N} \, \mathrm{d}P$$

$$\geqslant \lambda P(X_N^* \geqslant \lambda) + \int_{(X_N^* < \lambda)} X_N \, \mathrm{d}P,$$

which proves (2.5).

Now let $N \to \infty$ and note that $\{X^* \geqslant \lambda\} = \bigcup_N \{X_N^* \geqslant \lambda\}$ so that

$$P(X^* \geqslant \lambda) \leqslant \frac{1}{\lambda} \sup_n \mathbf{E}(X_n).$$

Since $|X|$ is a submartingale if X is a martingale, we have:

2.6.9. ***Corollary:*** If X is a martingale and $\lambda > 0$, then $P(\sup_n |X_n| \geqslant \lambda) \leqslant 1/\lambda \sup_n \|X_n\|_1$. Hence if X is also L^1-bounded, $X^* := \sup_n |X_n|$ is a.s. finite.

The random variable X_N^* shows us with hindsight where the largest 'returns' in our gambling occurred. In a submartingale (favourable game) the inequality shows how the increasing sequence $\mathbf{E}X_n$ provides an upper bound to the gambler's 'regret' at missed opportunities to stop – at least on average. Similar inequalities hold for supermartingales:

2.6.10. ***Theorem:*** Let $(X_n, \mathscr{F}_n)_{n \in N}$ be a supermartingale, $\lambda > 0$. Then we have

$$\lambda P\left(\sup_n X_n \geqslant \lambda\right) \leqslant \mathbf{E}(X_0) + \sup_n \mathbf{E}(X_n^-), \tag{2.7}$$

$$\lambda P\left(\inf_n X_n \leqslant -\lambda\right) \leqslant \sup_n \mathbf{E}(X_n^-), \tag{2.8}$$

and hence

$$\lambda P\left(\sup_n |X_n| \geqslant \lambda\right) \leqslant 3 \sup_n \|X_n\|_1. \tag{2.9}$$

Like the maximal inequality, (2.7) and (2.8) are immediate consequences of the optional sampling theorem (Theorem 2.4.7). We prove (2.7), leaving the (analogous) proof of (2.8) to the reader.

Define $T(\omega) = \inf\{n : X_n(\omega) \geqslant \lambda\}$, $T_k = T \wedge k$. The sequence (T_k) of stopping time is increasing, so by (Theorem 2.4.7) $\mathbf{E}(X_{T_k}) \leqslant \mathbf{E}(X_0)$. Now if $X_{T_k} < \lambda$,

$T_k = k$, hence we have, for each k,

$$\lambda P\left(\sup_{n \leqslant k} X_n \geqslant \lambda\right) + \int_{(\sup_{n \leqslant k} X_n < \lambda)} X_k \, dP \leqslant \mathbf{E}(X_{T_k}) \leqslant \mathbf{E}(X_0)$$

so that:

$$\lambda P\left(\sup_{n \leqslant k} X_n \geqslant \lambda\right) \leqslant \mathbf{E}(X_0) - \int_{(\sup_{n \leqslant k} X_n < \lambda)} (X_k^+ - X_k^-) \, dP$$

$$\leqslant \mathbf{E}(X_0) + \sup_{n \in N} \mathbf{E}(X_n^-).$$

Letting $k \to \infty$ yields (2.7).

We use the maximal inequality (2.5) to deduce a further result of Doob which is of fundamental importance in stochastic integration: if $\mathcal{M}^p$ denotes the vector space of L^p-bounded martingales, $1 \leqslant p < \infty$, we can define a *norm* on $\mathcal{M}^p$ by

$$||X||_p := \sup_n ||X_n||_p.$$

Now if $1 < p < \infty$, L^p is reflexive, so norm-bounded sets are relatively weakly compact, hence uniformly integrable (see Theorem 1.2.11). So $X_n \to X_\infty$ in L^p-norm and since $(|X_n|^p)$ is a submartingale, $||X||_p = ||X_\infty||_p$. Doob's inequality implies that this norm is actually *equivalent* to the norm on $\mathcal{M}^p$ defined by $||X|| = ||X^*||_p$.

2.6.11. ***Theorem:*** Let X be a positive submartingale and $1 < p < \infty$. Then $X^* \in L^p$ iff X is L^p-bounded, and in that case we have

$$||X^*||_p \leqslant \frac{p}{p-1} \sup_n ||X_n||_p. \tag{2.10}$$

Proof: If $X^* \in L^p$, then since $X_n \leqslant X^*$ a.s. for all $n \geqslant 0$, $\sup_n ||X_n||_p \leqslant ||X^*||_p$, so X is L^p-bounded.

Conversely, suppose X is L^p-bounded. Then $X_\infty = \lim_n X_n$ closes X as a submartingale and $X_n \to X_\infty$ in L^p-norm. In particular $\mathbf{E}(X_\infty | \mathscr{F}_n) \geqslant X_n$ for $n = 1, 2, \ldots, (N-1)$, where $N \in \mathbf{N}$ is arbitrary. Redefining the stopping time T_N in the proof of Theorem 2.6.8 by using X_∞ in place of X_N – note that T_N still has finite range: $\{0, 1, \ldots, N-1, \infty\}$ – we obtain

$$\lambda P(X_N^* \geqslant \lambda) \leqslant \int_{(X_N^* \geqslant \lambda)} X_\infty \, dP \leqslant \int_{(X^* \geqslant \lambda)} X_\infty \, dP \text{ for all } N.$$

But

$$P(X^* \geqslant \lambda) = \sup_n P(X_N^* \geqslant \lambda),$$

so we have, finally,

$$\lambda P(X^* \geqslant \lambda) \leqslant \int_{(X^* \geqslant \lambda)} X_\infty \, \mathrm{d}P. \tag{2.11}$$

To apply (2.11) we note first that by Fubini's theorem, if ϕ is any increasing right-continuous function $\mathbf{R}_+$ with $\phi(0)=0$, then

$$\mathbf{E}(\phi \circ X^*) = \int_\Omega \left(\int_0^{X^*(\omega)} \mathrm{d}\phi(\lambda) \right) \mathrm{d}P(\omega) = \int_\Omega \left(\int_0^\infty 1_{(X^* \geqslant \lambda)} \mathrm{d}\phi(\lambda) \right) \mathrm{d}P(\omega)$$

$$= \int_0^\infty \left(\int_\Omega 1_{(X^* \geqslant \lambda)} \mathrm{d}P \right) \mathrm{d}\phi(\lambda) = \int_0^\infty P(X^* \geqslant \lambda) \, \mathrm{d}\phi(\lambda)$$

Using this calculation with $\phi(\lambda) = \lambda^p$ and applying (2.11) we have

$$\|X^*\|_p^p = \int_0^\infty P(X^* \geqslant \lambda) \, \mathrm{d}(\lambda^p) \leqslant \int_0^\infty \left(\frac{1}{\lambda} \int_{(X^* \geqslant \lambda)} X_\infty \, \mathrm{d}P \right) \mathrm{d}(\lambda^p)$$

$$= \int_\Omega X_\infty(\omega) \left(\int_0^{X^*(\omega)} \frac{\mathrm{d}(\lambda^p)}{\lambda} \right) \mathrm{d}P(\omega) = \int_\Omega X_\infty(\omega) \left(\int_0^{X^*(\omega)} p\lambda^{p-2} \, \mathrm{d}\lambda \right) \mathrm{d}P(\omega)$$

$$= \frac{p}{p-1} \int_\Omega X_\infty \cdot X^{*p-1} \, \mathrm{d}P \leqslant \frac{p}{p-1} \|X_\infty\|_p \|X^{*p-1}\|_q,$$

where the final step follows by Hölder's inequality. But since

$$\frac{1}{p} + \frac{1}{q} = 1, \; \|X^{*p-1}\|_q = \left(\int_\Omega X^{*(p-1)q} \, \mathrm{d}P \right)^{1/q} = \|X^*\|_p^{p/q}$$

So we obtain

$$\|X^*\|_p \leqslant \frac{p}{p-1} \|X_\infty\|_p$$

upon dividing by $\|X^*\|_p^{p/q}$.

Note, however, that this argument is circular as it stands: e.g. in proving that $||X^*||_p$ is finite we have divided by $||X^*||_p^{p/q}$.

This problem can be overcome by replacing X^* by $Y_n = X^* \wedge n$, which still satisfies (2.11), and then letting n tend to infinity.

Applying Theorem 2.6.11 to $|X|$ if $X \in \mathcal{M}^p$ we see that the norms

$$||X||_p = \sup_n ||X_n||_p = ||X_\infty|| \text{ and } ||X|| = ||X^*||_p$$

are equivalent, since

$$||X||_p \leqslant ||X|| \leqslant \frac{p}{p-1} ||X||_p.$$

For $p=1$ there is no corresponding result and the subset $\mathcal{H}^1$ of uniformly integrable martingales in L^1 for which $X^* \in L^1$ has been studied extensively – see [49], [66].

2.6.12. ***Exercises:***

(1) Using inequality (2.5) on the submartingales (X_n^2), prove that if $X_n = Y_0 + Y_1 + \ldots + Y_n$, where $Y_0, Y_1, \ldots$ are independent, square-integrable random variables of mean zero, then

$$P\left(\max_{n \leqslant N} |X_n| \geqslant \lambda\right) \leqslant \frac{1}{\lambda^2} \mathbf{E}\left(\sum_{i=0}^{N} Y_i^2\right).$$

This inequality is due to Kolmogorov – see [8].

(2) (Harder – see [19] for references.) Apply the method of proof of Theorem 2.6.11 to the function $\phi(\lambda) = (\lambda - 1)^+$, to obtain

$$\mathbf{E}(X^*) \leqslant \frac{e}{e-1}\left(1 + \sup_n \mathbf{E}(X_n \log^+ X_n)\right).$$

This inequality has also been closely studied and again has a counterpart in ergodic theory.

Finally, we prove a general result on *norm-convergence* which will be used later. Note that any upwards filtering set T (i.e. if $s, t \in T$ there exists $u \in T$ with $s \leqslant u, t \leqslant u$) can serve as index set for martingales: given an upwards filtering set T and an increasing family $(\mathcal{F}_t)$ of sub-σ-fields of $\mathcal{F}$, we define an $(\mathcal{F}_t)$-*martingale* as an adapted family $(X_t)_T$ of L^1-functions such that $\mathbf{E}(X_t | \mathcal{F}_s) = X_s$ if $s, t \in T$, $s \leqslant t$. We then have the following general result:

2.6.13. ***Theorem:*** If $X=(X_t)_T$ is an $(\mathscr{F}_t)$-martingale and *either* X is uniformly integrable and $p=1$, *or* X is L^p bounded and $1<p<\infty$, then (X_t) converges in L^p-norm to some $X_\infty \in L^p$ which closes X, i.e.

$$\mathbf{E}(X_\infty|\mathscr{F}_t)=X_t \text{ for all } t\in T.$$

Proof: This result really belongs to functional analysis. We therefore employ the usual duality notation: if $f\in L^p$, $g\in L^q$, then $\langle f,g\rangle=\mathbf{E}(fg)$. Recall that the operator $\mathbf{E}(\cdot|\mathscr{F}_t)$ is strongly (and hence weakly) continuous on L^p, $1\leqslant p<\infty$.

Now as X is relatively weakly compact in L^p under either of the hypotheses of the theorem, there is a weak cluster point $X_\infty \in L^p$. By the weak continuity of $\mathbf{E}(\cdot|\mathscr{F}_t)$, $\mathbf{E}(X_\infty|\mathscr{F}_t)$ is a weak cluster point of $\{\mathbf{E}(X_s|\mathscr{F}_t)\}_s$. On the other hand, $\mathbf{E}(X_s|\mathscr{F}_t)=X_t$ for $s\geqslant t$, hence $\mathbf{E}(X_\infty|\mathscr{F}_t)=X_t$ for all $t\in \mathbf{T}$.

Furthermore, X_∞ is in the strongly closed convex hull of $(X_t)_T$ – since for convex sets the weak and strong closures coincide – so for given $\varepsilon>0$ we can find a convex combination $\sum_{i=1}^n \alpha_i x_{t_i}$ such that $\|X_\infty-\sum_{i=1}^n \alpha_i X_{t_i}\|_p<\varepsilon$. Choosing $s\in T$ which majorizes all the t_i we then have

$$\|X_\infty - X_s\|_p \leqslant \left\|X_\infty - \sum_{i=1}^n \alpha_i X_{t_i}\right\|_p + \left\|\mathbf{E}\left(\left(\sum_{i=1}^n \alpha_i X_{t_i} - X_\infty\right)\middle|\mathscr{F}_s\right)\right\|_p$$

since $X_s=\mathbf{E}(X_\infty|\mathscr{F}_s)$. But each X_{t_i} is $\mathscr{F}_s$-measurable and $\mathbf{E}(\cdot|\mathscr{F}_s)$ is an L^p-contraction, so the second term is no greater than the first and $\|X_\infty - X_s\|_p<2\varepsilon$ for all large enough $s\in T$. Hence (X_t) converges to X_∞ in L^p-norm, as required.

2.7. Convergence of vector-valued martingales

In this section, which may be omitted at a first reading, we consider processes which take their values in a Banach space $(B,\|\cdot\|)$. Some care is then needed with the definitions of L^p-spaces: it is natural to define $f:\Omega\to B$ to be *measurable* if $f^{-1}(C)\in\mathscr{F}$ for each Borel set $C\subseteq B$. Unfortunately, the set of such f need not even be closed under addition (see [78]). We therefore define $f:\Omega\to B$ to be *strongly measurable* if it is measurable and for each $\varepsilon>0$ there is a simple measurable function f_ε such that $P(\|f-f_\varepsilon\|>\varepsilon)<\varepsilon$. (It is not hard to see [78] that this is equivalent to f being *almost separably valued*, i.e. there is a separable closed subspace $B_0\subseteq B$ with $P(f\in B_0)=1$.) We then define $L_B^0 := L^0(\Omega,\mathscr{F},P,B)$ as the space of equivalence classes ($f\sim g$ iff $f=g$ a.s.) of strongly measurable $f:\Omega\to B$. The Banach spaces $L_B^p := \{f\in L_B^0 : \int_\Omega \|f\|^p \, dP<\infty\}$ are then well-defined for $p\geqslant 1$ and the simple measurable functions form a dense subspace $\mathscr{E}$ of L_B^p for $1\leqslant p\leqslant\infty$.

This allows us to define a conditional expectation operator: if $g=$

$\sum_{i=1}^{n}\alpha_i 1_{B_i}$ is a simple measurable function: $\Omega \to B$, and $\mathscr{G}$ is a sub-σ-field of $\mathscr{F}$, let $P_{\mathscr{G}}(g)=\sum_{i=1}^{n}\alpha_i \mathbf{E}(1_{B_i}|\mathscr{G})$. Clearly $P_{\mathscr{G}}(g)$ is $\mathscr{G}$-measurable, and in fact $P_{\mathscr{G}}(g)\in L_B^1(\Omega,\mathscr{G},P)$. The map $P_{\mathscr{G}}$ is a contraction: $\mathscr{E}\to L_B^1(\Omega,\mathscr{G},P)$ so extends to L_B^1 by continuity. It is easy to see that the extended map, which we denote by $\mathbf{E}_B(\cdot|\mathscr{G})$, has norm 1, leaves each $f\in L_B^1(\Omega,\mathscr{G},P)$ invariant and satisfies $\int_A \mathbf{E}_B(f|\mathscr{G})\mathrm{d}P=\int_A f\,\mathrm{d}P$ for all $f\in L_B^1$ and $A\in\mathscr{G}$. It is also clear that for each continuous affine map θ on L_B^1, $\mathbf{E}_B(\theta(f)|\mathscr{G})=\theta(\mathbf{E}_B(f|\mathscr{G}))$.

The only property of the conditional expectation operator which needs separate consideration is Jensen's inequality.

A real-valued function ϕ on a topological space X is *lower semi-continuous* if $\{x:\phi(x)\leqslant r\}$ is closed in X for all real r. We shall show that any lower semi-continuous convex function ϕ defined on a convex subset of a Banach space B can be approximated from below (on separable subsets) by a sequence of continuous affine functions. This allows us to state the following form of Jensen's inequality.

2.7.1. ***Theorem:*** If ϕ is a lower semi-continuous convex function on a non-empty closed convex subset C of the Banach space B, and $f\in L_B^1$ satisfies $\phi\circ f\in L^1$, then for any sub-σ-field $\mathscr{G}$ of $\mathscr{F}$

$$\phi(\mathbf{E}_B(f|\mathscr{G}))\leqslant \mathbf{E}(\phi\circ f|\mathscr{G})\text{a.s.}$$

Proof: Choose a separable subset $C_0\subseteq C$ containing almost all values of f and $\mathbf{E}(f|\mathscr{G})$. We prove below that there is a sequence (θ_i) of continuous affine functions such that $\phi(x)=\sup_i\theta_i(x)$ for all $x\in C_0$. But $\theta_i(\mathbf{E}_B(f|\mathscr{G}))=\mathbf{E}(\theta_i(f)|\mathscr{G})\leqslant \mathbf{E}(\phi(f)|\mathscr{G})$a.s. Hence also $\phi(\mathbf{E}_B(f)|\mathscr{G})\leqslant\mathbf{E}(\phi\circ f|\mathscr{G})$.

Applying this result with $\phi(x)=||x||$, or $\phi(x)=||x||^p$, we have

2.7.2. ***Corollary:*** The map $f\to\mathbf{E}_B(f|\mathscr{G})$ is a contraction on L_B^p, $1\leqslant p<\infty$.

Proof: Exercise.

To complete the argument we prove

2.7.3. ***Proposition:*** If ϕ is a lower semi-continuous convex function on a convex subset C of a *separable* Banach space, then there is a sequence (θ_i) of continuous affine functions with $\phi(x)=\sup_i\theta_i(x)$ on C.

Proof: Let $X=B\times\mathbf{R}$ with $||(x,r)||=||x||+|r|$, and consider the closed convex subset $S=\{(x,r):r\geqslant\phi(x)\}$. Then $X\backslash S$ is open, hence equals $\bigcup_{i=1}^{\infty}B_i$ for some sequence of open balls (B_i). Let $J=\{i:B_i\bigcap(C\times\mathbf{R})\neq\varnothing\}$. If $j\in J$, the Hahn–Banach theorem allows us to separate the open ball B_j from the closed convex set S: there exists $X_j^*=(x_j^*,\lambda_j)\in X^*$ such that

$$\sup\{x_j^*(x)+\lambda_j r:(x,r)\in B_j\}\leqslant\alpha_j\leqslant\inf\{x_j^*(x)+\lambda_j r:(x,r)\in S\} \tag{2.12}$$

If $(x,r)\in S$ and $s>r$, then $(x,s)\in S$. So if $\lambda_j<0$ we can choose r large enough, for fixed x, to contradict (2.12). Hence $\lambda_j\geqslant 0$. But if $(x,r)\in B_j\bigcap(C\times\mathbf{R})$, $x_j^*(x)+\lambda_j r<x_j^*(x)+\lambda_j\phi(x)$, since B_j is open, and $r<\phi(x)$, since $B_j\subset X\backslash S$. Hence $\lambda_j>0$. Now define $\theta_j(x)=\lambda_j^{-1}(\alpha_j-x_j^*(x))$. It is an easy exercise to check that θ_j has the required properties.

Now let $(\Omega,\mathscr{F},P,(\mathscr{F}_n),\mathbf{N})$ be a stochastic base and let B be a Banach space. A sequence $X=(X_n)\subseteq L_B^1$ is a *martingale* if $\mathbf{E}_B(X_{n+1}|\mathscr{F}_n)=X_n$ a.s. for all $n\geqslant 0$.

Jensen's inequality allows us to find (real) submartingales associated with a B-valued martingale X: the function $\phi(x)=||x||^p$ is convex and lower semi-continuous, hence if $X\subseteq L_B^1$ is a martingale, Theorem 2.7.1 implies that $||X_n||^p=\phi(X_n)=\phi(\mathbf{E}_B(X_{n+1}|\mathscr{F}_n))\leqslant\mathbf{E}(\phi\circ X_{n+1}|\mathscr{F}_n)=\mathbf{E}(||X_{n+1}||^p|\mathscr{F}_n)$, hence $(||X_n||^p)$ is a (non-negative) submartingale. Thus we can apply the maximal inequality (Corollary 2.6.9) to obtain, when $p=1$ in particular,

2.7.4. ***Theorem:*** If $X=(X_n)$ is a B-valued martingale and $\lambda>0$, then $\lambda P(\sup_n||X_n||\geqslant\lambda)\leqslant\sup_n\mathbf{E}(||X_n||)$.

The strategy for proving a.s. convergence theorems for B-valued martingales is the following: first we find a dense set in L_B^1 for which convergence is obvious. Then the maximal lemma (Theorem 2.7.4) is used together with the following general principle due to Banach to show that the set where convergence obtains is *closed*.

2.7.5. ***Theorem:*** Let (U_n) be a sequence of linear maps from a normed space E into $L_B^0(\Omega,\mathscr{F},P)$. Define $(U^*x)(\omega):=\sup_n||(U_nx)(\omega)||$ for $x\in E$, $\omega\in\Omega$. If for each $\varepsilon>0$ there is an $M>0$ such that $P(U^*x\geqslant M||x||_E)<\varepsilon$ for all $x\in E$, then the set $C:=\{x\in E:(U_nx)$ converges a.s.$\}$ is closed in E.

Proof: The statement $x\in C$ is equivalent to saying that $(Rx)(\omega)=0$ a.s., where $(Rx)(\omega):=\limsup_{m,n\to\infty}||(U_mx)(\omega)-(U_nx)(\omega)||$. Now clearly $Rx\leqslant 2U^*x$ a.s. by definition. Moreover, R is subadditive: $R(x+y)\leqslant Rx+Ry$ for $x,y\in E$, by definition of lim sup. Hence given $x\in\bar{C}$ and $\varepsilon>0$, find $y\in C$ with $||x-y||<\varepsilon/2M$ where M is as in the statement of the theorem. Then, since $y\in C$, $Rx\leqslant R(x-y)+Ry=R(x-y)\leqslant 2U^*(x-y)$, and so $P(Rx\geqslant\varepsilon)\leqslant P(U^*(x-y)\geqslant\varepsilon/2)\leqslant P(U^*(x-y)>M||x-y||)<\varepsilon$ by the choice of y and the hypothesis in the theorem. This holds for all $\varepsilon>0$, so $Rx=0$ a.s. Hence $x\in C$.

2.7.6. ***Lemma:*** If $\mathscr{F}=\sigma(\bigcup_n \mathscr{F}_n)$ and $1\leqslant p<\infty$, then the set $S=\bigcup_n L_B^p(\mathscr{F}_n)$ is dense in $L_B^p(\mathscr{F})$.

Proof: The measurable step-functions are dense in $L_B^p(\mathscr{F})$, so we need only show that each $g=\sum_{i=1}^m a_i 1_{B_i}$ can be approximated in L_B^p by functions in S. But this follows since each of the real-valued functions $1_{B_i}\in L^\infty(\mathscr{F})$ can be approximated pointwise by a sequence $(f_{n,i})_n$, where $f_{n,i}\in L^\infty(\mathscr{F}_n)\subseteq L^p(\mathscr{F}_n)$. Hence if we let $g_n=\sum_{i=1}^m \alpha_i f_{n,i}\in L_B^p(\mathscr{F}_n)$ for each n, we obtain $\|g_n-g\|_B\leqslant\sum_{i=1}^m\|\alpha_i\|\,|f_{n,i}-1_{B_i}|$, hence $\|g_n-g\|_p\leqslant\sum_{i=1}^m\|\alpha_i\|_B\|f_{n,i}-1_{B_i}\|_p$, which converges to zero as $n\to\infty$,. So $\bar{S}=L_B^p(\mathscr{F})$.

The lemma identifies the dense set we require for our principal convergence result:

2.7.7. ***Theorem:*** If the L_B^p-martingale (X_n) is closed by $Y\in L_B^p$, so that $X_n=\mathbf{E}_B(Y|\mathscr{F}_n)$ for all n, then $X_n\to Y$ a.s. and in L_B^p.

Proof: If $Y\in L_B^p(\mathscr{F}_n)$ we have $X_m=\mathbf{E}_B(Y|\mathscr{F}_m)=Y$ for $m\geqslant n$, so $X_m\to Y$ a.s. and in L_B^p. Hence this also holds for $Y\in S$. Norm-convergence is now trivial: if $Y\in L_B^p(\mathscr{F})$ and $\varepsilon>0$ we can find $Y_1\in S$ such that $\|Y-Y_1\|_p<\varepsilon/3$, by Lemma 2.7.6. Hence

$$\|Y-\mathbf{E}_B(Y|\mathscr{F}_n)\|_p\leqslant\|Y-Y_1\|_p+\|Y_1-\mathbf{E}_B(Y_1|\mathscr{F}_n)\|_p+\|\mathbf{E}_B(Y_1|\mathscr{F}_n)-\mathbf{E}_B(Y|\mathscr{F}_n)\|_p$$

$$\leqslant 2\|Y-Y_1\|_p+\|Y_1-\mathbf{E}_B(Y_1|\mathscr{F}_n)\|_p<\varepsilon$$

for large enough n, by Corollary 2.7.2 and the fact that $Y_1\in S$.

For a.s. convergence, use Theorem 2.7.5 with $E=L_B^p(\mathscr{F})$ and $U_nY=\mathbf{E}_B(Y|\mathscr{F}_n)$. The maximal inequality (Theorem 2.7.4) yields, for all $\lambda>0$,

$$\lambda P(U^*Y\geqslant\lambda)\leqslant\sup_n \mathbf{E}(\|\mathbf{E}_B(Y|\mathscr{F}_n)\|)\leqslant\mathbf{E}(\|Y\|)\leqslant\|Y\|_p,$$

where the penultimate inequality follows from Corollary 2.7.2 and the final one from Hölder's inequality (why?).

Given $\varepsilon>0$, let $\lambda=\|Y\|_p/\varepsilon$, then $P(U^*Y\geqslant 1/\varepsilon\|Y\|_p)<\varepsilon$ as required in Theorem 2.7.5. Hence the set $C=\{Y:\mathbf{E}_B(Y|\mathscr{F}_n)$ converges a.s.$\}$ is closed in L_B^p. But C contains the dense set S, hence $C=L_B^p$. The a.s. limit of $\mathbf{E}(Y|\mathscr{F}_n)$ must be Y, since we know that $X_n\to Y$ in L_B^p-norm.

The question of which martingales have the above form, i.e. are *closable* in L_B^p, can be answered as in the real case:

2.7.8. ***Theorem:*** If (X_n) is an L_B^p-martingale, $1\leqslant p<\infty$, then

(i) (X_n) is closable in L_B^p iff

(ii) (X_n) is relatively compact in L_B^p iff

(iii) (X_n) is weakly relatively compact in L_B^p.

(Note that the last is just uniform integrability if $p=1$, $B=\mathbf{R}$!)

Proof: If (X_n) is closable in L_B^p, it converges, hence is relatively compact and *a fortiori* weakly relatively compact. To prove that (iii) implies (i), let Y be a weak cluster point of (X_n), so that for $E\in\mathscr{F}$, $\int_E Y\,\mathrm{d}P$ is a weak cluster point of $\int_E X_n\,\mathrm{d}P$. If $E\in\mathscr{F}_n$ for some m, $\int_E X_n\,\mathrm{d}P=\int_E X_m\,\mathrm{d}P$ for $n\geqslant m$, so $\int_E X\,\mathrm{d}P=\int_E X_m\mathrm{d}P$. Now (i) follows.

In general Banach spaces, this is as far as we can go (although Theorem 2.7.8 clearly implies the characterisation for the real case given in Remark 2.6.7). The real impact of vector-valued martingales lies in their use to analyse the structure of various Banach spaces. We can define a Banach space B (or a closed absolutely convex subset C of B) to possess the *martingale convergence property* (MCP) if any B-valued (C-valued) uniformly integrable martingale (X_n) is closable in L_B^p. This property is shared, for example, by all reflexive Banach spaces (and, in fact, by every separable Banach space which is the dual of a Banach space). It is also equivalent to the *Radon–Nikodym property* (which says essentially that to each P-continuous B-valued measure μ of bounded variation, there corresponds $f\in L_B^1$ with $\mu(E)=\int_E f\,\mathrm{d}P$ for all $E\in\mathscr{F}$). This, in turn, can be characterised geometrically (see [41], [54]).

We must content ourselves here with showing that $L^1[0,1]$ does *not* have the MCP – the consequences of this result for the geometry of Banach spaces which contain copies of $L^1[0,1]$ are explored in [20], [54].

2.7.9. ***Theorem:*** If B has the MCP and (X_n) is a B-valued L_B^1-bounded martingale (so that $\sup_n\|X_n\|_1=K<\infty$), then (X_n) converges a.s. to an element of L_B^1.

Proof: This is proved by truncation of (X_n): fix $N>0$ and define a stopping time T by $T(\omega)=\inf(n:\|X_n\|\geqslant N)$. The stopped process X^T, where $X_n^T(\omega)=X_n(\omega)$ if $n<T(\omega)$ and $X_n^T(\omega)=X_{T(\omega)}(\omega)$ if $n\geqslant T(\omega)$, is again a martingale. (This may be proved directly or by extending the proof of Theorem 2.4.4.) We show that X^T is uniformly integrable: first define $Y=\|X_T\|$ on $\{T<\infty\}$ and 0 otherwise. Then $Y(\omega)=\lim_n X_n^T(\omega)$ on $\{T<\infty\}$ and $Y(\omega)\leqslant\liminf_{n\to\infty}\|X_n^T(\omega)\|$ on $\{T<\infty\}$. By Fatou's lemma $\mathbf{E}(Y)\leqslant\liminf_{n\to\infty}\mathbf{E}(\|X_n^T\|)$. Now $(\|X_n\|)$ is a submartingale, so optional sampling on the bounded stopping times $T\wedge n$ yields $\mathbf{E}(\|X_n^T\|)\leqslant\mathbf{E}(\|X_n\|)\leqslant K$, so $\mathbf{E}(Y)\leqslant K$ also. On the other hand, $\|X_n^T\|=Y$ on $\{T\leqslant n\}$ and $\|X_n^T\|=\|X_n\|\leqslant N$ on $\{T\leqslant n\}$, by definition of T. So $Y\vee N$ in L^1 is an upper bound for $||X^T||$, hence (X_n^T) is uniformly integrable.

Hence (X_n^T) converges a.s. to some Z in L_B^1. But on $\{T=\infty\}$, $X_n = X_n^T$ for all n, and $P(T<\infty) \leqslant P(\sup_n ||X_n|| \geqslant N) \leqslant K/N \to 0$ by the maximal inequality. So (X_n) converges to Z a.s. as required.

2.7.10. ***Corollary:*** $L^1[0,1]$ does not have the MCP.

Proof: Let $\mathscr{F}_n$ be the σ-field generated by the nth-order half-open dyadic intervals $J_i =](i-1)/2^n, i/2^n]$, where $1 \leqslant i \leqslant 2^n$ in $[0,1]$, and define $X_n(\omega) = 2^n \cdot 1_{J_i}$ for $\omega \in J_i$, $1 \leqslant i \leqslant 2^n$. Thus (X_n) takes its values in the unit ball of $L^1[0,1]$ and is a uniformly integrable martingale. But $||X_{n+1}(\omega) - X_n(\omega)|| = 1$ for all ω and all n, hence $(X_n(\omega))$ converges for no ω.

2.7.11. ***Exercises:*** The proofs of a.s. convergence theorems are usually quite subtle. Convince yourself that attempts to define a.s. convergence by means of a topology are doomed to failure by proving the following simple results (recall that $P(\Omega)=1$).

(i) Convergence a.s. is strictly stronger than convergence in probability (see section 0.1 for definitions).

(ii) If $f_n \to f$ a.s. implies $f_n \to f$ in some topology τ on L^0, then $f_n \to f$ in probability also implies $f_n \to f$ in τ.

2.8 ***Applications of the convergence theorems***

The martingale described in Example 2.3.3(3) allows us to construct Radon–Nikodym derivatives in a very natural way: recall that if $Q \ll P$ is a probability measure on $(\Omega, \mathscr{F})$ and $\mathscr{G} \subseteq \mathscr{F}$ is a separable sub-σ-field generated by the sequence $G_0, G_1, G_2, \ldots$ in $\mathscr{F}$, then $\mathscr{G}_n = \sigma(G_0, \ldots, G_n) n \geqslant 0$ defines an increasing sequence of sub-σ-fields of $\mathscr{G}$, each generated by a finite partition $\mathscr{P}_n$ of Ω. The random variables (X_n) defined by

$$X_n := \sum_{A \in \mathscr{P}_n} \frac{Q(A)}{P(A)} 1_A$$

define a $(\mathscr{G}_n)$-martingale and $Q(A) = \int_A X_n \, dP$ for all $A \in \mathscr{G}_n$, $n \geqslant 0$. We could extend this relation to $\mathscr{G}$ if we could find a $\mathscr{G}$-measurable L^1-function which closes (X_n) as a martingale. But this is easily done: (X_n) is a positive martingale, so converges a.s. to some $X_\infty \in L^1(\mathscr{G}, P)$ by Corollary 2.6.2. But by Corollary 2.6.9, (X_n) is even *uniformly integrable*: if

$$\lambda > 0, \; P(X_n \geqslant \lambda) \leqslant \frac{1}{\lambda},$$

as $\mathbf{E}(X_n) \leqslant 1$ for all n. Now

$$\int_{(X_n \geqslant \lambda)} X_n \, \mathrm{d}P = Q(X_n \geqslant \lambda)$$

since $(X_n \geqslant \lambda) \in \mathscr{G}_n$. But as $Q \ll P$, given $\varepsilon > 0$ there exists $\delta > 0$ such that $Q(A) < \varepsilon$ whenever $P(A) < \delta$. So taking $\lambda < 1/\delta$ we have

$$\int_{(X_n \geqslant \lambda)} X_n \, \mathrm{d}P < \varepsilon$$

uniformly in n.

Hence by Remark 2.6.7 $X_n \to X_\infty$ in L^1-norm and $X_n = \mathbf{E}(X_\infty | \mathscr{G}_n)$ for all $n \geqslant 0$. So if $A \in \bigcup_n \mathscr{G}_n$,

$$\int_A X_\infty \, \mathrm{d}P = \int_A X_n \, \mathrm{d}P = Q(A)$$

holds with appropriate choice of n. The relation $Q(A) = \int_A X_\infty \, \mathrm{d}P$ finally extends to all $A \in \mathscr{G} = \sigma(\bigcup_n \mathscr{G}_n)$ by the Monotone Class Theorem.

We have proved that if $Q \ll P$ then for any *separable* sub-σ-field $\mathscr{G}$ of $\mathscr{F}$ we can construct $X_{\mathscr{G}} \in L^1(\mathscr{G}, P)$ such that $Q(A) = \int_A X_{\mathscr{G}} \, \mathrm{d}P$ for all $A \in \mathscr{G}$. To find a single $X \in L^1$ satisfying $Q(A) = \int_A X \, \mathrm{d}P$ for all $A \in \mathscr{F}$ we can apply Theorem 2.6.13 to the martingale $(X_{\mathscr{G}})$ indexed by the separable sub-σ-fields of $\mathscr{F}$, ordered by inclusion. This index set is obviously upward filtering, and the martingale $(X_{\mathscr{G}})$ is uniformly integrable by the same argument as above. By Theorem 2.6.13 there is $X \in L^1$ with $\mathbf{E}(X|\mathscr{G}) = X_{\mathscr{G}}$ for each $\mathscr{G}$. Again, as above, $Q(A) = \int_A X \, \mathrm{d}P$ for all $A \in \mathscr{F}$. We have proved the following basic version of the Radon–Nikodym theorem:

2.8.1. ***Theorem:*** Let $Q \ll P$ be a second probability measure on $(\Omega, \mathscr{F})$. Then there is $X \in L^1(P)$ such that $Q(A) = \int_A X \, \mathrm{d}P$ for all $A \in \mathscr{F}$.

We write $X = \mathrm{d}Q/\mathrm{d}P$ and call this the *Radon–Nikodym* derivative of Q w.r.t. P.

The above methods suggest a considerable extension of Theorem 2.7.1, also due to Doob:

2.8.2. ***Theorem:*** Let $(\Omega, \mathscr{F})$ be *separable* and suppose $(T, \mathscr{T})$ is a second measurable space, where T indexes two measurable families of probability measures (P_t), (Q_t) on $(\Omega, \mathscr{F})$ such that $Q_t \ll P_t$ for each $t \in T$. Then there is a positive $\mathscr{T} \times \mathscr{F}$-measurable function X on $T \times \Omega$ such that for each $t \in T$

$$X(t,\cdot) = \frac{dQ_t}{dP_t}.$$

Proof: As in the proof of Theorem 2.8.1, now with $\mathscr{F}$ in place of $\mathscr{G}$, let

$$X_n(t,\omega) = \sum_{A \in \mathscr{P}_n} \frac{Q_t(A)}{P_t(A)} 1_A(\omega)$$

for each $(t,\omega) \in T \times \Omega$.

For each t, $(X_n(t,\cdot))_n$ converges P_t-a.s. to dQ_t/dP_t. Hence we set $X(t,\omega) = \lim_n X_n(t,\omega)$ when this limit is finite, and 0 otherwise. This proves the theorem.

2.8.3. ***Example:*** The martingales leading to Theorem 2.8.1 also furnish us with easy examples of martingales which converge a.s. but not in L^1-norm. Let $\Omega = [0,1[$, and let $\mathscr{F}_n$ be the σ-field generated by nth order dyadic intervals,

$$I_k = [\frac{k}{2^n}, \frac{k+1}{2^n}[, k = 1, \ldots, 2^n - 1.$$

Let P be Lebesgue measure, Q unit mass at 0. Then

$$X_n = \sum_{A \in \mathscr{P}_n} \frac{Q(A)}{P(A)} 1_A(\omega) = 2^n \cdot 1_{[0,1/2^n[}(\omega)$$

defines a positive martingale $X = (X_n)$ with $||X_n||_1 = 1$ for all n, while $X_n(\omega) \longrightarrow 0$ for all $\omega > 0$, hence P-a.s. But then (X_n) cannot converge in L^1-norm (the limit would have to be 0 – why?)
(*Exercise:* Show that this example is none other than 'double or nothing'!)

We have seen (Theorem 2.6.8) that if $Y \in L^1$, $\mathbf{E}(Y|\mathscr{F}_n) \longrightarrow Y$ in L^1-norm and a.s. Applying this to $Y = 1_A$, where $A \in \mathscr{F}$, yields $P(A|\mathscr{F}_n) \longrightarrow 1_A$ in L^1-norm and a.s. If $\mathscr{F}_n$ is generated by *independent* random variables $Y_1, \ldots, Y_n$ for each n, and A is in the *tail-σ-field*:

$$A \in \bigcap_{n=1}^{\infty} \sigma(Y_{n+1}, Y_{n+2}, \ldots),$$

then A is independent of $\mathscr{F}_n$ for each n, hence $P(A|\mathscr{F}_n) = P(A)$. So $P(A)$ is either 0 or 1. We have proved

2.8.4. ***Kolmogorov's 0–1 law:*** If (Y_n) is a sequence of independent random variables and A is in the tail σ-field then $P(A)$ is 0 or 1.

A further convergence theorem for conditional expectations is the following, where the random variable and the σ-field vary simultaneously:

2.8.5. ***Theorem (Hunt):*** Let $(\Omega, \mathscr{F}, P, (\mathscr{F}_n), \mathbf{N})$ be a stochastic base and write $\mathscr{F}_\infty$ for $\sigma(\bigcup_n \mathscr{F}_n)$. Suppose (X_n) is a sequence of random variables with $|X_n| \leqslant Y$ for all n, where $Y \in L^1$. Suppose also that $X_n \to X$ a.s. for some random variable X. Then as $n \to \infty$,

$$\mathbf{E}(X_n | \mathscr{F}_n) \to \mathbf{E}(X | \mathscr{F}_\infty) \text{ a.s.}$$

and in L^1-norm.

Proof: Dominated convergence ensures that $X_n \to X$ in L^1-norm. So

$$||\mathbf{E}(X_n|\mathscr{F}_n) - \mathbf{E}(X|\mathscr{F}_\infty)||_1 \leqslant ||\mathbf{E}(X_n - X|\mathscr{F}_n)||_1 + ||\mathbf{E}(X|\mathscr{F}_n) - \mathbf{E}(X|\mathscr{F}_\infty)||.$$

The first term vanishes as $n \to \infty$ since $\mathbf{E}(\cdot|\mathscr{F}_n)$ is an L^1-contraction, the second by Theorem 2.6.8. Hence L^1-norm convergence is proved.

For a.s. convergence, consider

$$U_m = \inf_{n \geqslant m} X_n, \quad V_m = \sup_{n \geqslant m} X_n.$$

Since $X_n \to X$ a.s. and in L^1-norm, $U_m - V_m \to 0$ a.s. and in L^1. Also, if $n \geqslant m$,

$$\mathbf{E}(U_m|\mathscr{F}_n) \leqslant \mathbf{E}(X_n|\mathscr{F}_n) \leqslant \mathbf{E}(V_m|\mathscr{F}_n),$$

as $\mathbf{E}(\cdot|\mathscr{F}_n)$ is positive. Letting $n \to \infty$,

$$\mathbf{E}(U_m|\mathscr{F}_n) \to \mathbf{E}(U_m|\mathscr{F}_\infty)$$

and

$$\mathbf{E}(V_m|\mathscr{F}_n) \to \mathbf{E}(V_m|\mathscr{F}_\infty).$$

So, in particular, these limits bound $\liminf \mathbf{E}(X_n|\mathscr{F}_n)$. For large m the difference between these bounds has arbitrarily small expectation

$$\mathbf{E}(\mathbf{E}(V_m - U_m|\mathscr{F}_\infty) = \mathbf{E}(V_m - U_m) = ||V_m - U_m||_1 \to 0.$$

So $\lim_n \mathbf{E}(X_n|\mathscr{F}_n)$ exists a.s. and must equal the L^1-limit.

The above applications of the convergence theorems are taken from [19].

There are many other applications in [12], [68], etc. In most cases these demand more background in probability theory than the above.

2.8.6. ***Exercise:*** Let $P=(p_n)$ be a probability measure on $\mathbf{N}$ – all subsets of $\mathbf{N}$ are assumed measurable – and let $\mathscr{F}_n$ be the σ-field generated by $\{0\}, \{1\}, \ldots, \{n-1\}, \{k : k \geqslant n\}$. Define $X=(X_n)$ by

$$X_n(i)=\begin{cases} 1/\sum_{k \geqslant n} p_k & \text{if } i \geqslant n \\ 0 & \text{if } i<n \end{cases}$$

Show that X is a martingale w.r.t. $(\mathscr{F}_n)$ and that X converges P-a.s. but not in $L^1(P)$-norm.

2.9. Decomposition theorems

Our second approach to the convergence theorems has yet to yield a result on supermartingales: we use a very simple but useful decomposition to achieve this. The continuous time analogue of this decomposition will be our main objective in Chapter 3 and is the key to stochastic integration in Chapter 4.

Suppose that $X=(X_n)$ is an adapted process. We split X into a martingale and a predictable process as follows:

$$M_0=X_0, \quad M_{n+1}=M_n+(X_{n+1}-\mathbf{E}(X_{n+1}|\mathscr{F}_n)),$$

$$A_0=0, \qquad A_{n+1}=A_n+(X_n-\mathbf{E}(X_{n+1}|\mathscr{F}_n)), \quad \text{if } n \geqslant 1.$$

It is clear that $A=(A_n)$ is predictable. To see that $M=(M_n)$ is a martingale, note that

$$\mathbf{E}(M_{n+1}-M_n|\mathscr{F}_n)=\mathbf{E}(X_{n+1}-\mathbf{E}(X_{n+1}|\mathscr{F}_n)|\mathscr{F}_n)=0.$$

Finally, $M_0-A_0=X_0$ and if $n \geqslant 1$, $X_{n+1}-X_n=(M_{n+1}-M_n)-(A_{n+1}-A_n)$, so by induction on n, $X_n=M_n-A_n$ for all n. This decomposition $X=M-A$ is unique up to additive constants: if $X=M'-A'$ also, then $M-M'=A'-A$ is a predictable martingale and hence is constant by Remark 2.3.2(3).

2.9.1. ***Definition:*** The decomposition $X=M-A$ is the *Doob decomposition* of the adapted process X.

When X is a supermartingale, $A_{n+1}-A_n=\mathbf{E}(X_{n+1}-X_n|\mathscr{F}_n) \geqslant 0$, hence in that case A is a predictable *increasing* process. Thus (A_n) converges a.s. to some random variable A_∞ if X is a supermartingale. This allows us to deduce Corollary 2.6.2 from the results of section 2.7 (taking $B=\mathbf{R}$) without using upcrossings:

2.9.2. ***Theorem:*** **If (X_n) is a supermartingale with $\sup_n \mathbf{E}(X_n^-)$ finite, then (X_n) converges a.s. to an integrable limit.**

Proof: X has Doob decomposition $X=M-A$. Since $\sup_n \mathbf{E}(X_n^-)=k<\infty$, X is actually bounded in L^1, as

$$\|X_n\|_1=\mathbf{E}(X_n)+2\mathbf{E}(X_n^-)\leqslant \mathbf{E}(X_0)+2k$$

for all $n\geqslant 0$.

Now $A_0=0$ and (A_n) increases, $M_n=X_n+A_n\geqslant X_n$ for all n, so $M_n^-\leqslant X_n^-$. Hence also

$$\|M_n\|_1=\mathbf{E}(M_n)+2\mathbf{E}(M_n^-)\leqslant \mathbf{E}(M_0)+2k,$$

so M is an L^1-bounded martingale and converges a.s. by Theorem 2.7.9, with $B=\mathbf{R}$. The a.s. convergence of (A_n) is clear and

$$\mathbf{E}(A_n)=\mathbf{E}(M_n)+\mathbf{E}(X_n),$$

so $(\mathbf{E}(A_n))_n$ is bounded. Hence by monotone convergence $A_\infty\in L^1$.

Moreover, given any predictable increasing process $A=(A_n)$ such that $A_\infty\in L^1$, we can associate a supermartingale with A: let $Z_n=\mathbf{E}(A_\infty-A_n|\mathscr{F}_n)$, then $Z=(Z_n)$ is a positive supermartingale, and $Z_n\to 0$ in L^1-norm by monotone convergence applied to (A_n).

Now suppose $X=(X_n)$ is a supermartingale with $\lim_n \mathbf{E}(X_n)>-\infty$. (The limit exists as $\mathbf{E}(X_n)$ decreases with increasing n.) Then the Doob decomposition $X=M-A$ has $A_\infty\in L^1$:

$$\mathbf{E}(X_n)=\mathbf{E}(M_n)-\mathbf{E}(A_n)=\mathbf{E}(X_0)-\mathbf{E}(A_n),$$

so $\{\mathbf{E}(A_n)\}_n$ is bounded, hence by monotone convergence $A_\infty\in L^1$.

Thus if we define $Z_n=\mathbf{E}(A_\infty-A_n|\mathscr{F}_n)$ as above and $Y_n=M_n-\mathbf{E}(A_\infty|\mathscr{F}_n)$, we obtain the new decomposition $X_n=Z_n+Y_n$, where $Z=(Z_n)$ is a positive supermartingale with $Z_n\to 0$ in L^1-norm, while $Y=(Y_n)$ is a martingale. The decomposition $X=Z+Y$ is unique with these properties. (Exercise: Prove this.)

2.9.3. ***Definition:*** A positive supermartingale Z with $Z_n\to 0$ in L^1-norm is called a *potential*. The above decomposition $X=Z+Y$ of the supermartingale X into the sum of a potential and a martingale is the *Riesz decomposition* of X.

The terminology stems from potential theory: a potential is a positive superharmonic function that dominates no positive harmonic function. Reading 'martingale' for 'harmonic function' we can see that the analogy holds here: if Y is a martingale and X is a supermartingale with $X_n \geqslant Y_n > 0$ for all n, then

$$\mathbf{E}(X_n) \geqslant \mathbf{E}(Y_n) = \mathbf{E}(Y_0) > 0,$$

hence $X_n \not\to 0$ in L^1-norm, so X is not a potential. Conversely if $\mathbf{E}(X_n) \not\to 0$ we can split the positive supermartingale X into the sum of a potential Z and a martingale Y. Then $Z_n \to 0$ in L^1-norm, so $Y_n > 0$ for all n, since otherwise $\mathbf{E}(X_n) \to 0$. Consequently a positive supermartingale is a potential iff it dominates no positive martingale.

The analogies and connections with potential theory are described in detail in [63]. For us, the Doob decomposition has its principal application in stochastic integration, because it allows us to define the 'quadratic variation' of a square-integrable martingale:

2.9.4. ***Definition:*** Suppose $X \in \mathcal{M}^2$, i.e. $X = (X_n)$ is a martingale with

$$\|X\|_2 := \sup_n \|X_n\|_2 < \infty.$$

Let $X^2 = M + A$ be the Doob decomposition of the submartingale X^2, so that M is a martingale and A a predictable increasing process with $A_0 \equiv 0$. We call A the (predictable) *quadratic variation* of X.

To explain the terminology, we prove a simple identity:

$$\text{for all } n, \quad A_{n+1} - A_n = \mathbf{E}((X_{n+1} - X_n)^2 | \mathcal{F}_n). \tag{2.13}$$

Proof: As A is predictable, $A_{n+1} - A_n = \mathbf{E}(A_{n+1} - A_n | \mathcal{F}_n) = \mathbf{E}(X_{n+1}^2 - M_{n+1} - X_n^2 + M_n | \mathcal{F}_n) = \mathbf{E}(X_{n+1}^2 - X_n^2 | \mathcal{F}_n)$, since M is a martingale. But since X is also a martingale,

$$\begin{aligned} \mathbf{E}((X_{n+1} - X_n)^2 | \mathcal{F}_n) &= \mathbf{E}((X_{n+1}^2 + X_n^2) | \mathcal{F}_n) - 2X_n \mathbf{E}(X_{n+1} | \mathcal{F}_n) \\ &= \mathbf{E}((X_{n+1}^2 - X_n^2) | \mathcal{F}_n), \end{aligned}$$

as required.

Writing $\Delta Y_n := Y_{n+1} - Y_n$ for any process $Y = (Y_n)$, we can write the martingale property as $\mathbf{E}(\Delta X_n | \mathcal{F}_n) = 0$. Hence (2.13) becomes

$$\Delta A_n = \mathbf{E}((\Delta X_n)^2 | \mathcal{F}_n),$$

which indicates that A plays the role of a 'conditional' quadratic variation.

Fixing $X \in \mathscr{M}^2$, we now define a space of 'square-integrable integrands' relative to X by writing $\ell^2(X) := \{Y: Y \text{ is an adapted process}, \mathbf{E}(\sum_n Y_n^2 \Delta A_n) < \infty\}$.

Define a seminorm on $\ell^2(X)$ by $\|Y\|_{\ell^2(X)} := (\mathbf{E}(\sum_n Y_n^2 \Delta A_n))^{1/2}$ and identify Y and Y' if

$$\sum_n (Y_n - Y'_n)^2 \Delta A_n = \text{a.s.}$$

The resulting quotient normed space will also be denoted by $\ell^2(X)$. Since each class has a representative with $\Delta A_n > 0$ for all n (why?) the measures Q_n defined by

$$\frac{\mathrm{d}Q_n}{\mathrm{d}P} = \Delta A_n$$

are all equivalent to P, so the norm on $\ell^2(X)$ is equivalent to the norm on the direct sum $\oplus_n L^2(Q_n)$. The proof that $\ell^2(X)$ is complete under this norm – and so is a Hilbert space – is identical with the classical proof of the completeness of the sequence space ℓ^2. (Exercise!)

We can define the 'stochastic integral' of $Y \in \ell^2(X)$ in terms of an *isometry* between the Hilbert spaces $\ell^2(X)$ and $L^2(P)$:

2.9.5. ***Definition:*** If $Y \in \ell^2(X)$, $X \in \mathscr{M}^2$, write $(Y \cdot X)_n = \sum_{m<n} Y_m \Delta X_m$.

This 'stochastic integral' is actually a martingale transform: for if $V_0 = 0$, $V_n = Y_{n-1}$ $(n \geqslant 1)$ $Y \cdot X$ is just the transform $V \cdot X$. Hence by Theorem 2.4.4, $Y \cdot X$ is again a martingale (and in $\mathscr{M}^2$ – check!).

We proceed to verify that the series $\sum_n Y_n \Delta X_n$ converges in $L^2(P)$ and that the map $Y \to Y \cdot X$ is an isometry.

Firstly, the series $\sum_n Y_n \Delta X_n$ converges in $L^2(P)$-norm, for in the sum

$$\mathbf{E}\left(\left(\sum_{n=k}^{k+p} Y_n \Delta X_n\right)^2\right) = \mathbf{E}\left(\sum_{m=k}^{k+p} \sum_{n=k}^{k+p} Y_n Y_m \Delta X_n \Delta X_m\right)$$

all crossterms must vanish as X is a martingale. (If $m<n$, Y_m, X_m, X_{m+1} are $\mathscr{F}_n$-measurable, then $\mathbf{E}(Y_n Y_m \Delta X_m \Delta X_n) = \mathbf{E}(Y_n Y_m \Delta X_m \mathbf{E}(\Delta X_n | \mathscr{F}_n)) = 0$.) Hence

$$\mathbf{E}\left(\left(\sum_{n=k}^{k+p} Y_n \Delta X_n\right)^2\right) = \mathbf{E}\left(\sum_{n=k}^{k+p} Y_n^2 (\Delta X_n)^2\right) = \mathbf{E}\left(\sum_{n=k}^{k+p} Y_n^2 \mathbf{E}((\Delta X_n)^2 | \mathscr{F}_n)\right)$$

$$= \mathbf{E}\left(\sum_{n=k}^{k+p} Y_n^2 \Delta A_n\right)$$

by (2.13). Now since $Y \in \ell^2(X)$, given $\varepsilon > 0$ we can find k_0 such that if $k \geqslant k_0$ and p are given,

$$\mathbf{E}\left(\sum_{n=k}^{k+p} Y_n^2 \Delta A_n\right) < \varepsilon,$$

hence also $\sum_n Y_n \Delta X_n$ converges in $L^2(P)$-norm. So in computing $\|Y \cdot X\|^2_{L^2(P)} = \mathbf{E}((\sum_n Y_n \Delta X_n)^2)$ we can again exchange $\mathbf{E}$ and $\sum_n$ to show that crossterms vanish. This means that

$$\|Y \cdot X\|^2_{L^2(P)} = \mathbf{E}\left(\sum_n Y_n^2 (\Delta X_n)^2\right) = \mathbf{E}\left(\sum_n Y_n^2 \Delta A_n\right) = \|Y\|^2_{\ell^2(X)}.$$

The map $Y \rightarrow Y \cdot X$ has a subspace H_X of $L^2(P)$ as its range. Given X_1 and X_2 in $\mathcal{M}^2$ it is easy to show by arguments like the above that H_{X_1} and H_{X_2} are orthogonal iff $(X_1 X_2)$ is also a martingale. The details are left to the reader (see also [68]). All these ideas have counterparts for continuous-time martingales and are basic to the study of stochastic integrals.

2.10. Optional sampling

In order to extend the optional sampling theorem (Theorem 2.4.7) to arbitrary stopping times we consider the convergence of *reverse supermartingales*: suppose $(\Omega, \mathcal{F}, P)$ is a probability space, and $(\mathcal{F}_n)$ is a *decreasing* sequence of sub-σ-fields of $\mathcal{F}$. If $X = (X_n)$ is a sequence of L^1 adapted to $(\mathcal{F}_n)$, then X is a reverse supermartingale if $\mathbf{E}(X_m | \mathcal{F}_n) \leqslant X_n$ for $m \leqslant n$. Note that X is a supermartingale (in the usual sense) for the stochastic base $(\Omega, \mathcal{F}, P, (\mathcal{F}_n), -\mathbf{N})$. These sequences have very simple convergence properties:

2.10.1. *Theorem:* If X is a reverse supermartingale, either $\lim_{n \to \infty} \mathbf{E}(X_n) = +\infty$ or (X_n) is uniformly integrable.

Proof: Since $\mathbf{E}(X_m)$ increases with m, the limit exists. Suppose it is finite, i.e. $\lim_n \mathbf{E}(X_n) = \alpha \in \mathbf{R}$. Write $Y_n = X_n - \mathbf{E}(X_0 | \mathcal{F}_n)$ for $n \geqslant 0$. By definition, $Y_n \geqslant 0$ and $(\mathbf{E}(X_0 | \mathcal{F}_n))_n$ is uniformly integrable. So in order to show that (X_n) is uniformly integrable we may assume $X_n \geqslant 0$. Given $\varepsilon > 0$ we can find $k \in \mathbf{N}$ such that $|\alpha - \mathbf{E}(X_n)| < \varepsilon/2$ for $n \geqslant k$. Now $\{X_0, \ldots, X_k\}$ is finite, hence uniformly integrable, and we need only consider X_n for $n \geqslant k$. If $\lambda \in \mathbf{R}$,

$$\int_{(X_n > \lambda)} X_n \, \mathrm{d}P = \mathbf{E}(X_n) - \int_{(X_n \leqslant \lambda)} X_n \, \mathrm{d}P \leqslant \mathbf{E}(X_n) - \int_{(X_n \leqslant \lambda)} X_k \, \mathrm{d}P$$

since $(X_n \leqslant \lambda) \in \mathcal{F}_n \subseteq \mathcal{F}_k$.

Since $\mathbf{E}(X_n)$ increases with n, $0 \leqslant \mathbf{E}(X_n) - \mathbf{E}(X_k) < \varepsilon/2$ if $n \geqslant k$, so

$$\int_{(X_n > \lambda)} X_n \, dP \leqslant \mathbf{E}(X_n) - \left(\mathbf{E}(X_k) - \int_{(X_n > \lambda)} X_k \, dP \right) < \frac{\varepsilon}{2} + \int_{(X_n > \lambda)} X_k \, dP.$$

But this is less than ε for large enough λ: since $X_n \geqslant 0$, we have

$$P(X_n > \lambda) \leqslant \frac{1}{\lambda} \mathbf{E}(X_n) \leqslant \frac{\alpha}{\lambda}.$$

So (X_n) is uniformly integrable.

2.10.2. ***Corollary:*** If X is a reverse supermartingale with $\sup_n \mathbf{E}(X_n)$ finite, then X is uniformly integrable and $X_n \to X_\infty \in L^1$ a.s. and in L^1-norm. X_∞ closes the supermartingale (X_n) and, if $X_0 \in L^p$, $X_n \to X_\infty$ in L^p-norm. If X is a reverse martingale, $X_\infty = \mathbf{E}(X_0 | \mathscr{F}_\infty)$, where $\mathscr{F}_\infty = \bigcap_n \mathscr{F}_n$.

Proof: Exercise – consider the upcrossings of $Y_{-n}, \ldots, Y_0$, where $Y_{-k} = X_k$.

2.10.3. ***Example (strong law of large numbers):*** Let (Y_n) be a sequence of independent identically distributed random variables with $Y_0 \in L^1$ and define

$$S_k = \sum_{n=0}^{k-1} Y_n \text{ for } k \geqslant 1.$$

Let $\mathscr{F}_n = \sigma(S_k : k \geqslant n)$ for $n \geqslant 0$. Then $X_n = \mathbf{E}(Y_0 | \mathscr{F}_n)$ defines a reverse martingale $(X_n, \mathscr{F}_n)$, which by Corollary 2.10.2 converges a.s. and in L^1-norm to $X_\infty = \mathbf{E}(Y_0 | \mathscr{F}_\infty)$, where $\mathscr{F}_\infty = \bigcap_n \mathscr{F}_n$ is the *tail σ-field* for (Y_n) – see 2.8.4. So by the 0–1 law X_∞ is a.s. constant and so $X_\infty = \mathbf{E}(X_\infty) = \mathbf{E}(Y_0)$. Now by symmetry $\mathbf{E}(Y_0 | \mathscr{F}_n) = \mathbf{E}(Y_i | \mathscr{F}_n)$ for $i \leqslant n-1$. So

$$\frac{1}{n} S_n = \frac{1}{n} \sum_{i=0}^{n-1} Y_i = \mathbf{E}\left(\frac{1}{n} \sum_{i=0}^{n-1} Y_i \middle| \mathscr{F}_n \right) = \mathbf{E}(Y_0 | \mathscr{F}_n) = X_n.$$

Hence we have proved that $(1/n) S_n \to \mathbf{E}(Y_0)$ a.s. (and in L^1-norm).

Turning now to optional sampling, i.e. the preservation of the (super-) martingale relation under arbitrary stopping times, we note that we cannot expect a definite result without some restrictions on the martingale.

Let $(\Omega, \mathscr{F}, P)$ be *coin-tossing space*, i.e. $\Omega = \{0, 1\}^{\mathbb{N}}$ with $\mathscr{F}$ the product σ-field and P the product of the symmetric distribution $P(0) = \frac{1}{2} = P(1)$ on each co-ordinate. Thus a point of Ω is a sequence $\omega = (\omega_k)$ of 0's and 1's. Let $\mathscr{F}_n = \sigma(\omega_1, \ldots, \omega_n : \omega \in \Omega)$ be generated by 'the first n tosses', i.e. by the projections onto the first n co-ordinates. Then $(\mathscr{F}_n)_{n\geqslant 1}$ is an increasing sequence of σ-fields generating $\mathscr{F}$. (Compare with Example 2.3.3(3).)

The random variables measuring the excess of the number of 'heads' over 'tails' are then a martingale relative to $(\mathscr{F}_n)$: we define

$$X_n(\omega)=\sum_{k=1}^{n}\omega_k-\left(n-\sum_{k=1}^{n}\omega_k\right)=2\sum_{k=1}^{n}\omega_k-n$$

for each $n\geqslant 1$, $\omega\in\Omega$. (This gives 'heads–tails in n tosses'.) Since $(X_{n+1}-X_n)(\omega)=2\omega_{n+1}-1$, $(X_{n+1}-X_n)$ is *independent* of $\mathscr{F}_n$, so $\mathbf{E}(X_{n+1}-X_n|\mathscr{F}_n)=\mathbf{E}(X_{n+1}-X_n)=0$ a.s. But here the martingale property is *not* preserved under stopping times: let $T(\omega)=\inf(n:X_n(\omega)=1)$ or $+\infty$ if the set is empty. This is clearly a stopping time and $\mathbf{E}(X_T)=1$. Thus

$$\mathbf{E}(X_1)=\tfrac{1}{2}(1)+\tfrac{1}{2}(-1)=0\neq\mathbf{E}(X_T),$$

contradicting the martingale property. Note that T is an *unbounded* stopping time.

2.10.4. ***Theorem (optional sampling):*** Let $X=(X_n)$ be a supermartingale closed by $Y\in L^1$ and suppose that S and T are stopping times with $S\leqslant T$. Then X_S and X_T are integrable and $\mathbf{E}(X_T|\mathscr{F}_S)\leqslant X_S$.

Proof: Write $Y_n=\mathbf{E}(Y|\mathscr{F}_n)$, $Z_n=X_n-Y_n$. Then (Y_n) is a uniformly integrable martingale and as Y closes X_n, $Z_n\geqslant 0$. (Z_n) is obviously a super-martingale (and closed by 0). We prove the assertions of the theorem separately for (Y_n) and (Z_n). In both cases we use the decreasing sequences of stopping times (S_n), (T_n), defined by $S_n(\omega)=S(\omega)$ on $\{S\leqslant n\}$ and $S_n(\omega)=+\infty$ on $\{S>n\}$, and similarly for (T_n).

For each $n\geqslant 0$, S_n has finite range and can be treated as a *bounded* stopping time (using $\{0,1,\dots,n+1\}$ instead of $\{0,1,\dots,n,+\infty\}$. So by Theorem 2.4.7, the martingales $Y_0,Y_1,\dots,Y_n,Y_\infty$, with $Y_\infty=Y$, has $Y_{S_n}=\mathbf{E}(Y|\mathscr{F}_{S_n})$ for each n. But $\mathscr{F}_S=\bigcap_n\mathscr{F}_{S_n}$, so by Corollary 2.10.2 the reverse martingale (Y_{S_n}) converges to $\mathbf{E}(Y_0|\mathscr{F}_S)$, which equals $\mathbf{E}(Y|\mathscr{F}_S)$ by the tower property of conditional expectations. Hence

$$\mathbf{E}(Y|\mathscr{F}_S)=\lim_n\mathbf{E}(Y|\mathscr{F}_{S_n}).$$

On the other hand, $Y_S=Y_{S_n}$ on $\{S\leqslant n\}\bigcup\{S=+\infty\}$, so $Y_{S_n}\to Y_S$ pointwise. So $Y_S=\mathbf{E}(Y|\mathscr{F}_S)$ and $Y_S\in L^1$. Now the tower property of conditional expectations yields

$$Y_S = \mathbf{E}(\mathbf{E}(Y|\mathscr{F}_T)|\mathscr{F}_S) = \mathbf{E}(Y_T|\mathscr{F}_S)$$

since $\mathscr{F}_S \subseteq \mathscr{F}_T$ when $S \leqslant T$.

Now consider the positive supermartingale (Z_n): we have $Z_{S_n} \geqslant 0$ for all n, as (Z_n) is closed by 0, and similarly for Z_{T_n}. Also $Z_{S_{n+1}} = Z_{S_n} = Z_S$ on $\{S \leqslant n\} \subseteq \{S \leqslant n+1\}$, while $Z_{S_{n+1}} \geqslant 0 = Z_{S_n}$ on the set $\{S = n+1\}$. So (Z_{S_n}), and similarly (Z_{T_n}), is an increasing sequence of positive random variables and so $Z_{S_n} \to Z_S$, $Z_{T_n} \to Z_T$ a.s. Now since by Theorem 2.4.7 (Z_{S_n}) and (Z_{T_n}) are supermartingales, $\mathbf{E}(Z_0)$ is an upper bound for their expectations, and by monotone convergence Z_S, Z_T are in L^1 and convergence takes place in L^1-norm. Moreover, for each fixed n, $S_n \leqslant T_n$, so

$$Z_{S_n} \geqslant \mathbf{E}(Z_{T_n}|\mathscr{F}_{S_n})$$

by Theorem 2.4.7.

Now let $W_n = \mathbf{E}(Z_{T_n}|\mathscr{F}_{S_n})$. Then $(W_n, \mathscr{F}_{S_n})$ is a reverse supermartingale, since on $\{T = n+1\}$, $T_n = +\infty$ and $T_{n+1} = n+1$, so that $\mathbf{E}(Z_{T_{n+1}} - Z_{T_n}|\mathscr{F}_{S_{n+1}}) \geqslant 0$ a.s. on $\{T = n+1\}$. Elsewhere this quantity is 0. So

$$W_{n+1} = \mathbf{E}(Z_{T_{n+1}}|\mathscr{F}_{S_{n+1}}) \geqslant \mathbf{E}(Z_{T_n}|\mathscr{F}_{S_{n+1}}) = \mathbf{E}(\mathbf{E}(Z_{T_n}|\mathscr{F}_{S_n})|\mathscr{F}_{S_{n+1}}) = \mathbf{E}(W_n|\mathscr{F}_{S_{n+1}}).$$

Now $(\mathscr{F}_{S_n})$ decreases to $\mathscr{F}_S$ and $\{\mathbf{E}(W_n)\}$ is bounded above by $\mathbf{E}(Z_0)$, so $W = \lim_n W_n$ exists a.s. If $A \in \mathscr{F}_S$, $\int_A W_n \, dP = \int_A Z_{T_n} \, dP \to \int_A W \, dP$. But $Z_{T_n} \to Z_T$ a.s. and $\mathbf{E}(Z_0)$ bounds $\{\mathbf{E}(Z_{T_n})\}$. Hence for all $A \in \mathscr{F}_S$, $\int_A W dP = \int_A Z_T dP$, so that $W = \mathbf{E}(Z_T|\mathscr{F}_S)$. Hence

$$Z_S = \lim_n Z_{S_n} \geqslant \lim \mathbf{E}(Z_{T_n}|\mathscr{F}_{S_n}) = \mathbf{E}(Z_T|\mathscr{F}_S).$$

2.10.5. ***Corollary:*** If X is a uniformly integrable martingale, then the family $\{X_T : T \text{ is a stopping time}\}$ is also uniformly integrable.

Proof: Since X is closed by X_∞, Theorem 2.10.4 applies and so

$$X_T = \mathbf{E}(X_\infty|\mathscr{F}_T)$$

for each T. But then (X_T) is uniformly integrable by Example 2.2.4.

This result is false in the continuous-time case, and the class of martingales for which it holds plays an important role in the analogue of the Doob decomposition.

2.10.6. ***Exercise:*** Prove that if $X \in L^1$ and T is a stopping time, then

$$\mathbf{E}(X|\mathscr{F}_T) = \mathbf{E}(X|\mathscr{F}_n) \text{ on } \{T = n\}.$$

Deduce that stopping the martingale $(\mathbf{E}(X|\mathscr{F}_n))_n$ at T amounts to conditioning X w.r.t. $\mathscr{F}_T$. Hence show that for *any* stopping times S and T, $\mathbf{E}(\cdot|\mathscr{F}_T)$ and $\mathbf{E}(\cdot|\mathscr{F}_S)$ commute, and their product is $\mathbf{E}(\cdot|\mathscr{F}_{S\wedge T})$.

2.11. Optimal stopping*

One of the most pressing 'practical' problems in gambling is knowing when to stop. Optional sampling already indicates that a 'fair' game remains fair if we choose to play only at certain times $n_1, n_2, \ldots$ (the values of our stopping time) whose choice involves no 'prescience'. More generally, one may ask for a stopping strategy T which optimises $\mathbf{E}(X_T)$, where X_n denotes the gambler's fortune at time n, whether or not the game is 'fair'. As usual, this will involve a stochastic base $(\Omega, \mathscr{F}, P, (\mathscr{F}_n), \mathbf{N})$, where $\mathscr{F}_n$ represents the information available at time n. Since the gambler wants to enjoy his fortune, $T(\omega)$ should be finite a.s.

Thus define a *stopping rule* T as a stopping time (relative to $(\mathscr{F}_n)$) which is a.s. finite. We seek an *optimal* stopping rule T, in other words we want $\mathbf{E}(X_T) = \sup_{S\in\mathscr{R}} \mathbf{E}(X_S)$, where $\mathscr{R}$ denotes the set of all stopping rules S. To ensure that the supremum is finite, we assume for simplicity that $X^* = \sup_n X_n \in L^1$. (The theory remains valid if we only assume $\sup_n X_n^+ \in L^1$ and $\mathbf{E}(X_S^-) < \infty$ for all $S \in \mathscr{R}$.)

If $X_0, X_1, \ldots, X_p$ is a *finite* adapted sequence, one may solve the problem by a 'dynamic programming', or *backward induction* principle: if we set $Z_p = X_p$ and for $n = 1, 2, \ldots, p$ we let $Z_{p-n} = \max(X_{p-n}, \mathbf{E}(Z_{p-n+1}|\mathscr{F}_{p-n}))$, then (Z_k) is a supermartingale, and by construction is the smallest supermartingale dominating (X_k). From this it is not hard to see (Exercise!) that $T_0 = \min(n: X_n = Z_n)$ defines the desired optimal stopping time. Details can be found in [12], [68].

Expressing (Z_n) in terms of the (X_n) directly poses a difficulty, since the family $\{X_T: T\in\mathscr{R}\}$ is uncountable, and the supremum of such a family need not be measurable. For example, if $A \subseteq [0,1]$, $\sup\{1_{\{a\}}: a\in A\} = 1_A$, which is only Borel-measurable if $A \in \mathscr{B}([0,1])$. The following construction resolves this difficulty.

2.11.1. ***Proposition:*** If L is a family of measurable functions $\Omega \to [-\infty, \infty]$ then there is a unique measurable function $g: \Omega \to [-\infty, \infty]$ such that

(i) $g \geqslant f$ for all $f \in L$, and

(ii) whenever h is a measurable function with $h \geqslant f$ for all $f \in L$, then $h \geqslant g$.

* The results in this section are not used in the rest of this book and may be omitted at a first reading.

We denoye g by ess sup(L), the *essential supremum* of L. There exists a sequence (f_n) in L with $\sup_n f_n = g$, and if L is upward filtering, (f_n) can be taken to be increasing, so that $g = \lim_n f_n$.

Sketch proof: (Details can be found in [68].) We may identify $[0,1]$ with $[-\infty,\infty]$ by means of an increasing bijection. For any countable family K in L let $f_K = \sup_K f$. Since f_K takes values in $[0,1]$, $\mathbf{E}(f_K)$ is well-defined. Let $\alpha = \sup \mathbf{E}(f_K)$, where K ranges through all countable sets $K \subseteq L$. Taking any sequence (K_n) of such sets with $\mathbf{E}(f_{K_n}) \to \alpha$, we see that the countable set $K^* = \bigcup_n K_n$ has $\mathbf{E}(f_{K^*}) = \alpha$. Then $g = f_{K^*}$ is the required function, since for given $f \in L$, $K' = K^* \bigcup \{f\}$ has $f_{K'} = g \vee f$, so that $\alpha \leqslant \mathbf{E}(f_{K'})$, which implies that $f \leqslant g$ a.s. Using K^* as the sequence (f_n) completes the proof. When L is upward filtering, replace (f_n) by (f'_n), where $f'_0 = f_0, f'_{n+1} \geqslant f'_n \vee f_{n+1}$, which exists in L for all n.

2.11.2: Now suppose we are given an adapted sequence (X_n) with $X^* = \sup_n X_n$ in L^1. Define the set $\mathscr{R}_n$ of stopping rules T such that $n \leqslant T < \infty$ a.s. Thus if the gambler plays at least until time n, his stopping strategy T has $\mathbf{E}(X_T | \mathscr{F}_n)$ as the best estimate at time n of his eventual fate. The optimal stopping problem hinges on the random variables (Z_n) defined by $Z_n = \text{ess sup}_{T \in \mathscr{R}_n} \mathbf{E}(X_T | \mathscr{F}_n)$. The sequence (Z_n) is called the *Snell envelope* of (X_n). It has the following properties:

2.11.3. *Theorem:* If $X = (X_n)$ is an adapted sequence with $X^* \in L^1$, the Snell envelope $\mathrm{Z} = (Z_n)$ satisfies

(i) $Z_n = \max(X_n, \mathbf{E}(Z_{n+1} | \mathscr{F}_n))$ for all $n \geqslant 0$,

(ii) $\mathbf{E}(Z_n) = \sup_{T \in \mathscr{R}_n} \mathbf{E}(X_T)$ for all $n \geqslant 0$.

If $X_n \geqslant 0$ for all n, (Z_n) is the smallest supermartingale dominating (X_n).

Proof: As the supremum of a sequence of $\mathscr{F}_n$-measurable functions, Z_n is $\mathscr{F}_n$-measurable, so Z is adapted. Since $X_n \leqslant Z_n \leqslant X^*$ a.s., $Z_n \in L^1$. Moreover, the family $(\mathbf{E}(X_T | \mathscr{F}_n))_{T \in \mathscr{R}_n}$ is upward filtering: this follows from the fact that if $T_1, T_2 \in \mathscr{R}_n$ and $A \in \mathscr{F}_n$ are given, then $T = T_1 1_A + T_2 1_{A^c}$ belongs to $\mathscr{R}_n$. Setting $A = \{\mathbf{E}(X_{T_2} | \mathscr{F}_n) < \mathbf{E}(X_{T_1} | \mathscr{F}_n)\}$ we have $\mathbf{E}(X_T | \mathscr{F}_n) = \max(\mathbf{E}(X_{T_1} | \mathscr{F}_n), \mathbf{E}(X_{T_2} | \mathscr{F}_n))$. Hence there is an increasing sequence (T_m) in $\mathscr{R}_n$ with $Z_n = \lim_m \mathbf{E}(X_{T_m} | \mathscr{F}_n)$. Applying $\mathbf{E}(\cdot | \mathscr{F}_{n-1})$ we have $\mathbf{E}(Z_n | \mathscr{F}_{n-1}) = \lim_m \mathbf{E}(X_{T_m} | \mathscr{F}_{n_1}) \leqslant Z_{n-1}$ since $\mathscr{R}_n \subseteq \mathscr{R}_{n-1}$. Thus Z is a supermartingale. Since $Z_n \geqslant X_n$ by construction (the constant stopping time n is in $\mathscr{R}_n$), we have shown that $Z_n \geqslant \max(X_n, \mathbf{E}(Z_{n+1} | \mathscr{F}_n)$. The reverse inequality follows if we set $X_T = X_n 1_{\{T=n\}} + X_{T \vee (n+1)} 1_{\{T>n\}}$ for $T \in \mathscr{R}_n$. Since $T \vee (n+1) \in \mathscr{R}_{n+1} \subseteq \mathscr{R}_n$ we have $Z_{n+1} \geqslant \mathbf{E}(X_{T \vee (n+1)} | \mathscr{F}_{n+1})$, so that

$$\mathbf{E}(X_T|\mathscr{F}_n)=X_n 1_{\{T=n\}}+1_{\{T>n\}}\mathbf{E}(X_{T\vee(n+1)}|\mathscr{F}_n)$$
$$\leqslant X_n 1_{\{T=n\}}+1_{\{T>n\}}\mathbf{E}(Z_{n+1}|\mathscr{F}_n)\leqslant\max(X_n,\mathbf{E}(Z_{n+1}|\mathscr{F}_n)).$$

This proves (i).

For (ii), it is clear that $\mathbf{E}(Z_n)\geqslant\mathbf{E}(X_T)$ for all $T\in\mathscr{R}_n$. On the other hand, $Z_n=\sup_n\mathbf{E}(X_{T_m}|\mathscr{F}_n)$, hence $\mathbf{E}(Z_n)\leqslant\sup_{\mathscr{R}_n}\mathbf{E}(X_T)$. Hence $\mathbf{E}(Z_n)=\sup_{\mathscr{R}_n}\mathbf{E}(X_T)$ for all $n\geqslant 0$.

If $X_n\geqslant 0$ for all n, so is Z_n and we have already shown that Z is a supermartingale. If $Y=(Y_n)$ is a positive supermartingale dominating X, then optional sampling gives $Y_n\geqslant\mathbf{E}(Y_T|\mathscr{F}_n)\geqslant\mathbf{E}(X_T|\mathscr{F}_n)$ for all $T\in\mathscr{R}_n$. Hence Y dominates Z also.

Not only is the Snell envelope a supermartingale, but it is a martingale until the first optimal stopping time if one exists. To see this we define the 'hitting time of the set $\{X=Z\}$' (cf. Chapter 3) by

$$H_0=\inf\{n: X_n=Z_n\} \text{ or } +\infty \text{ if } Z_n>X_n \text{ for all } n.$$

We shall show that $\sup\mathbf{E}(X_T)$, where T ranges over all stopping rules, is attained iff H_0 is a stopping rule. First we have:

2.11.4. ***Lemma:*** If $T\leqslant H_0$ the sequence $Z^T=(Z_{T\wedge n})$ is a martingale. If T is also a stopping rule, $\mathbf{E}(Z_T)\geqslant\mathbf{E}(Z_0)$.

Proof: If $T>n$, $H_0>n$, so $Z_n>X_n$ a.s. So by Theorem 2.11.3(i), we have $Z_n=\mathbf{E}(Z_{n+1}|\mathscr{F}_n)$ on $A=\{T>n\}$. Since $A\in\mathscr{F}_n$ we have $\mathbf{E}(Z_{T\wedge(n+1)}|\mathscr{F}_n)=Z_T 1_{A^c}+\mathbf{E}(Z_{n+1}|\mathscr{F}_n)1_A=Z_T 1_{A^c}+Z_n 1_A=Z_{T\wedge n}$, so $(Z_{T\wedge n})$ is a martingale. Hence $\mathbf{E}(Z_{T\wedge n})=\mathbf{E}(Z_0)$ for all n. But (Z_n) is dominated by $Y_n=\mathbf{E}(\sup_k X_k|\mathscr{F}_n)$, and we can apply Fatou's lemma to the non-negative sequence $(Y_n^T-Z_n^T)$ to obtain, if $T<\infty$ a.s., $\liminf_{n\to\infty}\mathbf{E}(Y_{T\wedge n}-Z_{T\wedge n})\geqslant\mathbf{E}(Y_T-Z_T)$. On the other hand, $\mathbf{E}(Y_{T\wedge n})=\mathbf{E}(Y_T)$ for all n, so $\mathbf{E}(Z_T)\geqslant\limsup_{n\to\infty}\mathbf{E}(Z_{T\wedge n})=\mathbf{E}(Z_0)$.

The lemma clarifies why H_0 must be optimal whenever it is a stopping rule: if H_0 is a.s. finite, $\mathbf{E}(X_{H_0})=\mathbf{E}(Z_{H_0})\geqslant\mathbf{E}(Z_0)$ by the lemma, but $\mathbf{E}(Z_0)=\sup_{T\in\mathscr{R}_0}\mathbf{E}(X_T)\geqslant\mathbf{E}(X_{H_0})$ if H_0 is a stopping rule (hence belongs to $\mathscr{R}_0$). So H_0 is optimal.

To see that H_0 is the *first* optimal stopping rule if one exists, we need the following 'optional sampling' result for Z:

2.11.5. ***Lemma:*** Let T_1 be a stopping rule, and denote by $\mathscr{R}_{T_1}$ the set of stopping rules $T\geqslant T_1$. Then $Z_{T_1}=\operatorname{ess\,sup}_{T\in\mathscr{R}_{T_1}}\mathbf{E}(X_T|\mathscr{F}_{T_1})$ and hence $\mathbf{E}(Z_{T_1})=\sup_{\mathscr{R}_{T_1}}\mathbf{E}(X_T)$.

Proof: $\mathscr{R}_{T_1}$ contains T_1, hence is non-empty. Write ${}_{T_1}Z = \text{ess sup}_{\mathscr{R}_{T_1}} \mathbf{E}(X_T|\mathscr{F}_{T_1})$. For $S \in \mathscr{R}_{T_1}$, we have $S \vee n \in \mathscr{R}_n$ and $S = S \vee n$ on $\{T_1 = n\}$. Hence $\mathbf{E}(X_S|\mathscr{F}_{T_1}) = \mathbf{E}(X_{S \vee n}|\mathscr{F}_n) \leqslant Z_n = Z_{T_1}$ on $\{T_1 = n\}$. This holds for all $n \in \mathbf{N}$ and $S \in \mathscr{R}_{T_1}$, so ${}_{T_1}Z \leqslant Z_{T_1}$ a.s. On the other hand if $S \in \mathscr{R}_n$, $S \vee T_1 \in \mathscr{R}_{T_1}$ and $S = S \vee T_1$ on $\{T = n\}$, hence $\mathbf{E}(X_S|\mathscr{F}_n) = \mathbf{E}(X_{S \vee T_1}|\mathscr{F}_{T_1}) \leqslant {}_{T_1}Z$ on $\{T_1 = n\}$ and therefore $Z_{T_1} \leqslant {}_{T_1}Z$ a.s. This proves the first assertion.

Finally, we have $\mathbf{E}(Z_{T_1}) \geqslant \mathbf{E}(X_T)$ for all $T \in \mathscr{R}_{T_1}$, while on the other hand, $\{\mathbf{E}(X_T|\mathscr{F}_{T_1}) : T \in \mathscr{R}_{T_1}\}$ is upward filtering and so $\mathbf{E}(Z_{T_1}) = \sup_m \mathbf{E}(X_{S_m}|\mathscr{F}_{T_1})$ for some sequence (S_m) in $\mathscr{R}_{T_1}$. Hence $\mathbf{E}(Z_{T_1}) \leqslant \sup_{\mathscr{R}_{T_1}} \mathbf{E}(X_T)$ also and we have equality as required.

Now suppose that T^* is an optimal stopping rule, so that $\mathbf{E}(X_{T^*}) = \sup_{\mathscr{R}_n} \mathbf{E}(X_T)$. Then by Lemma 2.11.5 we have $\mathbf{E}(Z_{T^*}) = \sup_{\mathscr{R}_{T^*}} \mathbf{E}(X_T) \leqslant \sup_{\mathscr{R}_0} \mathbf{E}(X_T) = \mathbf{E}(X_{T^*})$. But since $X_{T^*} \leqslant Z_{T^*}$ a.s. we must then have $X_{T^*} = Z_{T^*}$ a.s. By definition of H_0 we conclude that $H_0 \leqslant T^*$, so that H_0 is also a stopping rule and is the first optimal one, by the remark preceding Lemma 2.11.5. To summarise:

2.11.6. ***Theorem:*** Let (X_n) be an adapted sequence with $\sup_n X_n \in L_1$. Then $\sup_{\mathscr{R}_0} \mathbf{E}(X_T)$, where $\mathscr{R}_0$ is the set of all stopping rules, is attained iff

$$H_0 = \begin{cases} \inf\{n : X_n = Z_n\} \\ +\infty \text{ if } Z_n > X_n \text{ for all } n \end{cases}$$

is a stopping rule. In that case H_0 is the first optimal stopping rule for (X_n).

2.11.7. ***Exercise:*** Suppose we are given a stopping rule T. Let $\mathscr{R}_T$ be the set of all stopping rules dominating T and define the stopping time $H_T = \inf\{n \geqslant T : X_n = Z_n\}$, where $\inf \varnothing = +\infty$. Show that $\sup_{S \in \mathscr{R}_T} \mathbf{E}(X_S)$ is attained iff H_T is a stopping rule, and that in that case H_T is the smallest optimal stopping rule in $\mathscr{R}_T$. (*Hint:* show that if $S \leqslant H_T$, then with $A = \{T \leqslant n < S\}$ we have $Z_n = \mathbf{E}(Z_{n+1}|\mathscr{F}_n)$ and hence $\mathbf{E}(Z_{(S \wedge n) \vee T}) = \mathbf{E}(Z_T)$ follows by induction on $n \in \mathbf{N}$. As in Lemma 2.11.4, $\mathbf{E}(Z_S) = \mathbf{E}(Z_T)$. The rest is as before.)

2.11.8. ***Remark:*** One can also ask for the *largest* optimal stopping rule to find a characterisation of optimal rules. The idea here is to discover when Z^S ceases to be a martingale, where $S \in \mathscr{R}_0$ is given. The Doob decomposition (Definition 2.9.1) $Z = M - B$ of Z into a martingale M and a predictable increasing process B with $B_0 = 0$ yields the following answer: let $K_0 = \sup\{n \geqslant 0 : \; B_n = 0\}$. Since B is predictable $\{K_0 = k\} = \bigcap_{m \leqslant k} [B_m = 0] \bigcap \{B_{k+1} > 0\} \in \mathscr{F}_k$ for all $k \in \mathbf{N}$, hence K_0 is a stopping time.

This 'exit time' of B from $\{0\}$ thus gives us the last time where $\mathbf{E}(B_{K_0})=\mathbf{E}(B_0)=0$. We have

2.11.9. ***Lemma:*** $H_0 \leqslant K_0$ a.s.

Proof: If $S \leqslant H_0$ is a stopping rule, Lemma 2.11.4 ensures that $\mathbf{E}(Z_{S\wedge n})=\mathbf{E}(Z_0)$ for all $n\in\mathbf{N}$. Writing $Z=M-B$ for the Doob decomposition of Z, we have $\mathbf{E}(M_{S\wedge n})=\mathbf{E}(M_0)$ by Theorem 2.4.7, since $S\wedge n$ is bounded. So $\mathbf{E}(B_{S\wedge n})=\mathbf{E}(B_0)=0$, hence $B_{S\wedge n}=0$ a.s. This implies that $S \leqslant K_0$ a.s. Now if $S_n=H_0\wedge n$ for $n\in\mathbf{N}$, we have $S_n\leqslant H_0$, hence $S_n\leqslant K_0$, hence also $H_0\leqslant K_0$ a.s.

This result, together with Lemma 2.11.4, allows us to characterise all optimal stopping rules. More generally, if $T\in\mathscr{R}_0$ is given, we can characterise all optimal rules in $\mathscr{R}_T$ in terms of the stopping time $K_T=\sup\{n\geqslant T: B_n=B_T\}$. The idea for this characterisation for continuous-time processes is due to Bismut and Skalli [5]. The corresponding discrete-time result is due to N.J.Cutland, and is contained, for the case $T=0$, in [22].

First consider the set $A_m=\{T=m\}$ for $m\geqslant 0$: we let $X'_n=X_{m+n}$, $\mathscr{F}'_n=\mathscr{F}_{m+n}$. Then it is clear that $Z'_n=Z_{m+n}$ relates the Snell envelopes of these processes, and

$$B'_{n+1}-B'_n=Z'_n-\mathbf{E}(Z'_{n+1}|\mathscr{F}'_n)=Z_{m+n}-\mathbf{E}(Z_{m+n+1}|\mathscr{F}_{m+n})=B_{m+n+1}-B_{m+n}.$$

Hence *on* A_m we have

(i) $H_T=\inf\{n\geqslant m: X_n=Z_n\}=m+\inf\{k: X_{m+k}=Z_{m+k}\}=$ $m+\inf\{n: X'_n=Z'_n\}=m+H'_0$;

(ii) $K_T=\sup\{n\geqslant m: B_n-B_m=0\}=m+\sup\{n\geqslant 0: B'_n=B'_0\}=m+K'_0$;

where H'_0 and K'_0 are defined in the obvious way. Consequently we have proved:

2.11.10. ***Lemma:*** $H_T\leqslant K_T$ a.s. for all $T\in\mathscr{R}_0$. This allows us to characterise optimal stopping rules in $\mathscr{R}_0$:

2.11.11. ***Theorem:***

(i) A stopping rule $T\in\mathscr{R}_0$ is optimal iff $X_T=Z_T$ a.s. and $T\leqslant K_0$.

(ii) If K_0 is a.s. finite, then it is the largest optimal stopping rule.

Proof: If T is optimal in $\mathscr{R}_0$, then it is optimal in $\mathscr{R}_T$, so $\mathbf{E}(X_T)=\mathbf{E}(Z_0)=\mathbf{E}(Z_T)=\mathbf{E}(Z_{T\wedge n})$ by Lemma 2.11.5. But $X_T\leqslant Z_T$ a.s., hence $X_T=Z_T$ a.s. Since $Z=M-B$ and $\mathbf{E}(M_{T\wedge n})=\mathbf{E}(M_0)$ by Theorem 2.4.7, we also have $\mathbf{E}(B_{T\wedge n})=\mathbf{E}(B_0)=0$, so that $T\leqslant K_0$ a.s.

Conversely, if $T \in \mathscr{R}_0$ satisfies $T \leqslant K_0$ and $Z_T = X_T$, then $\mathbf{E}(M_{T\wedge n}) = \mathbf{E}(M_0)$, while $\mathbf{E}(B_{T\wedge n}) = 0$. Hence $\mathbf{E}(Z_{T\wedge n}) = \mathbf{E}(Z_0)$. As in Lemma 2.11.4, with $Y = \sup_m X_m$, $Y_n = \mathbf{E}(Y|\mathscr{F}_n)$ and $U_n = (Y_{T\wedge n} - Z_{T\wedge n}) \geqslant 0$, we can apply Fatou's lemma and Theorem 2.6.6 to obtain $\mathbf{E}(Z_T) \geqslant \lim_n \sup \mathbf{E}(Z_{T\wedge n}) = \mathbf{E}(Z_0)$. On the other hand $\mathbf{E}(Z_T) \leqslant \mathbf{E}(Z_0)$ by Lemma 2.11.5, and $X_T = Z_T$, hence $\mathbf{E}(X_T) = \mathbf{E}(Z_T) = \mathbf{E}(Z_0) = \sup_{\mathscr{R}_0} \mathbf{E}(X_S)$, so T is optimal.

To see that K_0 is optimal we need only show that $X_{K_0} = Z_{K_0}$ and apply (i). Now by Lemma 2.11.10, $H_T \leqslant K_T$ a.s. for any $T \in \mathscr{R}_0$. If K_0 is a.s. finite, we can deduce that $H_{K_0} \leqslant K_{K_0} = K_0$ and since $K_0 \leqslant H_{K_0}$ by definition, we have shown that $K_0 = H_{K_0}$. Hence $X_{K_0} = Z_{K_0}$ as required.

2.11.12: The extension of Theorem 2.11.11 to a characterisation of optimal stopping times in $\mathscr{R}_T$, for arbitrary $T \in \mathscr{R}_0$, is complicated by the fact that we cannot use optional sampling directly on the martingale M if T is unbounded. However, since the range of T is countable, we can deal separately with X and Z on each of the sets $A_m = \{T = m\}$. Thus, define a new process $Y = (Y_n)$ by

$$Y_n = \begin{cases} X_n 1_{A_m} & \text{for } n \geqslant m \\ 0 & \text{for } n < m \end{cases}.$$

Then Y is adapted and the stopping time

$$S_m = \begin{cases} H_T & \text{on } A_m \\ m & \text{on } A_m^c \end{cases}$$

is optimal in the following sense: $\mathbf{E}(Y_{S_m}) = \sup_{S \in \mathscr{R}_m} \mathbf{E}(Y_S)$. This follows easily from Exercise 2.11.7.

Now let $S \in \mathscr{R}_T$ be optimal, so that $\mathbf{E}(X_S) = \sup_{T' \in \mathscr{R}_T} \mathbf{E}(X_{T'})$. Fix $m \in \mathbf{N}$ and define $Y = (Y_n)$ as above. Also let

$$\bar{S} = \begin{cases} S & \text{on } A_m \\ m & \text{on } A_m^c \end{cases}.$$

We claim that this is an optimal stopping time for Y in $\mathscr{R}_m$. For if $\mathbf{E}(Y_{\bar{S}}) < \mathbf{E}(W_{\bar{S}})$, where W denotes the Snell envelope of Y, then $\mathbf{E}(X_S 1_{A_m}) < \mathbf{E}(Z_S 1_{A_m})$, which implies that $\mathbf{E}(X_S) < \mathbf{E}(Z_S)$ since $X_n \leqslant Z_n$ for all n. Hence S is not optimal. But optimal stopping rules $S \in \mathscr{R}_m$ satisfy $m \leqslant S \leqslant K_m$ and $X_S = Z_S$, as is easily seen from the characterisation in Theorem 2.11.11. Hence on A_m we have $T \leqslant S \leqslant K_T$ and $X_S = Z_S$. Conversely, if a stopping rule $S \in \mathscr{R}_T$ satisfies $S \leqslant K_T$ and $X_S = Z_S$, then by the above construction, with fixed A_m, we have $\mathbf{E}(X_S 1_{A_m}) = \mathbf{E}(Z_S 1_{A_m}) = \mathbf{E}(Z_T 1_{A_m})$. Hence also

$\mathbf{E}(X_S 1_{\bigcup_1^k A_m}) = \mathbf{E}(Z_T 1_{\bigcup_1^k A_m})$ for all $k \in \mathbf{N}$, and since X_S and Z_T are in L^1, we have $\mathbf{E}(X_S) = \mathbf{E}(Z_T)$. This again implies that S is optimal in $\mathscr{R}_T$. Thus we have

2.11.13. ***Theorem:*** Let $T \in \mathscr{R}_0$ be given. Then

(i) $S \in \mathscr{R}_T$ is optimal iff $X_S = Z_S$ and $S \leqslant K_T$.

(ii) If K_T is a.s. finite, then it is the largest optimal stopping rule in $\mathscr{R}_T$.

Proof: (i) was proved above. (ii) follows as in Theorem 2.11.11, with K_T replacing K_0.

2.11.14. ***Remark:*** For any $\varepsilon > 0$, one can find explicit *ε-optimal* stopping rules, i.e. T such that $\mathbf{E}(X_T) + \varepsilon \geqslant \sup_{S \in \mathscr{R}_0} \mathbf{E}(X_S)$, by means of the Snell envelope Z of X: define $T_\varepsilon = \inf\{n \geqslant 0: X_n \geqslant Z_n - \varepsilon\}$. Then $T_\varepsilon \leqslant H_0$, so $(Z_{T_\varepsilon \wedge n})_n$ is a martingale by Lemma 2.11.4. Moreover, $\sup_n \|Z_{T_\varepsilon \wedge n}\|_1 \leqslant \|\sup_n X_n\|_1$ so by Theorem 2.6.1 $(Z_{T_\varepsilon \wedge n})$ converges a.s., so that $\lim_n Z_n$ is finite on the set $\{T_\varepsilon = +\infty\}$. By definition of T_ε, $Z_n \geqslant X_n + \varepsilon$ on this set, so that $\lim_n Z_n \geqslant \limsup_n X_n + \varepsilon$ on $\{T_\varepsilon = +\infty\}$. On the other hand, $\limsup_n Z_n = \limsup_n X_n$ if $\limsup_n Z_n < \infty$: for we have $X_S \leqslant \sup_{m \geqslant k} X_m$ for $S \in \mathscr{R}_k$, so if $n \geqslant k$, $Z_n \leqslant \mathbf{E}(\sup_{m \geqslant k} X_m | \mathscr{F}_n)$. But $\sup_{m \geqslant k} X_m \in L^1(\mathscr{F}_\infty)$, so $\limsup_{n \to \infty} Z_n \leqslant \sup_{m > k} X_m$ by Theorem 2.6.6. Letting $m \to \infty$ we have $\limsup_n Z_n \geqslant \limsup_n X_n$. As $X_n \leqslant Z_n$ the reverse inequality is obvious. Hence we have $P(T_\varepsilon = +\infty) = 0$, so T_ε is a stopping rule. Finally $\mathbf{E}(X_{T_\varepsilon}) + \varepsilon \geqslant \mathbf{E}(Z_{T_\varepsilon}) \geqslant \mathbf{E}(Z_0)$ by Lemma 2.11.4, so T_ε is ε-optimal.

3

Continuous-time martingales

The interplay between analysis and probability theory which creates 'stochastic analysis' begins in earnest when the parameter set or *time set* $\mathbf{T}$, by which we index the families of random variables to be considered, is an interval contained in $[0, \infty]$. For simplicity we shall take $\mathbf{T}$ to be $[0, \infty[$, unless explicitly stated otherwise. We again fix a complete probability space $(\Omega,\mathscr{F},P)$.

3.1. Stochastic processes and stochastic bases

A (*stochastic*) *process* is a map $X:\mathbf{T}\times\Omega\to\mathbf{R}$ such that the maps $\omega\to X(t,\omega)$ are measurable (from $(\Omega,\mathscr{F})$ to $(\mathbf{R},\mathscr{B}(\mathbf{R}))$ for all $t\in\mathbf{T}$. Equivalently, X may be considered as a family $X=(X_t)$ of random variables *indexed* by $\mathbf{T}$. (We only consider real-valued processes, i.e. with *state space* $(\mathbf{R},\mathscr{B}(\mathbf{R}))$. The definition can be extended by using an arbitrary measurable space instead.)

The process X is *measurable* if the map X is measurable for the product σ-field $\mathscr{B}(\mathbf{T})\times\mathscr{F}$, where $\mathscr{B}(\mathbf{T})$ is the Borel σ-field on $\mathbf{T}$. Note that this was automatic in the discrete case (when $\mathbf{T}=\mathbf{N}$). In continuous time the *paths* $t\to X(t,\omega)$, that is, the 'time evolution' of $X(t,\omega)$ for fixed $\omega\in\Omega$, become the principal objects of study. We therefore make the following identification: two processes X and Y are said to be *indistinguishable* if for P-almost all $\omega\in\Omega$ the corresponding paths of X and Y coincide as functions of t. We shall regard X and Y as the same process in this case, so that the basic object is actually an equivalence class of stochastic processes. A process in the class of the zero process is called *evanescent*. More generally, we say that the process X has a certain property (e.g. continuity) iff P-almost all paths of X have that property.

One may also regard X as a *random function*, that is, as a map from $\mathbf{T}$ into the space $L^0(\mathscr{F})$ of (P-equivalence classes of) measurable functions $(\Omega,\mathscr{F})\to(\mathbf{R},\mathscr{B}(\mathbf{R}))$. For each $t\in T$, $X_t:=X(t)$ is a function in $L^0(\mathscr{F})$. (Again we suppress the fact that L^0 consists of P-equivalence classes of functions.) Now if Y is another random function and for each $t\in T$, X_t and Y_t define the same function in L^0, i.e. $X_t = Y_t$ P-a.s., then Y is a *modification* of X. In the discrete case any two modifications are indistinguishable, as the index set, $\mathbf{N}$, is countable. In the present set-up this no longer holds in general. However, if X and Y are *right-continuous* processes (almost all *paths* are right-continuous!) we may restrict attention to rationals in $\mathbf{T}$, so if Y is a modification of X, X and Y are indistinguishable also. The same remarks hold for left-continuous processes. We shall see that in the case of (super-) martingales we can restrict our attention to such processes.

In order to model the time-evolution of our 'knowledge' of a process $X=(X_t)$ we need measurability conditions that vary with t. So define a *filtration* $(\mathscr{F}_t)_{t\in\mathbf{T}}$ as an increasing family ($\mathscr{F}_s \subseteq \mathscr{F}_t$ if $s \leqslant t$ in $\mathbf{T}$) of sub-σ-fields of $\mathscr{F}$. We shall say that $(\Omega,\mathscr{F},P,(\mathscr{F}_t),\mathbf{T})$ is a *stochastic base* if the probability space $(\Omega,\mathscr{F},P)$ is complete and if the filtration $(\mathscr{F}_t)$ indexed by $\mathbf{T}$ satisfies the following ('usual' [19]) conditions:

(i) $\mathscr{F}_0$ contains all P-null sets.

(ii) $(\mathscr{F}_t)$ is *right-continuous*: $\mathscr{F}_t = \mathscr{F}_{t+} := \bigcap_{t<s} \mathscr{F}_s$ for all $t\in\mathbf{T}$.

A process X is *adapted to* $(\mathscr{F}_t)$ if for each $t\in\mathbf{T}$, X_t is $\mathscr{F}_t$-measurable. Since all P-null sets belong to each $\mathscr{F}_t$, any modification of an adapted process is also adapted to $(\mathscr{F}_t)$.

Again, the condition (ii) is no real restriction if the process X is adapted to $(\mathscr{F}_t)$ and has right-continuous paths: for then X_t is $\mathscr{F}_{t+}$-measurable, so X is adapted to the filtration $(\mathscr{F}_{t+})$, which certainly satisfies (ii). For a further discussion of the 'usual conditions' see [83; Ch. 1, pp. 27–9, Ch. 2, pp. 35–40].

3.2. Progressive processes and stopping times

Now fix a stochastic base $(\Omega,\mathscr{F},P,(\mathscr{F}_t),\mathbf{T})$. Let X be a measurable process. Measurability relative to the filtration is then defined as follows: X is *progressive* (or progressively measurable w.r.t. $(\mathscr{F}_t)$) if for each $t\in T$, the map $(s,\omega)\mapsto X(s,\omega)$, defined on $[0,t]\times\Omega$, is measurable for the product $\mathscr{B}_t \times \mathscr{F}_t$. Here $\mathscr{B}_t$ denotes the Borel σ-field on $[0,t]$.

It is easy to see that every progressive process is adapted.

Exercise: Prove this, noting that $\mathscr{B}_t \times \mathscr{F}_t$ is generated by 'rectangles'.

The converse is false, however: let $\Omega = \mathbf{T} = \mathbf{R}^+$, with the probability measure P defined, for example, as $P(A) = \int_A e^{-x}\,dx$, and with all the σ-fields $\mathscr{F}_t$ being generated by the points of $\mathbf{R}^+$. Now define $X = 1_D$, where $D = \{(t,t): t\in\mathbf{R}^+\}$ is the *diagonal* of $\mathbf{T}\times\Omega$. Then for *each* $t\in\mathbf{T}$,

$$\omega \to X_t(\omega) = 1_D(t,\omega) = \begin{cases} 1 \text{ if } \omega = t \\ 0 \text{ if } \omega \neq t \end{cases}$$

is $\mathscr{F}_t$-measurable, as $\{\omega : X_t(\omega) > 0\} = \{t\} \in \mathscr{F}_t$. However, the map $X : (t,\omega) \mapsto 1_D(t,\omega)$ is not $\mathscr{B}_t \times \mathscr{F}_t$-measurable:

$$\{(s,\omega) : X(s,\omega) > 0,\ s \in [0,t]\} = \{(s,s) : 0 \leqslant s \leqslant t\}$$

has the interval $[0,t]$ as its projection onto Ω, and $[0,t] \notin \mathscr{F}_t$, being uncountable.

So we need additional conditions on X to obtain the converse. Again, right-continuity of the paths is sufficient:

3.2.1. ***Proposition:*** If X is adapted and right-continuous then X is progressive.

Proof: Fix $t \in \mathbf{T}$, and divide $[0,t]$ into dyadic intervals of order n. Define

$$X^n(s,\omega) = X(k/2^n, \omega) \text{ for } (k-1)/2^n \leqslant s < k/2^n,$$

where $k/2^n$ ranges from $1/2^n$ to $[2^n t]/2^n$. Also let $X^{(n)}(t,\omega) = X(t,\omega)$ for all $n \in \mathbf{N}$. The process $X^{(n)}$ is then a simple process on $[0,t] \times \Omega$, hence is measurable for $\mathscr{B}_t \times \mathscr{F}_t$. (Note that it need not be adapted!) Since almost all paths $s \to X(s,\omega)$ are right continuous, there is a P-null set N such that for $\omega \notin N$ we have $\lim_{u \downarrow s} X(u,\omega) = X(s,\omega)$ for all $s \in [0,t]$. Hence $X^{(n)}(s,\omega) \mapsto X(s,\omega)$ for all $s \in [0,t]$ and $\omega \notin N$. But since $X^{(n)}$ is $\mathscr{B}_t \times \mathscr{F}_t$-measurable for each n, so is X, i.e. X is progressive. (*Exercise:* Prove that each right-continuous process is measurable.)

An important property of progressive measurability is that it is preserved under stopping. To see this we first define stopping times.

3.2.2. ***Definition:*** A random variable $S : \Omega \to [0,\infty]$ is a *stopping time* for the filtration $(\mathscr{F}_t)$ if $\{S \leqslant t\} \in \mathscr{F}_t$ for each $t \in \mathbf{T}$. The natural σ-field of 'events that occur by time S' is given by

$$\mathscr{F}_S = \{A \in \mathscr{F} : A \cap \{S \leqslant t\} \in \mathscr{F}_t \text{ for each } t \in \mathbf{T}\}.$$

Since we assume that the filtration $(\mathscr{F}_t)$ is right-continuous, it is actually enough to show that $\{S < t\} \in \mathscr{F}_t$ for all $t \in \mathbf{T}$: for, in that case,

$$\text{if } t \in \mathbf{T},\ \{S \leqslant t\} = \bigcap_n \{S < t + 1/n\} \in \bigcap_n \mathscr{F}_{t+1/n} = \mathscr{F}_{t+} = \mathscr{F}_t.$$

As in Exercise 2.4.2, we leave some simple properties of stopping times to the reader:

3.2.3. ***Exercises:*** Suppose that S and T are stopping times. Prove the following assertions:

(i) $S \vee T$, $S \wedge T$ and $S+t$ $(t \in \mathbf{T})$ are stopping times.

(ii) If $A \in \mathscr{F}_S$ then $A \bigcap \{S \leqslant T\} \in \mathscr{F}_T$.

(iii) If $S \leqslant T$ then $\mathscr{F}_S \subseteq \mathscr{F}_T$.

(iv) $\{S<T\}$, $\{S=T\}$, $\{S>T\}$ belong to both $\mathscr{F}_S$ and $\mathscr{F}_T$.

Trivially S is an $\mathscr{F}_S$-measurable random variable: if $A=\{S \leqslant t\}$, then $A \bigcap \{S \leqslant t\} \in \mathscr{F}_t$, hence $A \in \mathscr{F}_S$.

We can stop a process X by a stopping time S as follows: define X_S as the random variable $\omega \rightarrow X_{S(\omega)}(\omega)$ on $\{S<\infty\}$, then the *stopped process* X^S is given by $X_t^S(\omega) := X_{S \wedge t}(\omega)$ on $\{S<\infty\}$. Also set $X_S 1_{\{S<\infty\}}=0$ on $\{S=\infty\}$. This allows us to define X_S consistently, whether or not X_∞ is well-defined.

3.2.4. ***Proposition:*** If S is a stopping time and X is progressive, then X_S is $\mathscr{F}_S$-measurable and X^S is progressive.

Proof: We must show that for any Borel set $B \subseteq \mathbf{R}$ and $t \in \mathbf{T}$, $\{X_S \in B\} \bigcap \{S \leqslant t\}$ belongs to $\mathscr{F}_t$. Now this set equals $\{X_{S \wedge t} \in B\} \bigcap \{S \leqslant t\}$ so we need only show that X^S is progressive (when, in particular, $\{X_{S \wedge t} \in B\} \in \mathscr{F}_t$). Now $S \wedge t \leqslant t$, so by Exercise 3.2.3(iii) we know that $S \wedge t$ is $\mathscr{F}_t$-measurable. Hence $(s,\omega) \rightarrow (S(\omega) \wedge t, \omega)$ is $\mathscr{B}_t \times \mathscr{F}_t$-measurable and so is X, being progressive. So $(s,\omega) \rightarrow X(S(\omega) \wedge t, \omega)$ is $\mathscr{B}_t \times \mathscr{F}_t$-measurable, hence X^S is progressive.

3.2.5. ***Example (first hitting time):*** The 'first time' that a process X hits a Borel set $B \subseteq \mathbf{R}$ is defined as $D'_B(\omega) = \inf\{s>0: X_s \in B\}$, where $\inf \varnothing = +\infty$. In general it is very difficult to show that this is a stopping time: it can be false if $\mathscr{F}$ is not complete, see [19; Vol. I]. For now, we shall assume the following measure-theoretic result without proof (see Theorem 3.5.1 and the supplement to this chapter for a proof).

Let $(\Omega, \mathscr{G}, P)$ be a complete probability space and let $A \in \mathscr{B}(\mathbf{R}^+) \times \mathscr{G}$, where $\mathscr{B}(\mathbf{R}^+)$ denotes the Borel sets on $\mathbf{R}^+$. Then the *projection* $\Pi_\Omega(A) = \{\omega : (t,\omega) \in A \text{ for some } t \in \mathbf{R}^+\}$ is in $\mathscr{G}$.

We apply this to a *progressive* process X as follows: let $B \subseteq \mathbf{R}$ be a Borel set, $s \in \mathbf{T}$, and define $A_s = \{(u,\omega): X_{u \wedge s}(\omega) \in B\}$.

As X is progressive, the map $(u,\omega) \rightarrow X(u \wedge s, \omega)$ is $\mathscr{B}_s \times \mathscr{F}_s$-measurable, hence also $\mathscr{B}(\mathbf{R}^+) \times \mathscr{F}_s$-measurable. So $\Pi_\Omega(A_s) \in \mathscr{F}_s$ by the above result. However, since for $t \in \mathbf{T}$, $\inf(s>0\colon X_s(\omega) \in B < t$ iff $X_u(\omega) \in \mathscr{B}$ for some $u<t$ (and hence for some $n \in \mathbf{N}$, $u<t-1/n$), we have $\{D'_B<t\} = \Pi_\Omega(\bigcup_n A_{t-1/n})$. As $\bigcup_n A_{t-1/n} \in \mathscr{B}(\mathbf{R}^+) \times \mathscr{F}_t$, we have $\{D'_B<t\} \in \mathscr{F}_t$, so D'_B is a stopping time.

3.2.6. ***Exercise:*** Prove directly that D'_B is a stopping time if B is *open* and X has right-continuous paths.

3.3. Martingales: regularity properties and convergence

As always we fix a stochastic base $(\Omega,\mathscr{F},P,(\mathscr{F}_t),\mathbf{T})$.

3.3.1. ***Definition:*** A martingale X is a process with $X_t \in L^1$ for all $t \in \mathbf{T}$ and $X_t = \mathbf{E}(X_s|\mathscr{F}_t)$ for $s \geqslant t$ in $\mathbf{T}$. (X is thus automatically adapted to $(\mathscr{F}_t)$.) A *supermartingale* X is an adapted process with $X_t \in L^1$ and $X_t \geqslant \mathbf{E}(X_s|\mathscr{F}_t)$ for $s \geqslant t$. X is a submartingale if $-X$ is a supermartingale.

The following properties are immediate (cf. Properties 2.3.4):

(i) If X and Y are supermartingales, so is $X \wedge Y$.

(ii) A supermartingale X is a martingale iff $t \to \mathbf{E}(X_t)$ is constant.

(iii) The set of $(\mathscr{F}_t)$-martingales is a vector space.

(iv) If X is a (sub-)martingale and $\phi: \mathbf{R} \to \mathbf{R}$ is (increasing and) convex, then $\phi \circ X$ is a submartingale.

We now extend the basic inequalities of Chapter 2. Our first results hold without any restrictions on the filtration $(\mathscr{F}_t)$, but deal only with functions defined on a countable dense subset D in $\mathbf{T}$.

3.3.2. ***Proposition:*** Let $X = (X_t)_{t \in \mathbf{R}^+}$ be a supermartingale and D a countable dense subset of $\mathbf{T}$. The following inequalities are valid:

(i) for $\lambda > 0$, $$\lambda P\left(\sup_{t \in D}|X_t| \geqslant \lambda\right) \leqslant 3 \sup_{t \in \mathbf{T}} \mathbf{E}(|X_t|), \tag{3.1}$$

(ii) if X is an L^p-martingale,
$$\| \sup_{t \in D}|X_t| \|_p \leqslant \frac{p}{p-1} \sup_{t \in D} \|X_t\|_p \text{ for } 1 < p < \infty, \tag{3.2}$$

(iii) if for a compact interval $I \subseteq \mathbf{T}$, and $a < b$ in $\mathbf{R}$, $U_a^b(D \cap I)(\omega)$ denotes the number of upcrossings of $[a,b]$ by $X(t,\omega)$ as t increases through the countable set $D \cap I$, then
$$\mathbf{E}(U_a^b(D \cap I)) \leqslant \frac{1}{b-a} \sup_{t \in I} \mathbf{E}(X_t - a)^-. \tag{3.3}$$

Proof: If we replace D by a finite subset F, the inequalities follow from Theorems 2.6.10, 2.6.11 and Exercises 2.5.3 respectively. The result follows by taking the sup on the left over the class of finite subsets of D (or $D \cap I$).

3.3.3. ***Exercise:*** Show that if X is a right-continuous supermartingale, then almost all paths of X are bounded on each compact interval in $\mathbf{T}$.

3.3.4. ***Theorem:*** Let X be a supermartingale, and let D be a countable dense subset of $\mathbf{T}$. Then the restriction of D of $s \to X_s(\omega)$ has both left and right limits at all $t \in \mathbf{T}$ for almost all $\omega \in \Omega$.

Proof: The set $M_{n,a,b} = \{\omega : U_a^b(X, D \cap [0,n])(\omega) = +\infty\}$ of paths which upcross $[a,b]$ infinitely often in $D \cap [0,n]$ has measure 0 by (3.3). Hence so does $M = \bigcup_{\substack{a<b \in \mathbf{Q} \\ n \in \mathbf{N}}} M_{b,a,b}$. Thus if $\omega \notin M$ then

$$\limsup_{\substack{s \to t \\ s<t \\ s \in D}} X_s(\omega) = \liminf_{\substack{s \to t \\ s<t \\ s \in D}} X_s(\omega) = X_{t-}(\omega)$$

by definition of upper and lower limits, since D is dense in $\mathbf{T}$.

Similarly for $X_{t+}(\omega)$ if $\omega \notin M$.

On M we can set $X_{t-} = X_{t+} = 0$. Then X_{t-} and X_{t+} are random variables defined on $\mathbf{T}$. Note that for a complete filtration $(\mathscr{F}_t)$, X_{t+} is $\mathscr{F}_{t+}$-measurable and X_{t-} is $\mathscr{F}_{t-}$-measurable. With these hypotheses the supermartingale relation extends to X_{t+} and X_{t-}:

3.3.5. ***Lemma:*** With the above assumptions

$$X_t \geqslant \mathbf{E}(X_{t+} | \mathscr{F}_t) \text{ and } X_{t-} \geqslant \mathbf{E}(X_t | \mathscr{F}_{t-}) \text{ a.s.} \tag{3.4}$$

Proof: Let $t_n \downarrow t$ in D $(t_n > t)$, then $\mathscr{F}_{t_{n+1}} \subseteq \mathscr{F}_{t_n}$ and hence $X_{t_{n+1}} \geqslant \mathbf{E}(X_{t_n} | \mathscr{F}_{t_{n+1}})$, so $(X_{t_n}, \mathscr{F}_{t_n})$ is a reverse supermartingale and $\{\mathbf{E}(X_{t_n})\}$ is bounded above by $\mathbf{E}(X_t)$, so $(X_{t_n}, \mathscr{F}_{t_n})$ is uniformly integrable by Theorem 2.10.1. Hence if $A \in \mathscr{F}_t$ we have

$$\int_A X_t \, \mathrm{d}P \geqslant \int_A X_{t_n} \, \mathrm{d}P, \text{ and as } t_n \downarrow t, \int_A X_t \, \mathrm{d}P \geqslant \int_A X_{t+} \, \mathrm{d}P$$

since (X_{t_n}) is uniformly integrable. Hence $X_t \geqslant \mathbf{E}(X_{t+} | \mathscr{F}_t)$. On the other hand, if $s_n \uparrow t$ in D $(s_n < t)$ we can use the positive supermartingale $Y_n = X_{s_n} - \mathbf{E}(X_t | \mathscr{F}_{s_n})$ instead: we know $Y_n \to X_{t-} - \mathbf{E}(X_t | \mathscr{F}_{t-})$ and the limit must be non-negative.

These results enable us to determine under what conditions a supermartingale has a right-continuous modification:

3.3.6. ***Theorem:*** If $(\mathscr{F}_t)$ is a right-continuous filtration, then the supermartingale $(X_t, \mathscr{F}_t : t \in \mathbf{T})$ has a right-continuous modification if and only if $t \to \mathbf{E}(X_t)$ is right-continuous on $\mathbf{T}$.

Proof: Since $\mathcal{F}_t = \mathcal{F}_{t+}$, the first part of (3.4) reduces to $X_t \geqslant X_{t+}$, as X_{t+} is $\mathcal{F}_{t+}$-measurable. We obtain equality a.s. (and hence a right-continuous modification) iff the random variables have equal expectation. As in the lemma, if $t_n > t$ and $t_n \to t$, $(X_{t_n})_n$ is uniformly integrable, hence $\mathbf{E}(X_{t_n}) \to \mathbf{E}(X_{t+})$. But $s \to \mathbf{E}(X_s)$ is right-continuous at t iff $\mathbf{E}(X_{t_n}) \to \mathbf{E}(X_t)$. This is therefore equivalent to $X_t = X_{t+}$, and hence X_{t+} is a right-continuous modification of X. Conversely, if (Y_t) is a right-continuous modification of (X_t), then $\mathbf{E}(Y_t) = \mathbf{E}(X_t)$ for all $t \in \mathbf{T}$, hence $t \to \mathbf{E}(X_t)$ is right-continuous.

3.3.7. ***Remarks:***

(1) To see that the modification in Theorem 3.3.6 is in fact a supermartingale, we need only show that $(X_{t+}, \mathcal{F}_{t+})$ is a supermartingale: but if $s_n \downarrow s$, $t_n \downarrow t$, $s < s_n < t < t_n$ are sequences as above, producing uniformly integrable reverse supermartingales $(X_{s_n}, \mathcal{F}_{s_n})$, $(X_{t_n}, \mathcal{F}_{t_n})$, then for $A \in \mathcal{F}_{s+}$, $\int_A X_{s_n} \, dP \geqslant \int_A X_{t_n} \, dP$ and since these families are uniformly integrable, taking limit s as $n \to \infty$ gives $\int_A X_{s+} \, dP \geqslant \int_A X_{t+} \, dP$.

(2) The right-continuous supermartingale $(X_{t+}, \mathcal{F}_{t+})$ has left limits a.s. when restricted to $\mathbf{T} \cap \mathbf{Q}$. By right-continuity this means that almost all paths $t \to X_{t+}(\omega)$ have left limits for all $t \in \mathbf{T}$. Hence if $(\mathcal{F}_t)$ is a right-continuous filtration and the map $t \to \mathbf{E}(X_t)$ is right-continuous for the supermartingale (X_t), then (X_t) has a right-continuous modification which has left limits a.s. (P).

(3) In particular, since $t \to \mathbf{E}(X_t)$ is *constant* for martingales, any martingale has a right-continuous modification with left limits. Define a *cadlag* process as: right-continuous and with left limits a.s. (French: *cadlag* = *continu à droite, limites à gauche.*)

We shall now restrict our attention to right-continuous (in fact cadlag) supermartingales. Since we assume that the filtration $(\mathcal{F}_t)$ is right-continuous, the above remarks indicate that in practice this is no real restriction. The basic convergence theorems are now easy to extend:

3.3.8. ***Theorem:*** Let $X = (X_t)_{t \in \mathbf{T}}$ be a right-continuous supermartingale.

(i) If $\sup_{t \in \mathbf{T}} \mathbf{E}(X_t^-) < +\infty$, then (X_t) converges to an integrable limit X_∞ a.s. as $t \leftarrow \infty$.

(ii) If $X_t \geqslant 0$ for all t, X_∞ *closes* X, i.e. $X_t \geqslant \mathbf{E}(X_\infty | \mathcal{F}_t)$, $t \in \mathbf{T}$.

(iii) If X is a uniformly integrable martingale, then $X_t \to X_\infty$ in L^1-norm and $X_t = \mathbf{E}(X_\infty | \mathcal{F}_t)$ for all $t \in \mathbf{T}$. Conversely if $X_t = \mathbf{E}(X_\infty | \mathcal{F}_t)$ for all t, X is uniformly integrable.

Proof: (i) Since (X_t) is right-continuous,

$$U_a^b(\mathbf{Q} \cap [0, n]) = U_a^b([0, n]) \leqslant \frac{1}{b-a} \sup_t \mathbf{E}(X_t - a)^-$$

for all $n\in\mathbf{N}$. Thus the set $M_{a,b}$ where X upcrosses $[a,b]$ infinitely often has measure zero. Now set

$$M=\bigcup_{\substack{a,b\in\mathbf{Q}\\ a<b}} M_{a,b}$$

as usual, then $X_\infty(\omega)=\liminf_{t\to\infty}X_t(\omega)=\limsup_{t\to\infty}X_t(\omega)$ whenever $\omega\notin M$. The integrability of X_∞ is proved as in the discrete case. The proofs of (ii) and (iii) are also identical to those of the discrete case.

Finally, to prove an optional sampling theorem, we again need *closable* supermartingales: this means that (X_t) is right-continuous and for some $Y\in L^1$, $X_t\geqslant\mathbf{E}(Y|\mathscr{F}_t)$ for all $t\in\mathbf{T}$. If T is any stopping time, we then set $X_{T(\omega)}(\omega)=Y(\omega)$ on $\{T=\infty\}$.

3.3.9. ***Theorem:*** Let X be a closable supermartingale and let $S\leqslant T$ be stopping times. Then X_S and X_T are integrable and $X_S\geqslant\mathbf{E}(X_T|\mathscr{F}_S)$.

Proof: Write $D_n=\{k/2^n:k\in\mathbf{N}\}$ and apply Theorem 2.10.4 to the discrete supermartingale $(X_t)_{t\in D_n}$: set $S^{(n)}(\omega)=k/2^n$ if $S(\omega)=t$ for $t\in[(k-1)/2^n,k/2^n[$. Then $S^{(n)}\downarrow S$ and similarly we can find $T^{(n)}\downarrow T$. The countably valued stopping times $S^{(n)}$, $T^{(n)}$ satisfy the conditions of Theorem 2.10.4, so $X_{S^{(n)}}\geqslant\mathbf{E}(X_{T^{(n)}}|\mathscr{F}_{S^{(n)}})$ for each n. On the other hand $(X_{S^{(n)}})$ is a reverse supermartingale, since $(S^{(n)})$ decreases with increasing n. Also $\{\mathbf{E}(X_{S^{(n)}})\}$ is bounded above by $\mathbf{E}(X_0)$, so $(X_{S^{(n)}})$ is uniformly integrable. Hence $X_{S^{(n)}}\to X_S$ in L^1-norm and similarly $X_{T^{(n)}}\to X_T$. Now as $\int_A X_{T^{(n)}}\mathrm{d}P\leqslant\int_A X_{S^{(n)}}\mathrm{d}P$ for all $A\in\mathscr{F}_{S^{(n)}}$ and $\mathscr{F}_S\subseteq\mathscr{F}_{S^{(n)}}$, the inequality holds for all $A\in\mathscr{F}_S$, and as $X_{S^{(n)}}\to X_S$ in L^1, $\int_A X_{S^{(n)}}\mathrm{d}P\to\int_A X_S\mathrm{d}P$ and similarly $\int_A X_{T^{(n)}}\mathrm{d}P\to\int_A X_T\mathrm{d}P$. Hence $\int_A X_T\mathrm{d}P\leqslant\int_A X_S\mathrm{d}P$ for all $A\in\mathscr{F}_S$.

3.3.10. ***Exercises:***

(1) Show that if X is a closable supermartingale and T is a stopping time, then the stopped process X^T, where $X_t^T=X_{T\wedge t}$ is again a right-continuous supermartingale.

(2) Let T be a non-negative random variable, and suppose the random variable Y is independent of T and is never zero. Define

$$X_t(\omega)=\begin{cases}0 & \text{if } t<T(\omega)\\ Y(\omega) & \text{if } t\geqslant T(\omega)\end{cases}$$

and let $\mathscr{F}_t=\sigma(X_s:s\leqslant t)$ for each $t\in\mathbf{T}$. Prove that:

(i) $(\mathscr{F}_t)$ is a right-continuous filtration.

(ii) T is an $(\mathscr{F}_t)$-stopping time.

(iii) $\mathscr{F}_T=\sigma(T,Y)$.

3.4. Predictable stopping times and predictable processes

The success of martingale transforms (Definition 2.4.3) depended upon the fact that the 'multiplier process' $V=(V_n)$ in the transform $Z=V\cdot X$ was predictable, i.e. V_n was $\mathscr{F}_{n-1}$-measurable. It is not immediately obvious how we should define a continuous-time analogue of this concept. Clearly we want to describe the statement that a random variable X_t is 'known just before time t'. If the paths $t \rightarrow X_t(\omega)$ are *left*-continuous on $\mathbf{T}$, this requirement is met. In fact, one may characterise the *predictable σ-field* $\Sigma_p \subset \mathscr{B}(\mathbf{T}) \times \mathscr{F}$ as the σ-field generated by all left-continuous adapted processes. If we restrict our attention to predictable $Y=(Y_t)$ in attempting to define a stochastic integral $\int Y \mathrm{d}X$ for a right-continuous martingale, the above remark at least indicates how this restriction overcomes the problem of a 'common jump' of integrand and integrator which occurs in the elementary theory of Stieltjes integrals.

3.4.1. It is convenient and illuminating to describe Σ_p and other sub-σ-fields of $\mathscr{B}(\mathbf{T}) \times \mathscr{F}$ in terms of stochastic intervals: given two stopping times S and T with $S \leqslant T$, define the *stochastic interval* $]\!]S,T]\!] = \{(t,\omega) \in \mathbf{T} \times \Omega : S(\omega) < t \leqslant T(\omega)\}$. $[\![S,T[\![$, $]\!]S,T[\![$ and $[\![S,T]\!]$ are defined similarly. Note that $[\![T]\!] := [\![T,T]\!]$ is the *graph* $\{(t,\omega) \in \mathbf{T} \times \Omega : T(\omega) = t\}$ of T. (All these sets are subsets of $\mathbf{T} \times \Omega$, whether T is finite or not.) In order to describe the σ-fields generated by these collections of 'intervals' in $\mathbf{T} \times \Omega$ we define the following classes of stopping times:

3.4.2. ***Definition:***

(i) A stopping time T is *predictable* if there is an increasing sequence (T_n) of stopping times such that $T = \lim_n T_n$ and $T_n < T$ on $\{T > 0\}$ for all n. We say (T_n) *announces* T.

(ii) A stopping time S is *totally inaccessible* if $[\![T]\!] \cap [\![S]\!]$ is an evanescent set for every predictable stopping time T.

(iii) A stopping time S is *accessible* if $[\![S]\!]$ is contained up to evanescent sets in the union of the graphs of a sequence of predictable stopping times (S_n).

3.4.3. ***Remarks:***

(1) Predictable stopping times can be defined independently of an announcing sequence – see the remark following Th. 77 in Ch. III of [19].

(2) The sequence (T_n) announcing a predictable stopping time represents an approximation of T from below. Approximation *from above* is trivial for any stopping time T: let $T_n = T + 1/n$, then (T_n) decreases to T a.s. and $T_n > T$

for all n on $\{T<\infty\}$. Moreover, each T_n is then predictable: the sequence $S_{m,n}=T+(1/n)(1-1/m)$, $m=1,2,\dots$ announces T_n.

(3) It is clear that $S\wedge T$ is predictable if both S and T are.

(4) Predictable stopping times are always accessible. If T is totally inaccessible, it must be strictly positive a.s.: for if $\{T=0\}=A$ we can define $S(\omega)=0$ on A and $S(\omega)=+\infty$ on $\Omega\setminus A$. This is a predictable stopping time (e.g. let $S_n=0$ on A and $S_n=n$ on $\Omega\setminus A$) and $[\![S]\!]\bigcap[\![T]\!]\supseteq\{0\}\times A$ is not evanescent if $P(A)>0$.

(5) If T is both accessible and totally inaccessible, then $T=\infty$ a.s.: the set $\{T<\infty\}=\{\omega\in\Omega:(t,\omega)\in[\![T]\!]$ for some $t\in[0,\infty[\}$ is P-null since $[\![T]\!]$ is evanescent by hypothesis.

This suggests that an arbitrary stopping time may be split into an accessible and a totally inaccessible part. Before we can prove this we need to discuss the σ-field containing the 'strict past' of T:

3.4.4. ***Definition:*** The σ-field $\mathscr{F}_{T-}$ of events *strictly prior* to the stopping time T is the σ-field generated by $\mathscr{F}_0$ and all sets of the form $A\bigcap\{t<T\}$, for $t\in\mathbf{T}$ and $A\in\mathscr{F}_t$.

To see that $\mathscr{F}_{T-}\subseteq\mathscr{F}_T$ we must check that the generators of $\mathscr{F}_{T-}$ belong to $\mathscr{F}_T$. This is trivial for $\mathscr{F}_0$, using the constant stopping time $0\leqslant T$. Now if $A\in\mathscr{F}_t$ for $t\in\mathbf{T}$, then for all $s\in\mathbf{T}$ we have $A\bigcap\{t<T\}\bigcap\{T\leqslant s\}\in\mathscr{F}_s$: the set is empty if $s\leqslant t$, and if $s>t\{t<T\}=\{T\leqslant t\}^c\in\mathscr{F}_t\subseteq\mathscr{F}_s$. Hence $A\bigcap\{t<T\}\in\mathscr{F}_T$.

Note further than $\{t<T\}=\Omega\bigcap\{t<T\}\in\mathscr{F}_{T-}$, so that $\{T\leqslant t\}\in\mathscr{F}_{T-}$ and hence T is $\mathscr{F}_{T-}$-measurable.

3.4.5. ***Proposition:*** Let S and T be stopping times. Then we have:

(i) if $S\leqslant T$ then $\mathscr{F}_{S-}\subseteq\mathscr{F}_{T-}$;

(ii) if $A\in\mathscr{F}_S$ then $A\bigcap\{S<T\}\in\mathscr{F}_{T-}$;

(iii) if $A\in\mathscr{F}$, $A\bigcap\{T=\infty\}\in\mathscr{F}_{T-}$;

(iv) if $S\leqslant T$ and $S<T$ on $\{0<T<\infty\}$ then $\mathscr{F}_S\subseteq\mathscr{F}_{T-}$.

Proof: (i) If $S\leqslant T$, $(A\bigcap\{t<S\})\bigcap\{t<T\}=A\bigcap\{t<S\}\in\mathscr{F}_t$ whenever $A\in\mathscr{F}_t$. So $\mathscr{F}_{T-}$ contains the generators of $\mathscr{F}_{S-}$.

(ii) If $S(\omega)<T(\omega)$ we can find rational r with $S(\omega)\leqslant r<T(\omega)$. Hence $A\bigcap\{S<T\}=\bigcup_{r\in\mathbf{Q}^+}(A\bigcap\{S\leqslant r\})\bigcap\{r<T\})$, and as $A\in\mathscr{F}_S$, $B=A\bigcap\{S\leqslant r\}\in\mathscr{F}_r$, so $B\bigcap\{r<T\}$ is a generator of $\mathscr{F}_{T-}$.

(iii) For $A\in\mathscr{F}_t$, $A\bigcap\{T=\infty\}=\bigcap_n A\bigcap\{T>t+n\})$ and $A\in\mathscr{F}_{t+n}$, so $A\bigcap\{T=\infty\}$ is a countable intersection of generators of $\mathscr{F}_{T-}$. Now the sets for which $A\bigcap\{T=\infty\}\in\mathscr{F}_{T-}$ form a σ-field containing $\bigcup_{t\geqslant 0}\mathscr{F}_t$, which therefore contains $\mathscr{F}$.

(iv) Write $A \in \mathscr{F}_S$ as $A = (A \cap \{T=0\}) \cup (A \cap \{S<T\}) \cup (A \cap \{T=\infty\})$. Then $A \cap \{T=0\} = A \cap \{S=0\} \cap \{T=0\} \in \mathscr{F}_0$ and $A \cap \{S<T\} \in \mathscr{F}_{T-}$ by (ii), while $A \cap \{T=\infty\} \in \mathscr{F}_{T-}$ by (iii). Hence $A \in \mathscr{F}_{T-}$.

3.4.6. ***Exercises:***

(1) Show that in Exercise 3.3.10(2), $\mathscr{F}_{T-} = \sigma(T)$.

(2) Let (T_n) be an increasing sequence of stopping times with $T = \lim_n T_n$. Show that $\mathscr{F}_{T-} = \sigma(\bigcup_n \mathscr{F}_{T_n -})$. Show further that if $T_n < T$ on $\{0 < T < \infty\}$, then $\mathscr{F}_{T-} = \sigma(\bigcup_n \mathscr{F}_{T_n})$.

3.4.7. ***Definition:*** For any stopping time T the *restriction* T_A of T to a set $A \in \mathscr{F}$ is defined as $T_A(\omega) = T(\omega)$ if $\omega \in A$ and $T_A(\omega) = \infty$ if $\omega \notin A$.

Since $\{T_A \leqslant t\} = A \cap \{T \leqslant t\}$, T_A is a stopping time iff $A \in \mathscr{F}_T$. Furthermore if T is accessible (respectively, totally inaccessible) then so is T_A, since $[\![T_A]\!] \subseteq [\![T]\!]$.

We can now prove the promised decomposition theorem:

3.4.8. ***Theorem:*** For any stopping time T the set $\{T < \infty\}$ can be partitioned essentially uniquely into two sets A and B of $\mathscr{F}_{T-}$ such that T_A is accessible and T_B is totally inaccessible.

Proof: Let (S_n) be any increasing sequence of stopping times bounded above by T. Define $S = \lim_n S_n$, then S is a stopping time and $S \leqslant T$. Then (S_n) announces T on the set $L(S_n) = \{S = T, S_n < T \text{ for all } n\}$, thus the restriction T_L of T to $L = L(S_n) \cup \{T=0\}$ is predictable. On the other hand, $L(S_n) = (\bigcap_n \{S_n < T\}) \setminus \{S < T\}$ belongs to $\mathscr{F}_{T-}$ by Proposition 3.4.5(ii). Since $\{T=0\} \in \mathscr{F}_0 \subseteq \mathscr{F}_{T-}$, $L \in \mathscr{F}_{T-}$ also. If we let f be (a representative of) the essential supremum of the family $\{1_{L(S_n)} + 1_{\{T=0\}} : (S_n)$ increases and is bounded above by $T\}$, then f is $\mathscr{F}_{T-}$-measurable, so $A = \{f > 0\} \in \mathscr{F}_{T-}$. Then A contains $\{T=0\}$ and $\{T=\infty\}$, and $[\![T_A]\!]$ is contained in a countable union of graphs of predictable stopping times T_L. Hence T_A is accessible. With $B = \Omega \setminus A \in \mathscr{F}_T$, the stopping time T_B is totally inaccessible, since for any predictable S announced by (S_n) we can set $S'_n = S_n \wedge T$, so that $L(S'_n)$ contains $\{S = T\}$ and hence $[\![S]\!] \cap [\![T_B]\!]$ is evanescent. The uniqueness follows from the fact that any restriction T_C of an accessible stopping time is accessible (since $[\![T]\!]$ contains $[\![T_C]\!]$) and similarly for totally inaccessible times. So if there were two decompositions of T associated with sets A, B, A' and B', $T_{A \cap B'}$, would be both accessible and totally inaccessible, hence a.s. infinite. Hence if U is accessible, V totally inaccessible and $U \wedge V = T$, $U \vee V = +\infty$ then $U = T_A$, $V = T_B$ a.s.

We call T_A the *accessible part*, T_B the *totally inaccessible part* of T.

Predictable stopping times need not be preserved under restrictions. To obtain a characterisation of the class of sets which do preserve predictable stopping times, we first prove that the lattice of predictable stopping times is closed under increasing sequences and 'stationary' decreasing sequences. This implies that the above class of sets is a monotone class.

3.4.9. *Proposition:*

(i) If (T_n) is an increasing sequence of predictable stopping times then $T = \lim_n T_n$ is predictable.

(ii) If (T_n) is a stationary decreasing sequence of predictable stopping times (so that for each $\omega \in \Omega$ there exists $n = n(\omega)$ with $T_{n+k}(\omega) = T_n(\omega)$ for all k) then $T = \lim_n T_n$ is predictable.

Proof: (i) Suppose $(T_{n,p})_p$ announces T_n. Let $S_k = \max_{p,n \leqslant k} T_{n,p}$, then (S_k) increases to T, and on $\{T > 0\} = \bigcup_n \{T_n > 0\}$ we have $S_k < \max(T_1, \ldots, T_k) < T$, so (S_k) announces T.

(ii) On $\{T = \infty\}$ there is nothing to prove. So suppose $T < \infty$ a.s. and let $(T_{n,p})_p$ announce T_n. Since $T_{n,p} \to T_n$ a.s. we can select a (diagonal) subsequence $(S_{n,p})$ converging to T_n 'rapidly in measure', e.g. if

$$d(s,t) = \frac{|s-t|}{1+|s-t|}$$

for $s, t \in \mathbf{T}$, $d(t, \infty) = d(\infty, t) = 1$, $d(\infty, \infty) = 0$, we can choose $(S_{n,p})$ such that $P(d(S_{n,p}, T_n) > 2^{-p}) \leqslant 2^{-(n+p)}$ for all $p \geqslant 1$. Put $S_p = \inf_n S_{n,p}$, so that (S_p) increases with p, and since $T(\omega) = T_n(\omega)$ for large enough n, while $S_{n,p} < T_n$ on $\{T_n > 0\}$, we have $S_p < T$ on $\{T > 0\} = \bigcap_n \{T_n > 0\}$. But then (S_p) announces T, as $T = \lim_n S_p$: to see this, let $S = \lim_p S_p$. For each $p \in \mathbf{N}$,

$$P(d(S,T) > 2^{-p}) \leqslant P(d(S_p, T) > 2^{-p})$$

$$\leqslant \sum_n P(d(S_{n,p}, T) > 2^{-p}) \leqslant \sum_n P(d(S_{n,p}, T_n) > 2^{-p}) \leqslant \sum_n 2^{-n-p}$$

$$= 2^{-p} \to 0$$

as $p \to \infty$. Hence $P(S < T) = 0$, so $S = T$ a.s.

3.4.10. ***Exercise:*** Prove Proposition 3.4.9, with 'accessible' replacing 'predictable'. (Hint: for (i) first show that if $A = \{T_n < T \text{ for all } n\}$ then $R_n = (T_n)_A \wedge n$ defines a sequence announcing T_A.)

3.4.11. ***Theorem:*** Let T be a predictable stopping time and $A \in \mathscr{F}_T$. Then T_A is predictable iff $A \in \mathscr{F}_{T-}$.

Proof: Suppose T_A is announced by (S_n). We can write A as follows:

$$A=\{T_A \leqslant T\} \setminus (A^c \cap \{T=\infty\}) = \left(\bigcap_n \{S_n < T\}\right) \setminus (A^c \cap \{T=\infty\}).$$

By Proposition 3.4.5(ii), $\{S_n < T\} = \Omega \cap \{S_n < T\} \in \mathscr{F}_{T-}$ and by 3.4.5(iii) $A^c \cap \{T=\infty\} \in \mathscr{F}_{T-}$, hence $A \in \mathscr{F}_{T-}$.

Conversely, the family of sets $A \in \mathscr{F}_T$ for which T_A and T_{A^c} are predictable is a σ-field, by Proposition 3.4.9. So we may restrict ourselves to generators of $\mathscr{F}_{T-}$. By Exercise 3.4.6(ii) we know that $\bigcup_n \mathscr{F}_{T_n}$ generates $\mathscr{F}_{T-}$ whenever (T_n) announces T. But if $A \in \mathscr{F}_{T_m}$, T_A is announced by (S_n), where $S_n(\omega) = (T_{m+n})_A(\omega) \wedge n$. So T_A is predictable for all $A \in \bigcup_n \mathscr{F}_{T_n}$, hence for all $A \in \mathscr{F}_{T-}$.

3.4.12. ***Remark:***

Similar arguments show that if S is predictable and $A \in \mathscr{F}_{S-}$ then $A \cap \{S < T\} \in \mathscr{F}_{T-}$ for any stopping time T. So if S and T are both predictable, then the restriction of T to $A = \{S < T\}$ is also predictable.

We are now in a position to define σ-fields on $\mathbf{T} \times \Omega$ generated by classes of stochastic intervals:

3.4.13. ***Definition:*** Let S and T be stopping times, $S \leqslant T$. We identify the σ-fields generated on $\mathbf{T} \times \Omega$ by the class of stochastic intervals of the forms $[\![S, T[\![$ and by the evanescent sets to obtain:

(i) the *predictable σ-field* Σ_p when S, T run through the class $\mathscr{T}_p$ of predictable stopping times;

(ii) the *accessible σ-field* Σ_a when S, T run through the class $\mathscr{T}_a$ of accessible stopping times;

(iii) the *optional σ-field* Σ_o when S, T run through the class $\mathscr{T}$ of all stopping times.

(The explicit inclusion of the evanescent sets in these σ-fields is not usually made in the literature. Since we wish to characterise these σ-fields as those generated by certain classes of processes, it is necessary to include the evanescent sets: our 'processes' are themselves equivalence classes, with the identification of indistinguishable processes – see Section 3.1. While the sections $A_\omega = \{t : (t, \omega) \in A\}$ of an evanescent set A are empty for almost all $\omega \in \Omega$, these sections need not be Borel-measurable for the exceptional ω. For a stochastic interval, on the other hand, all such sections are intervals in $\mathbf{T}$. See also [19; Ch. IV p. 61 *et seq.*].)

3.4.14. ***Remarks:***

(1) $\Sigma_p \subseteq \Sigma_a \subseteq \Sigma_o$. Moreover, if Σ_π is the σ-field generated by all

progressive processes (i.e. $A \in \Sigma_\pi$ iff the process $(t,\omega) \to 1_A(t,\omega)$ is progressive), then $\Sigma_0 \subseteq \Sigma_\pi$: for any stopping time T,

$$1_{[\![T,\infty[\![}(t,\omega) = \begin{cases} 1 \text{ if } T(\omega) \leqslant t \\ 0 \text{ if } T(\omega) > t \end{cases}$$

is adapted since $\{T \leqslant t\} \in \mathscr{F}_t$, and the paths $t \to 1_{[\![T,\infty[\![}(t,\omega)$ are right-continuous for all ω. So $[\![T,\infty[\![\in \Sigma_\pi$ by Proposition 3.2.1, and hence $[\![S,T[\![\in \Sigma_\pi$ for all S,T with $S \leqslant T$.

(2) Note that $[\![S,T]\!] = \bigcap_n [\![S, T+1/n[\![$, while $[\![T]\!] = \bigcap_n [\![T, T+1/n[\![$. Hence Σ_0, Σ_a, Σ_p can be defined using $[\![S,T]\!]$ instead of $[\![S,T[\![$. Moreover, the restriction 0_A of the zero stopping time to $A \in \mathscr{F}_0$ is predictable, and since for any T, $T+1/n$ is predictable, $[\![0,T]\!] \in \Sigma_p$ for all T. Now if S is any stopping time less than T, $S+1/n$ is predictable, so $[\![S+1/n,\infty[\![\in \Sigma_p$. Hence $]\!]S,T]\!] = (\bigcap_n [\![S+1/n,\infty[\![) \bigcap [\![0,T]\!] \in \Sigma_p$. On the other hand, let Σ be generated by sets $[\![0_A]\!]$, $]\!]S,T]\!]$, $S<T$ stopping times, let R be a predictable stopping time, and let $A = \{R=0\} \in \mathscr{F}_0$, while (R_n) announces R. Then $[\![R,\infty[\![= [\![0_A]\!] \bigcup (\bigcap_n]\!]R_n,\infty[\![)$, hence $\Sigma_p \subseteq \Sigma$. Since we proved that $\Sigma \subseteq \Sigma_p$, these σ-fields are equal.

(3) This means that Σ_p is generated by sets of the form $\{0\} \times A$, $A \in \mathscr{F}_0$, and $]s,t] \times B$, $s<t$ in $\mathbf{T}$, $B \in \mathscr{F}_s$. We need only observe that any stopping time T is approximated from above by simple (i.e. finite range) stopping times, e.g.

$$T_n = \sum_{k=1}^{n} \frac{k}{n} t \cdot 1_{\{(k-1)t/n \leqslant T < kt/n\}}.$$

Since processes of the form $1_{\{0\} \times A}$, $1_{]s,t] \times B}$ $(A \in \mathscr{F}_0, B \in \mathscr{F}_s)$ are left-continuous, we have shown that the left-continuous adapted processes generate a σ-field containing Σ_p. But, conversely, any left-continuous adapted process X can be approximated by

$$X^{(n)} = X_0 1_{\{0\} \times \Omega} + \sum_{k \geqslant 1} X_{k/n} 1_{]k/n,(k+1/n]},$$

which is predictable. In fact $X(t,\omega) = \lim_n X^{(n)}(t,\omega)$ a.s. (P) for *all* $t \in \mathbf{T}$, by left-continuity of X. Hence X is predictable. We have proved

3.4.15. ***Theorem:*** Σ_p is the σ-field generated by the family of all left-continuous adapted processes.

3.4.16. ***Corollary:*** Σ_p is generated by the continuous adapted processes.

Proof: Let S be a stopping time and define $X_t(\omega) = t - S(\omega) \wedge t$. Then $t \to X_t(\omega)$ is continuous for each $\omega \in \Omega$, and $(X>0) =]\!]S,\infty[\![$. Similarly, $Y_t = t1_A + 1_{A^c}$ is adapted and has continuous paths, while $\{0\} \times A = (Y=0)$.

Hence the generators of Σ_p belong to the σ-field generated by the continuous adapted processes.

3.4.17. ***Theorem:*** Σ_o is generated by the family of adapted cadlag processes.

Proof: Since the processes $1_{[\![S,T[\![}$, for $S \leqslant T$ arbitrary stopping times, are cadlag (i.e. are right-continuous and have left limits a.s.) and generate Σ_o, we need only prove that all adapted cadlag processes are Σ_o-measurable. So let $X=(X_t)$ be adapted, cadlag. For each $k \in \mathbf{N}$ define a sequence $(T_n^k)_n$ of stopping times:

$$T_1^k \equiv 0,\ T_{n+1}^k(\omega) = \inf\{t > T_n^k(\omega): |X_t(\omega) - X_{T_n^k}(\omega)| \geqslant 1/k\}$$

where, as usual, we take $\inf \varnothing = +\infty$. Since X and $X_{T_n^k} 1_{[\![T_n^k, \infty[\![}$ are progressive, the set

$$B_{n+1}^k =]\!]T_n^k, \infty[\![\bigcap \{(t,\omega): |X_t(\omega) - X_{T_n^k}(\omega) 1_{[\![T_n^k, \infty[\![}(t,\omega)| \geqslant 1/k\}$$

is progressive. Now the first hitting time of $[\![1, \infty[\![$ by the progressive process $1_{B_{n+1}^k}$ is T_{n+1}^k, which shows (by Example 3.2.5) that T_{n+1}^k is a stopping time. The right-continuity of X also guarantees that $|X_{T_{n+1}^k} - X_{T_n^k}| \geqslant 1/k$ on $\{T_{n+1}^k < \infty\}$. Because $t \rightarrow X_t(\omega)$ has left limits at each t for almost all $\omega \in \Omega$, so that $(X_t(\omega))$ has no oscillatory discontinuities, $(T_n^k)_n$ will increase to $+\infty$ a.s. for each k. So let $X^k = \sum_{n \geqslant 1} X_{T_n^k} 1_{[\![T_n^k, T_{n+1}^k[\![}$, then X^k is optional and, by right-continuity of X, (X^k) converges to X up to evanescent sets. Hence X is Σ_o-measurable.

One can also use transfinite induction to show that each right-continuous process is optional – see [49].

We shall now call Σ_o-measurable processes *optional processes*, the Σ_a-measurable processes *accessible processes* and the Σ_p-measurable processes *predictable processes*. They are not all that far apart:

3.4.18. ***Theorem:*** Let X be an optional process. Then there is a predictable process Y such that the set $\{(t,\omega): X(t,\omega) \neq Y(t,\omega)\}$ is contained in the union of the graphs of a sequence of stopping times.

Proof: The class of bounded optional processes for which the assertion holds is an algebra closed under monotone convergence (we may assume that the associated predictable processes are bounded – truncate if necessary). We can therefore apply the Monotone Class Theorem to the class of indicators of the stochastic intervals $[\![S, T[\![$, S, T stopping times, whose associated predictable processes are the indicators of intervals

$]\!]S,T]\!]$. So the theorem holds for a class of processes generating Σ_o, hence for all optional processes.

3.4.19. *Examples:*

(1) Let $[\![S,T[\![$ be a stochastic interval and let Y be $\mathscr{F}_S$-measurable. Then $X_t(\omega)=Y(\omega)1_{[\![S,T[\![}(t,\omega)$ is optional: to see this, suppose first that $Y=1_A$, where $A\in\mathscr{F}_S$. Then $X_t(\omega)=X(t,\omega)=1_{[\![S_A,T_A[\![}(t,\omega)$, hence is optional. Hence the result holds for $\mathscr{F}_S$-simple functions and thus for all $\mathscr{F}_S$-measurable Y.

(2) Similarly, if Z is $\mathscr{F}_{S-}$-measurable, and S, T are predictable stopping times, then $X_t(\omega)=Z(\omega)1_{[\![S,T[\![}(t,\omega)$ is predictable: again let $A\in\mathscr{F}_{S-}$, $Z=1_A$. By Theorem 3.4.11, S_A, T_A are predictable, so $[\![S_A,T_A[\![\in\Sigma_p$. The result follows as in (i).

(3) If Z is $\mathscr{F}_S$-measurable, $X_t(\omega)=Z(\omega)1_{]\!]S,T]\!]}(t,\omega)$ is predictable for any stopping times S, T: for any $A\in\mathscr{F}_S$ we also have $]\!]S_A,T_A]\!]\in\Sigma_p$, by Remark 3.4.14(2).

(4) Thus restricting a random variable to stochastic intervals yields optional processes. But if $Y\in L^1$, the martingale $M_t=\mathbf{E}(Y|\mathscr{F}_t)$ also yields an optional process: by Theorem 3.3.6, (M_t) has a modification with right-continuous paths, and since it is adapted, this process is optional. Furthermore, if $N_t=\mathbf{E}(Y|\mathscr{F}_{t-})$, then there is a modification of (N_t) with left-continuous paths, so (N_t) gives rise to a predictable process, by Theorem 3.4.15.

These examples will be of importance in discussing the optional and predictable projections of a measurable process.

3.5. Cross-sections and projections

The discussion of the first hitting time of a set by a process required the following result on the projections of measurable sets, proved in the supplement to this chapter:

3.5.1. *Theorem:* Let $(\Omega,\mathscr{F},P)$ be a complete probability space and let $A\in B(\mathbf{T})\times\mathscr{F}$, where $B(\mathbf{T})$ is the Borel σ-field on $\mathbf{T}$. Then the projection $\Pi_\Omega(A)=\{\omega:(t,\omega)\in A \text{ for some } t\in\mathbf{T}\}$ belongs to $\mathscr{F}$.

This enabled us to show that for any *progressive* process X the first hitting time of a Borel set in $\mathbf{R}$ is a stopping time. Adopting a slightly different emphasis, we can also define the *début* of any set $A\subseteq\mathbf{T}\times\Omega$ as $D_A(\omega)=\inf\{t\in\mathbf{T}: (t,\omega)\in A\}$. This is a stopping time for $A\in\Sigma_\pi$: given $u>0$, $A\cap[\![0,u]\!]$ is $\mathscr{B}([0,u])\times\mathscr{F}_u$-measurable and for $t\in\mathbf{T}$, $\{D_A<t\}=\Pi_\Omega(A\cap[\![0,t[\![)$ then belongs to $\mathscr{F}_t$. Since the first hitting time of $B\subset\mathbf{R}$ is $A\cap]\!]0,\infty[\![$, where $A=\{(t,\omega): X_t(\omega)\in B\}$, this provides an alternative proof of the result proved in Example 3.2.5.

3.5.2. *Exercise:* Define the *section* of $A \subseteq \mathbf{T} \times \Omega$ by $\omega \in \Omega$ as $A(\omega) = \{t \in \mathbf{T} : (t,\omega) \in A\}$. Then $D_A(\omega) = \inf\{t \in \mathbf{T} : [0,t] \cap A(\omega) \neq \emptyset\}$. Define the *n-début* $D_A^{(n)}(\omega) = \inf\{t \in \mathbf{T} : [0,t] \cap A(\omega)$ has at least n points$\}$. Show that $D_A^{(n)}$ is a stopping time for each $n \in \mathbf{N}$, if $A \in \Sigma_\pi$.

The above theorem can be refined to show that the projection onto Ω of a measurable set $B \subseteq \mathbf{T} \times \Omega$ has the form $\{Z < \infty\}$ for some random variable Z whose graph is contained in B. The proof (which requires capacitability theory) is again given in the supplement. We can state the result as follows:

3.5.3. ***First cross-section theorem:*** If $(\Omega, \mathscr{F}, P)$ is a probability space and $B \in \mathscr{B}(\mathbf{T}) \times \mathscr{F}$, then there is a random variable $Z : \Omega \to [0,\infty]$ such that

(i) if $Z(\omega) < \infty$ then $(Z(\omega), \omega) \in B$,
(ii) $P(Z < \infty) = P(D_B < \infty)$.

Thus, although Z need not be a stopping time, $[\![Z]\!] \subset B$ by (i) and up to P-null sets, $\Pi_\Omega(B) = \{D_B < \infty\} = \{Z < \infty\}$ by (ii). Therefore Z is almost surely a cross-section of B.

Our main interest in this result lies in finding to what extent it is possible to replace Z by a (predictable) stopping time T when B is taken from $\Sigma_o[\Sigma_p]$. This is in fact possible, provided we weaken (ii) to $P(T < \infty) \geqslant P(\Pi_\Omega(B)) - \varepsilon$ for given $\varepsilon > 0$. The proof now uses properties of Σ_o, Σ_p and of débuts, but we leave it until the supplement.

3.5.4. ***Second cross-section theorem:*** Let $(\Omega, \mathscr{F}, P, (\mathscr{F}_t), \mathbf{T})$ be a stochastic base and let $B \in \Sigma_o$. For any given $\varepsilon > 0$ there is a stopping time T such that

(i) $[\![T]\!] \subseteq B$,
(ii) $P(T < \infty) \geqslant P(\Pi_\Omega(B)) - \varepsilon$.

If B is accessible (predictable) we can choose T to be accessible (predictable).

This result has far-reaching consequences for processes and stopping times:

3.5.5. *Corollary:* Two optional [predictable] processes X and Y are indistinguishable iff $X_T = Y_T$ for all [predictable] stopping times T.

Proof: The set $B = \{(t,\omega) : X_t(\omega) \neq Y_t(\omega)\}$ belongs to $\Sigma_o[\Sigma_p]$, so by Theorem 3.5.4 we can find a [predictable] stopping time T such that $[\![T]\!] \subseteq B$ and $P(T < \infty) \geqslant P(\Pi_\Omega(B)) - \varepsilon$, for given $\varepsilon > 0$. So if $P(\Pi_\Omega(B)) > 0$ we can take $P(T < \infty) > 0$. But then there exists $t \in \mathbf{T}$ for which $X_{T \wedge t} \neq Y_{T \wedge t}$ on a set of positive measure in Ω, since $\{(t,\omega) : T(\omega) = t\} \subseteq \{(t,\omega) : X_t(\omega) \neq Y_t(\omega)\}$. But

X and Y are indistinguishable iff $P(\Pi_\Omega(B))=0$, hence iff $X_T=Y_T$ for all [predictable] T.

3.5.6. ***Corollary:*** A random variable $T:\Omega\to[0,\infty]$ is a [predictable, accessible] stopping time iff $[\![T]\!]\in[\Sigma_p,\Sigma_a]\Sigma_o$.

Proof: Note that $[\![T]\!]=\bigcap_n[\![T,T+1/n[\![$, so the necessity is obvious. Also, if $[\![T]\!]\in\Sigma_o$, then T is a stopping time, as $\Sigma_o\subseteq\Sigma_\pi$ and $T(\omega)=D_{[\![T]\!]}(\omega)=\inf\{t:T(\omega)=t\}$, so that Theorem 3.5.1 applies. If $[\![T]\!]\in\Sigma_a$ (respectively, Σ_p) we can apply Theorem 3.5.4 with $\varepsilon=2^{-n}$ successively for each $n\in\mathbf{N}$ to obtain a sequence (T_n) of accessible (respectively, predictable) stopping times with $[\![T_n]\!]\subseteq[\![T]\!]$ and $P(T_n<\infty)\geqslant P(T<\infty)-2^{-n}$. Using $T_1\wedge T_2\wedge\ldots\wedge T_n$ instead of T_n, if necessary, we may assume that (T_n) is decreasing. Then $T=\lim_n T_n$ and since $[\![T_n]\!]\subset[\![T]\!]$ we can always find $n=n(\omega)$ for $\omega\in\Omega$ such that $T_{n(\omega)}(\omega)=T(\omega)$. Hence, by Proposition 3.4.9(ii), T is accessible (respectively, predictable). Note that the 'stationarity' of the sequence (T_n) is only needed in the predictable case.

3.5.7. ***Exercise:*** Show that if X and Y are bounded optional [accessible, predictable] processes and $\mathbf{E}(X_T1_{\{T<\infty\}})=\mathbf{E}(Y_T1_{\{T<\infty\}})$ for all [accessible, predictable] stopping times T, then X and Y are indistinguishable. (Hint: use Theorem 3.5.4 on $\{X<Y\}$ and $\{Y<X\}$.)

The cross-section theorems also permit an exhaustion argument which enables us to 'cut down' on the graphs of stopping times containing sets in Σ_o, Σ_a, Σ_p:

3.5.8. ***Corollary:*** If $A\in\Sigma_o$ (respectively, Σ_a, Σ_p) is contained in the union of a sequence of graphs of stopping times, then there exists a disjoint sequence of graphs of (accessible, predictable) stopping times (T_n) whose union equals A.

Proof: If $A\subseteq\bigcup_n[\![S_n]\!]$, let $T_n=D_{B_n}$, where $B_n=(A\setminus\bigcup_{k<n}[\![S_k]\!])\bigcap[\![S_n]\!]$. The B_n are disjoint, $A=\bigcup_n B_n=\bigcup_n[\![T_n]\!]$ and T_n is a stopping time if $B_n\in\Sigma_o\subseteq\Sigma_\pi$. Thus the result is proved for $A\in\Sigma_o$. If A is accessible and (T_n) is as above, decompose T_n into its accessible part U_n and totally inaccessible part V_n by Theorem 3.4.8. Then $B=\bigcup_n[\![V_n]\!]=A\setminus\bigcup_n[\![U_n]\!]$ is accessible. Now if $P(\Pi_\Omega(B))>0$ we can find an accessible stopping time V such that $P(V<\infty)>0$ and $[\![V]\!]\subseteq B$ by Theorem 3.5.4. If $[\![V]\!]\subseteq\bigcup_n[\![R_n]\!]$ for predictable R_n, we know that $[\![V_n]\!]\bigcap[\![R_m]\!]$ is evanescent for all m, n. Hence so is $B\bigcap[\![R_m]\!]$, and also $[\![V]\!]\bigcap[\![R_m]\!]$. But this is impossible, so B is evanescent. Hence $A=\bigcup_n[\![U_n]\!]$. If A is predictable, then it is accessible, so $A=\bigcup_n[\![U_n]\!]$

again. Now for each n, $[\![U_n]\!] \subseteq \bigcup_m [\![R_{m,n}]\!]$ for some sequence of predictable $R_{m,n}$. Hence A is also contained in a sequence of graphs of predictable stopping times, R_n say. Now define a predictable stopping time T_n via $B_n = (A \setminus \bigcup_{k<n} [\![R_k]\!]) \bigcap [\![R_n]\!]$ as above and note that these T_n have disjoint graphs with union A.

We can now characterise predictable *processes* in terms of $\mathscr{F}_{T-}$:

3.5.9. ***Theorem:*** An accessible process X is predictable iff for each predictable stopping time T, $X_T 1_{\{T<\infty\}}$ is $\mathscr{F}_{T-}$-measurable.

Proof: The collection $\mathscr{C}$ of processes for which $X_T 1_{\{T<\infty\}}$ is $\mathscr{F}_{T-}$-measurable forms a monotone class, hence we need only consider X of the form $1_{[\![0_A]\!]}$ or $1_{]\!]U,V]\!]}$, where U, V are stopping times with $U \leqslant V$. But $1_{[\![0_A]\!]} \cdot 1_{\{T<\infty\}} = 1_{A \bigcap \{T=0\}}$ is $\mathscr{F}_{T-}$-measurable.

On the other hand, let $X = 1_{]\!]U,V]\!]} = 1_{]\!]U,\infty[\![} - 1_{]\!]V,\infty[\![}$. Then it is clear that $X_T(\omega) = X(T(\omega), \omega) = 1_{\{U<T\}}(\omega)$, so that by Proposition 3.4.5(ii) $X_T 1_{\{T<\infty\}}$ is $\mathscr{F}_{T-}$-measurable. (Note that this shows that for a predictable process X, $X_T 1_{\{T<\infty\}}$ is $\mathscr{F}_{T-}$-measurable for *any* stopping time T.)

Now suppose X is an accessible process such that $X_T 1_{\{T<\infty\}}$ is $\mathscr{F}_{T-}$-measurable for all predictable T. Since X is optional there is a predictable process Y such that $\{X \neq Y\} = A$ is contained in the union of the graphs of a sequence of stopping times (Theorem 3.4.18). But X and Y are both accessible, so $A \in \Sigma_a$. Thus by Corollary 3.5.8 we can find a sequence (S_n) of predictable stopping times with disjoint graphs whose union contains A: we have $A = \bigcup_n [\![T_A]\!]$ for accessible T_n, and each $[\![T_A]\!]$ is contained in a countable union of graphs of predictable stopping times, which we can again reduce to a disjoint union by Corollary 3.5.8. Now the same holds for A. The random variable $X_{S_n} 1_{\{S_n<\infty\}}$ is then $\mathscr{F}_{S_n-}$-measurable by hypothesis and $(X_{S_n} 1_{\{S_n<\infty\}}) 1_{[\![S_n]\!]} = X_{S_n} 1_{[\![S_n]\!]}$ is therefore predictable: to see this proceed as in Examples 3.4.18 and recall that S_n is predictable: so that if $X_{S_n} 1_{\{S_n<\infty\}}$ is replaced by 1_B for some $B \in \mathscr{F}_{S_n-}$, $1_B . 1_{[\![S_n]\!]} = [\![(S_n)_B]\!] \in \Sigma_p$ by Theorem 3.4.11 and Corollary 3.5.6. Finally, as $C = \bigcap_n [\![S_n]\!]^c \in \Sigma_p$, $Y . 1_C \sum_n X_{S_n} 1_{[\![S_n]\!]} = X$ is predictable.

This characterisation of predictable processes allows us to extend the criterion for indistinguishability contained in Exercise 3.5.7 to a means of ordering processes:

3.5.10. ***Theorem:*** Let X and Y be finite optional [accessible, predictable]. Then the following are equivalent:

(i) $X \geqslant Y$ except on an evanescent set.

(ii) If for an arbitrary [accessible, predictable] stopping time T, the random variables $X_T 1_{\{T<\infty\}}$, $Y_T 1_{\{T<\infty\}}$ are in L^1, then

$$\mathbf{E}(X_T 1_{\{T<\infty\}}) \geqslant \mathbf{E}(Y_T 1_{\{T<\infty\}})$$

Proof: It is clear that (i) implies (ii). If (ii) holds but $\{X<Y\}$ is not evanescent, then the second cross-section theorem yields an [accessible, predictable] stopping time such that $P(T<\infty)>0$ and $[\![T]\!] \subseteq \{X<Y\}$, so that $X_T 1_{\{T<\infty\}} < Y_T 1_{\{T<\infty\}}$. Now we can find $k \in \mathbf{R}$ such that $B = \{T<\infty\} \cap \{|X_T| \leqslant k\} \cap \{|Y_T| \leqslant k\}$ has positive probability. Then $B \in \mathscr{F}_T$, and if X and Y are predictable, $B \in \mathscr{F}_{T-}$ by Theorem 3.5.9. Hence T_B is an [accessible, predictable] stopping time for which $\mathbf{E}(X_{T_B} 1_{\{T_B<\infty\}}) < \mathbf{E}(Y_{T_B} 1_{\{T_B<\infty\}})$, contradicting (ii).

We now define *projection maps* for stochastic processes: these will resemble conditional expectation operators and transform bounded measurable processes into optional, respectively predictable, processes. We call a measurable process $X=(X_t)$ *bounded* if $\sup_t |X_t| \in L^\infty$, and denote the vector space of such processes by $\mathscr{B}^\infty(\Sigma)$, where $\Sigma = \mathscr{B} \times \mathscr{F}$. Similarly we write $\mathscr{B}^\infty(\Sigma_o)$, $\mathscr{B}^\infty(\Sigma_p)$ for the spaces of bounded optional, respectively bounded predictable, processes.

3.5.11. ***Projection theorem:*** There are unique linear order-preserving projections Π_o: $\mathscr{B}^\infty(\Sigma) \to \mathscr{B}^\infty(\Sigma_o)$ and Π_p: $\mathscr{B}^\infty(\Sigma) \to \mathscr{B}^\infty(\Sigma_p)$ such that for any $X \in \mathscr{B}^\infty(\Sigma)$ and

(i) for all stopping times T, $(\Pi_o X)_T 1_{\{T<\infty\}} = \mathbf{E}(X_T 1_{\{T<\infty\}} | \mathscr{F}_T)$,

(ii) for all predictable stopping times T,

$$(\Pi_p X)_T 1_{\{T<\infty\}} = \mathbf{E}(X_T 1_{\{T<\infty\}} | \mathscr{F}_{T-}).$$

We call $\Pi_o X$ the *optional projection* and $\Pi_p X$ the *predictable projection* of X.

Proof: If $\mathscr{H}_o$, respectively $\mathscr{H}_p$, denotes the class of $X \in \mathscr{B}^\infty(\Sigma)$ for which (i) or (ii) holds, then $\mathscr{H}_o$ and $\mathscr{H}_p$ are vector spaces containing the constants, and since (i) implies that $\mathbf{E}((\Pi_o X)_T 1_{\{T<\infty\}}) = \mathbf{E}(X_T 1_{\{T<\infty\}})$ for all T, while (ii) gives $\mathbf{E}((\Pi_p X)_T 1_{\{T<\infty\}}) = \mathbf{E}(X_T 1_{\{T<\infty\}})$ for all predictable T, Exercise 3.5.7 guarantees the uniqueness of $\Pi_o X$, $\Pi_p X$ and Theorem 3.5.10 ensures that Π_o, Π_p are order-preserving. Moreover, $\mathscr{H}_o$ and $\mathscr{H}_p$ are closed under limits of increasing uniformly bounded sequences: if (X^n) increases to X, $(\Pi_o X^n)$ and $(\Pi_p X^n)$ increase and so does $\mathbf{E}((\Pi_o X^n)_T 1_{\{T<\infty\}}) = \mathbf{E}(X^n_T 1_{\{T<\infty\}}) = \mathbf{E}((\Pi_p X^n)_T 1_{\{T<\infty\}})$, so $\Pi_o X = \liminf_{n\to\infty} (\Pi_o X^n)$, $\Pi_p X = \liminf_{n\to\infty} (\Pi_p X^n)$ yield $X \in \mathscr{H}_o$, $(\mathscr{H}_p)$.

Hence $\mathscr{H}_o$ and $\mathscr{H}_p$ satisfy the Monotone Class Theorem 0.1.5. To show that each equals $\mathscr{B}^\infty(\Sigma)$ we must find a uniformly bounded family of measurable processes, closed under multiplication and generating Σ, contained in $\mathscr{H}_o$ and $\mathscr{H}_p$. In each case we take the class of processes of the form $X = Z1_{[\![r,s]\!]}$, where $Z \in L^\infty(\mathscr{F})$ and $r, s \in \mathbf{T}$. We consider each case in turn.

(i) Let $Y = (Y_t)$ be a right-continuous modification of the martingale $(\mathbf{E}(Z|\mathscr{F}_t))_t$. Y can be chosen to have left limits a.s., so is optional by Theorem 3.4.17. Set $\Pi_o X = Y.1_{[\![r,s]\!]}$, then for any stopping time T, $\mathbf{E}(X_T 1_{\{T<\infty\}}|\mathscr{F}_T) = 1_{\{r\leqslant T\leqslant s\}}.\mathbf{E}(Z|\mathscr{F}_T) = 1_{\{r\leqslant T\leqslant s\}} Y_T$ by optional sampling (Theorem 3.3.9). Hence $(\Pi_o X)_T 1_{\{T<\infty\}} = \mathbf{E}(X_T 1_{\{T<\infty\}}|\mathscr{F}_T)$ for all stopping times T.

(ii) Let $Y_t = \mathbf{E}(Z|\mathscr{F}_t)$ as in (i) and let T be a predictable stopping time, announced by (T_n). Since $\mathscr{F}_{T-} = \sigma(\bigcup_n \mathscr{F}_{T_n})$ by Exercise 3.4.6(2), we can define a uniformly integrable martingale $V = (V_n)$ converging to $\mathbf{E}(Z|\mathscr{F}_{T-}) = Y_{T-}$ by setting $V_n = \mathbf{E}(Z|\mathscr{F}_{T_n})$. Now as $W = (W_t) = (\mathbf{E}(Z|\mathscr{F}_{t-}))$ has a.s. left-continuous paths, W is predictable and $W_T = Y_{T-} = \mathbf{E}(Z|\mathscr{F}_{T-})$. Set $\Pi_p X = W.1_{[\![r,s]\!]}$, then for T announced by (T_n) we have, since T and Y_{T-} are $\mathscr{F}_{T-}$-measurable,

$$\mathbf{E}(X_T 1_{\{T<\infty\}}|\mathscr{F}_{T-}) = 1_{\{r\leqslant T\leqslant s\}}.\mathbf{E}(Z|\mathscr{F}_{T-}) = 1_{\{r\leqslant T\leqslant s\}} W_T$$

$$= (\Pi_p X)_T.1_{\{T<\infty\}}.$$

The following corollaries are immediate:

3.5.12. ***Corollary:*** Let $X \in \mathscr{B}^\infty(\Sigma)$ and let T be a stopping time. Then

(i) $\mathbf{E}(X_T 1_{\{T<\infty\}}) = \mathbf{E}((\Pi_o X)_T 1_{\{T<\infty\}})$.

(ii) If T is predictable, $\mathbf{E}(X_T 1_{\{T<\infty\}}) = \mathbf{E}((\Pi_p X)_T 1_{\{T<\infty\}})$.

3.5.13. ***Corollary:***

(i) For all $X \in \mathscr{B}^\infty(\Sigma)$, $\Pi_p X = \Pi_p(\Pi_o X)$.

(ii) If $X \in \mathscr{B}^\infty(\Sigma_o)$, then $\Pi_o X = X$.

(iii) If $X \in \mathscr{B}^\infty(\Sigma_p)$, then $\Pi_p X = X$.

(iv) If X is positive, but not necessarily bounded, we can define $\Pi_o X$, respectively $\Pi_p X$, as the limit of the increasing sequence of optional, respectively predictable, projections of the sequence $(X_n) \subseteq \mathscr{B}^\infty(\Sigma)$, $X_n = X \wedge n$.

3.5.14. ***Corollary:***

(i) If $X \in \mathscr{B}^\infty(\Sigma)$ and $Y \in \mathscr{B}^\infty(\Sigma_o)$, then $\Pi_o(XY) = Y(\Pi_o X)$.

(ii) If $X \in \mathscr{B}^\infty(\Sigma)$ and $Y \in \mathscr{B}^\infty(\Sigma_p)$, then $\Pi_p(XY) = Y(\Pi_p X)$.

Proof: (i) Y_T and $(\Pi_o X)_T 1_{\{T<\infty\}}$ are $\mathscr{F}_T$-measurable by Proposition 3.2.4, since optional processes are progressive. So $\Pi_o(XY)_T 1_{\{T<\infty\}} = \mathbf{E}((XY)_T 1_{\{T<\infty\}}|\mathscr{F}_T) = Y_T \mathbf{E}(X_T 1_{\{T<\infty\}}|\mathscr{F}_T) = Y_T(\Pi_o X)_T 1_{\{T<\infty\}}$ for all T. Since Y and $(\Pi_o X)$ are optional $Y.(\Pi_o X) = \Pi_o(XY)$ by the uniqueness of Π_o.

(ii) $(\Pi_p X)_T 1_{\{T<\infty\}}$ is $\mathscr{F}_{T-}$-measurable if T is predictable, by Proposition 3.5.9, and similarly for $Y_T . 1_{\{T<\infty\}}$. Hence $\Pi_p(XY)_T 1_{\{T<\infty\}} = \mathbf{E}((XY)_T 1_{\{T<\infty\}}|\mathscr{F}_{T-}) = Y_T(\Pi_p X)_T 1_{\{T<\infty\}}$ and the result again follows by uniqueness of Π_p.

3.5.15. *Remarks:*

(1) The identities of Corollary 3.5.12 can also be used to deduce those of Theorem 3.5.11: if $A \in \mathscr{F}_T$ apply Corollary 3.5.12(i) to the stopping time T_A, while if T is predictable and $A \in \mathscr{F}_{T-}$ then T_A is predictable by Theorem 3.4.11 and we can apply Corollary 3.5.12(ii) to T_A.

However, we cannot make the same deductions by restricting to *finite* stopping times: if $\mathbf{E}(X_T) = \mathbf{E}((\Pi_o X)_T)$ for all finite T, where X is a bounded, right-continuous martingale with $X_0 = 0$, then $\Pi_p X = X$, while $\mathbf{E}(X_T) = 0$ for all finite T, by optional sampling. So we cannot deduce that $(\Pi_o X)_T = \mathbf{E}(X_T|\mathscr{F}_T)$.

(2) One can also define a projection $\Pi_a : \mathscr{B}^\infty(\Sigma) \to \mathscr{B}^\infty(\Sigma_a)$. In fact, we would require $\Pi_a X = \Pi_a(\Pi_o X)$, so we may restrict attention to $\mathscr{B}^\infty(\Sigma_o)$, and hence to processes $X = 1_{[\![S,\infty]\!]}$, where S is a stopping time. Let (S_A, S_B) be the decomposition of S into accessible and totally inaccessible parts by Theorem 3.4.8, and let $C = [\![S_A, \infty[\![\bigcup]\!]S_B, \infty[\![$. Then $C \in \Sigma_a$, and for all accessible T, $[\![S_B]\!] \bigcup [\![T]\!]$ is evanescent, so $C \bigcap [\![T]\!] = [\![S, \infty[\![\bigcap [\![T]\!]$ for all such T and so $\mathbf{E}((1_C)_T 1_{\{T<\infty\}}) = \mathbf{E}(X_T 1_{\{T<\infty\}})$. Hence we can take $\Pi_a X = 1_C$.

(3) Given any $X \in \mathscr{B}^\infty(\Sigma)$, the set $\{(t,\omega) : (\Pi_o X)(t,\omega) \neq (\Pi_p X)(t,\omega)\}$ is contained in the union of the graphs of a sequence of stopping times, thus giving rather more precise information than Theorem 3.4.18: by the Monotone Class Theorem we can assume $X = Z.1_{[\![r,s]\!]}$. Then $(\Pi_p X)_t = Y_{t-}.1_{\{r \leq t \leq s\}}$ and $(\Pi_o X)_t = Y_t . 1_{\{r \leq t \leq s\}}$, where (Y_t) is a right-continuous modification of the martingale $(\mathbf{E}(X|\mathscr{F}_t))_t$. So the set where $\Pi_o X$ and $\Pi_p X$ differ is described in terms of the *jumps* of Y. Our claim then follows from the following:

3.5.16. *Definition:* If $X = (X_t)$ is adapted and cadlag, we say that X *charges* the stopping time T if $P(X_T \neq X_{T-}) \bigcap \{T < \infty\} > 0$ and a sequence (T_n) of stopping times *exhausts the jumps* of X if:

(i) $X_{T_n} \neq X_{T_n -}$ a.s. on $\{T_n < \infty\}$ (we say X has a *jump* at T_n) for all n;

(ii) the graphs of (T_n) are disjoint;

(iii) X charges no other stopping time, i.e. if $[\![T]\!]$ is disjoint from $\bigcup_n [\![T_n]\!]$, X does not charge T.

3.5.17. ***Theorem:*** If X is adapted and cadlag there is a sequence (T_n) of stopping times exhausting the jumps of X. If X is accessible (predictable), the T_n are accessible (predictable).

Proof: Write Y_t for X_{t-}, so $Y=(Y_t)$ is predictable. The set $A=\{X\neq Y\}=\bigcup_k A_k$, where $A_k=\{(t,\omega):|X_t(\omega)-Y_t(\omega)|>1/k\}$. Now since X has no oscillatory discontinuities, the sections $A_k(\omega)$ are free of accumulation points, hence A_k is the union of the graphs of its n-débuts (see Exercise 3.5.2). So A is contained in a countable union of graphs of stopping times, and the result follows from Corollary 3.5.8.

3.5.18. ***Corollary:*** X is accessible iff it does not charge any totally inaccessible stopping time.

Proof: The necessity is clear. Conversely if all the T_n in Theorem 3.5.17 are accessible, $A=\bigcap_n [\![T_n]\!]^c \in \Sigma_a$ and again let $Y_t=X_{t-}$. The process Y is predictable, so $1_A . Y$ and $X_{T_n} . 1_{[\![T_n]\!]}$ are accessible. But by construction $X=1_A . Y+\sum_n X_{T_n} . 1_{[\![T_n]\!]}$.

3.6. Potentials and increasing processes

Recall that a discrete-time supermartingale has a unique decomposition as the difference of a martingale and an increasing process (Definition 2.9.1). In the continuous-time case we shall derive an analogous result, and, as before, the increasing process will be predictable (although the terminology is now applied to a rather more subtle concept). We shall also demand that the paths of our processes are sufficiently regular, and consequently make the following definitions:

3.6.1. ***Definition:*** Let $(\Omega,\mathscr{F},P,(\mathscr{F}_t),\mathbf{T})$ be a stochastic base.

(i) A measurable process $A=(A_t)$ is an *increasing process* provided
   (a) $A_0=0$ a.s.,
   (b) $A_t \in L^1$ for all $t\in\mathbf{T}$,
   (c) almost all paths $t\to A_t(\omega)$ are increasing and right-continuous,
   (d) A is adapted to $(\mathscr{F}_t)$.

   Note that $A_\infty(\omega):=\lim_{t\to\infty} A_t(\omega)$ exists for almost all $\omega\in\Omega$. We call A an *integrable* increasing process if $A_\infty \in L^1$.

(ii) A positive right-continuous supermartingale $X=(X_t)$ is called a *potential* if $X_t\to 0$ in L^1-norm as $t\to\infty$. (Cf. Definition 2.9.3.)

Integrable increasing processes generate potentials in a natural way: let

$A=(A_t)$ be an integrable increasing process and let $X=(X_t)$ be a right-continuous modification of $\mathbf{E}(A_\infty|\mathscr{F}_t)-A_t=\mathbf{E}(A_\infty-A_t|\mathscr{F}_t)$. It is clear that X is a potential. We call X the *potential generated by* A.

On the other hand, let $Y=(Y_t)$ be a uniformly integrable potential such that $Y=M-A$ for some martingale M and increasing process A. We see that $\{\mathbf{E}(A_t)\}$ is bounded, since $\mathbf{E}(Y_t)\to 0$ and $\{\mathbf{E}(M_t)\}_t$ is constant. Hence $A_\infty \in L^1$, and so $M=Y+A$ is uniformly integrable, and $M_t=\mathbf{E}(M_\infty|\mathscr{F}_t)$ for all $t\in\mathbf{T}$. But $Y_\infty=0$ by hypothesis, so $M_\infty=A_\infty$, and so Y is in fact the potential generated by A.

These potentials satisfy a strong optional sampling theorem:

3.6.2. ***Theorem:*** Let X be the potential generated by an integrable increasing process A. Then
(i) for each stopping time T, $X_T=\mathbf{E}(A_\infty|\mathscr{F}_T)-A_T$;
(ii) the family $\{X_T: T \text{ is a stopping time}\}$ is uniformly integrable.

Proof: (i) follows from optional sampling applied to the martingale M, i.e. $M_T=\mathbf{E}(A_\infty|\mathscr{F}_T)$ and $X_T=M_T-A_T$ by definition.

To prove (ii), note that since $X_T=M_T-A_T$ and M is uniformly integrable, $M_T=\mathbf{E}(A_\infty|\mathscr{F}_T)$ for all stopping times T. Also $A_T\leqslant A_\infty\in L^1$, so the M_T and the A_T form uniformly integrable families as T ranges over the set of all stopping times.

For a general supermartingale X the existence of X_∞ is not guaranteed, but we can make the following

3.6.3. ***Definition:*** A supermartingale X is in *class* (D) if X is right-continuous and the family $\{X_T: T \text{ is a finite stopping time}\}$ is uniformly integrable. The potentials considered in Theorem 3.6.2 are thus in class (D).

We saw (Definition 2.9.5) that in discrete time, all uniformly integrable supermartingales are in class (D). This is not true in continuous time – see [44] for an example based on Brownian motion in $\mathbf{R}^3$. However, all uniformly integrable martingales are in class (D), by optional sampling: we have $X_T=\mathbf{E}(X_\infty|\mathscr{F}_T)$ for all stopping times T, so $\{X_T: T \text{ is a finite stopping time}\}$ is uniformly integrable, by Proposition 2.2.4. It is an easy exercise to show that all positive submartingales which are closed by an L^1-function also belong to class (D). More important for our purposes is the following result:

3.6.4. ***Proposition:*** Let $X=(X_t)$ be a right-continuous supermartingale. Then there exists a sequence (T_n) of stopping times, increasing to $+\infty$, such that the stopped processes X^{T_n} belong to class (D).

Proof: Define $T_n = n \wedge \inf\{t : |X_t| \geqslant n\}$, then $T_n \uparrow +\infty$. The supermartingale $(X_{t\wedge n})_t$ is closed by X_n and $T_n \leqslant n$. So by optional sampling $X_{T_n} \in L^1$, and $|X_S^{T_n}| = |X_{T_n \wedge S}| \leqslant |X_{T_n}| \vee n$ for all finite stopping times S. So the family $\{X_S^{T_n} : S \text{ is a finite stopping time}\}$ is uniformly integrable.

Finally, Theorem 3.6.2 shows that any potential which is generated by an integrable increasing process belongs to class (*D*). We shall prove that, conversely, each class (*D*) potential is generated by a unique *predictable* integrable increasing process. Using a 'Riesz decomposition theorem' and Proposition 3.6.4, we shall be able to extend this result to an analogue of the Doob decomposition for all right-continuous supermartingales.

3.6.5. ***Definition:*** A measurable process $A = (A_t)$ is a *raw increasing process* if almost all paths $t \to A_t(\omega)$ are finite, positive, right-continuous increasing functions of t, and each $A_t \in L^1$. If A_∞, defined by $A_\infty(\omega) = \lim_{t\to\infty} A_t(\omega)$ a.s., is also in L^1, we say that A is *integrable.*

(Note that if, further, $A_0 = 0$ a.s. and A is adapted, then A is an increasing process in the sense of Definition 3.6.1.)

For each fixed ω we have a Lebesgue–Stieltjes measure induced on **T** by the increasing right-continuous function $A_\cdot$: we denote this measure by $\mathrm{d}A_\cdot(\omega)$. Omitting the requirement $A_0 = 0$ a.s. allows this measure to have mass at 0. For a consistent formulation we must also adopt the convention that $A_{0-} = 0$ a.s. and augment the filtration $(\mathscr{F}_t)$ by a σ-field $\mathscr{F}_{0-} \subseteq \mathscr{F}_0$ containing all P-null sets.

The measure $\mathrm{d}A_\cdot(\omega)$ enables us to integrate processes. More precisely, let A be an integrable raw increasing process. We write $X \in L^1(A)$ if the measurable process X satisfies $\mathbf{E}(\int_{[0,\infty[} |X_s| \mathrm{d}A_s) < \infty$. The map $\omega \to \int_0^t X_s(\omega) \mathrm{d}A_s(\omega)$ is measurable by Fubini's theorem, hence $Y_t = \int_0^t X_s \mathrm{d}A_s$ defines a process $Y = (Y_t)$. It is easy to see that Y has a unique optional (predictable) modification whenever X and A are both optional (predictable).

3.6.6. ***Remark:*** The above definitions (and much of what follows) may easily be extended to processes of *integrable variation*: a right-continuous process $V = (V_t)$ in L^1 has integrable variation if $\mathbf{E}(\int_0^\infty |\mathrm{d}V_s|) < \infty$, where $|\mathrm{d}V_\cdot|$ denotes the total variation of the Lebesgue–Stieltjes measure associated with V. Any such process can be written essentially uniquely as the difference of two integrable raw increasing processes.

3.6.7. ***Definition:***

(i) Let A be an integrable raw increasing process. A induces a measure

μ_A on $\mathscr{B}(\mathbf{T})\times\mathscr{F}$ – we can describe this measure as a linear functional on $L^1(A)$ as follows: for X in $L^1(A)$ let $\mu_A(X)=\mathbf{E}(\int_{[0,\infty[}X_s\,\mathrm{d}A_s)$. Clearly μ_A is positive and $\mu_A(\mathbf{T}\times\Omega)=\mathbf{E}(A_\infty)<\infty$. Moreover, if X is indistinguishable from the zero process, then the set of paths for which $\int_0^\infty X_s(\omega)\mathrm{d}A_s(\omega)\neq 0$ has P-measure 0. So $\mu_A(X)=0$.

(ii) A *stochastic measure* μ is a positive bounded measure μ on a sub-σ-field of $\mathscr{B}(\mathbf{T})\times\mathscr{F}$ which vanishes on evanescent sets.

We have just seen that any integrable raw increasing process A induces a stochastic measure on $\mathscr{B}(\mathbf{T})\times\mathscr{F}$. The converse also holds:

3.6.8. ***Theorem:***

(i) Let μ be a stochastic measure on $\mathbf{T}\times\Omega$. Then there exists an integrable raw increasing process A, unique up to indistinguishability, such that $\mu=\mu_A$.

(ii) The process A is optional if and only if $\mu(X)=\mu(\Pi_o X)$ and predictable if and only if $\mu(X)=\mu(\Pi_p X)$ for all bounded measurable processes X.

Proof: (i) For each $t\in\mathbf{T}$, $F\in\mathscr{F}$, let $\mu_t(F)=\mu([0,t]\times F)$. Then μ_t is a finite positive measure on $\mathscr{F}$, and if $P(F)=0$ then $[0,t]\times F$ is evanescent, hence $\mu_t(F)=0$. So $\mu_t\ll P$. Define $\hat{A}_t=\mathrm{d}\mu_t/\mathrm{d}P$. Then $\hat{A}_t\in L^1_+$ and if $s\leqslant t$, $\hat{A}_s\leqslant\hat{A}_t$. To obtain a process with right-continuous paths, set

$$\bar{A}_t=\sup_{\substack{s\leqslant t\\ s\in\mathbf{Q}}}\hat{A}_s$$

and finally $A_t=\hat{A}_{t+}$. Then $\bar{A}_t\leqslant\hat{A}_t\leqslant A_t$ and as the map $t\to\mu_t(X)$ is right-continuous, A_t is still a density of μ_t relative to P. (A_t) has right-continuous increasing paths by construction and each $A_t\in L^1$ by dominated convergence. Hence

$$\mu_A([0,t]\times F)=\mathbf{E}\left(\int_0^t 1_F\,\mathrm{d}A_s\right)=\mathbf{E}(1_F A_t)=\mu([0,t]\times F)$$

for all $t\in\mathbf{T}$, $F\in\mathscr{F}$ and since such sets generate $\mathscr{B}(\mathbf{T})\times\mathscr{F}$, $\mu=\mu_A$.

To prove (ii) we need a simple concept that is useful in many contexts (see [19; Ch. VI, pp. 54–6] for a detailed discussion); (ii) will follow from Theorems 3.6.11 and 3.6.14.

3.6.9. ***Definition:*** Let $A=(A_t)$ be an increasing process (and hence optional!). The *time change* $C=(C_t)$ *associated with* A is given by

$$C_t(\omega)=\inf\{s:A_s(\omega)>t\} \quad \text{for } t\in\mathbf{T},\ \omega\in\Omega.$$

As A is right-continuous and increasing, $C_{t-}=\lim_{u\uparrow t}C_u=\inf\{s:A_s\geqslant t\}$. Both C_t and C_{t-} are stopping times, being the débuts of the optional sets $A^{-1}(]t,\infty[)$ and $A^{-1}([t,\infty[)$ respectively – see Corollary 3.5.6. (Here $A^{-1}(B):=\{(s,\omega):A_s(\omega)\in B\}$ for any Borel set $B\subseteq\mathbf{R}$.) Moreover, if A is predictable, the set $H=A^{-1}([t,\infty[)$ is in Σ_p and $[\![C_{t-}]\!]=\{(u,\omega):\inf\{s:A_s(\omega)\geqslant t\}=u\}\subseteq\{(u,\omega):A_u(\omega)\geqslant t\}=H$. So $[\![C_{t-}]\!]=H\setminus]\!]C_{t-},\infty[\![\in\Sigma_p$, hence C_{t-} is predictable.

The idea behind these definitions lies in a result due to Lebesgue, which allows one to reduce Stieltjes integrals to Lebesgue integrals: note that almost all paths $\alpha:t\to A_t(\omega)$ are right-continuous increasing functions of t (even if A is only a raw increasing process). So consider a general right-continuous increasing function $\alpha:\mathbf{R}^+\to\mathbf{R}^+$, and describe its reflection in the diagonal by $\gamma(t)=\inf(s:\alpha(s)>t)$. Jump discontinuities of α correspond to intervals of constancy of γ and vice versa. Moreover, $\gamma(t)<\infty$ if and only if $t<\lim_{s\to\infty}\alpha(s)$.

It is clear that γ is again right-continuous and increasing. Hence γ has left-limits $\gamma_-(s)=\gamma(s-)=\inf(t:\alpha(t)\geqslant s)=\sup(t:\alpha(t)<s)$. (We define $\alpha(0-)=0$, $\alpha(\infty)=\lim_{s\to\infty}\alpha(s)$).

The reader may verify the following easy consequences of these definitions:

(i) $\alpha(\gamma(s))\geqslant\alpha(\gamma_-(s))\geqslant s$, $\alpha_-(\gamma_-(s))\leqslant\alpha_-(\gamma(s))\leqslant s$.

(ii) $\alpha(s)=\inf(t:\gamma(t)>s)=\sup(t:\gamma(t)\leqslant s)$ for t in $\mathbf{R}$.

(iii) $\gamma_-(s)\leqslant t$ if and only if $s\leqslant\alpha(t)$.

Lebesgue's result can then be stated in the following form:

3.6.10. ***Proposition:*** If α is as above and f is a positive Borel function on $\mathbf{R}^+$, then

$$\int_0^\infty f(t)\,\mathrm{d}\alpha(t)=\int_{\alpha(0)}^\infty f(\gamma_-(s))1_{\{\gamma_-(s)<\infty\}}\,\mathrm{d}s=\int_{\alpha(0)}^\infty f(\gamma(s))1_{\{\gamma(s)<\infty\}}\,\mathrm{d}s \tag{3.4}$$

Proof: The last two integrals are equal since γ has only countably many discontinuities, so $\gamma=\gamma_-$ off a set of zero measure. To prove that $\int_0^\infty f(t)\mathrm{d}\alpha(t)=\int_{\alpha(0)}^\infty f(\gamma(s))1_{\{\gamma(s)<\infty\}}\mathrm{d}s$ we may assume that $f=1_{[0,s]}$ for some $s\in\mathbf{R}^+$. Then $\int_0^\infty f(t)\mathrm{d}\alpha(t)=\alpha(s)-\alpha(0)$ and $\int_{\alpha(0)}^\infty f(\gamma(s))1_{\{\gamma(s)<\infty\}}\mathrm{d}s$ is the length of the *interval* $\{t\leqslant\alpha(0):\gamma(t)\leqslant s\}$. But $\sup\{t:\gamma(t)\leqslant s\}=\inf\{t:\gamma(t)>s\}=\alpha(s)$, so the interval has length $\alpha(s)-\alpha(0)$, as required.

We now apply these results to an increasing process A to compare $\mu_A(X)$ and $\mu_A(\Pi_o X)$ for any bounded measurable process X:

3.6.11. ***Theorem:*** Let A be an integrable increasing process and X a bounded measurable process. Then $\mu_A(X)=\mu_A(\Pi_o X)$. Conversely, if A is an integrable raw increasing process and $\mu_A(X)=\mu_A(\Pi_o X)$ for all $X\in B^\infty(\Sigma)$, then A is optional.

Proof: Let $C=(C_t)$ be the time change associated with $A=(A_t)$. We may also assume that X is positive. Apply Proposition 3.6.10 to the paths $t\to X_t(\omega)$, $t\to A_t(\omega)$ for each $\omega\in\Omega$. Then we have, as $A_0=0$,

$$\mu_A(X)=\mathbf{E}\left(\int_0^\infty X_t(\omega)\,\mathrm{d}A_t(\omega)\right)=\mathbf{E}\left(\int_0^\infty X_{C_t}.1_{\{C_t<\infty\}}\,\mathrm{d}t\right)=\int_0^\infty \mathbf{E}(X_{C_t}.1_{\{C_t<\infty\}})\mathrm{d}t,$$

where the final equality again follows by Fubini's theorem. Similarly,

$$\mu_A(\Pi_o X)=\int_0^\infty \mathbf{E}((\Pi_o X)_{C_t}1_{\{C_t<\infty\}})\mathrm{d}t.$$

But by definition of Π_o (Corollary 3.5.11), the integrands are the same in both cases, as C_t is a stopping time. So $\mu_A(X)=\mu_A(\Pi_o X)$.

Conversely, given an integrable raw increasing process A and $t>0$, let f in L^∞ satisfy $\mathbf{E}(f|\mathscr{F}_t)=0$, and define $X=f.1_{[\![0,t]\!]}$. Then $\Pi_o X=M.1_{[\![0,t]\!]}$, where (M_s) is a right-continuous modification of the martingale $(\mathbf{E}(f|\mathscr{F}_s))_s$, as in the proof of Theorem 3.5.11. But $\mathbf{E}(f|\mathscr{F}_s)=\mathbf{E}(\mathbf{E}(f|\mathscr{F}_t)|\mathscr{F}_s)=0$ for $s\leqslant t$, so $\Pi_o X=0$. Hence also $0=\mu_A(\Pi_o X)=\mu_A(X)=\mathbf{E}(\int_0^t f\mathrm{d}A_s)=\mathbf{E}(f.A_t)$. Now this means that A_t is $\mathscr{F}_t$-measurable: for we chose f orthogonal to $L^1(\mathscr{F}_t)$, i.e. $\mathbf{E}(fg)=0$ for all g in $L^1(\mathscr{F}_t)$, and then showed that $\mathbf{E}(fA_t)=0$ also. By Corollary 1.1.10 A_t lies in the closed subspace $L^1(\mathscr{F}_t)$.

3.6.12. ***Corollary:*** Let A be an increasing process and let T be a stopping time. Suppose that X and Y are measurable processes with $\Pi_o X=\Pi_o Y$. Then

$$\mathbf{E}\left(\int_0^T X_t\mathrm{d}A_t\right)=\mathbf{E}\left(\int_0^T Y_t\mathrm{d}A_t\right).$$

Proof: If $T=+\infty$ a.s. the result follows at once from Theorem 3.6.11. But if T is any stopping time, set $B_t=A_{T\wedge t}$ and apply Theorem 3.6.11 to B.

3.6.13. ***Corollary:*** Let A be an increasing process and let M be a positive right-continuous uniformly integrable martingale. Then for any stopping time T, $\mathbf{E}(\int_0^T M_t \, dA_t) = \mathbf{E}(M_T A_T)$.

Proof: Let $X = M . 1_{[\![0,T]\!]}$, $Y = M_T \cdot 1_{[\![0,T]\!]}$. If S is a stopping time,

$$\mathbf{E}(X_S 1_{\{S<\infty\}}) = \mathbf{E}(M_{S(\omega)}(\omega) . 1_{[\![0,T]\!]}(S(\omega),\omega))$$

$$= \mathbf{E}(M_S 1_{\{S \leqslant T\}} 1_{\{S<\infty\}}) = \mathbf{E}(M_T 1_{\{S \leqslant T\}} 1_{\{S<\infty\}}) = \mathbf{E}(Y_S 1_{\{S<\infty\}}),$$

where we have used optional sampling on the martingale M, since $\{S \leqslant T\}$, $\{S<\infty\}$ belong to $\mathscr{F}_S$. Hence $\Pi_o X = \Pi_o Y$ and, by Corollary 3.6.12, $\mathbf{E}(\int_0^T M_t dA_t) = \mathbf{E}(\int_0^T M_T dA_t) = \mathbf{E}(M_T A_T)$ as required.

Exactly the same analysis can be used to compare $\mu_A(X)$ and $\mu_A(\Pi_p X)$ if A is predictable:

3.6.14. ***Theorem:*** Let A be a predictable integrable increasing process and let X be a bounded measurable process. Then $\mu_A(X) = \mu_A(\Pi_p X)$. Conversely, if A is an integrable raw increasing process and $\mu_A(X) = \mu_A(\Pi_p X)$ for all $X \in B^\infty(\Sigma)$, then A is predictable.

Proof: If A is predictable, the stopping times C_{t-} arising from the time change $C = (C_t)$ associated with A are predictable. Hence $\mu_A(X) = \mathbf{E}(\int_0^\infty X_s dA_s) = \mathbf{E}(\int_0^\infty X_{C_{s-}} 1_{\{C_{s-}<\infty\}} ds)$ by Proposition 3.6.10, and similarly $\mu_A(\Pi_p X) = \mathbf{E}(\int_0^\infty (\Pi_p X)_{C_{s-}} 1_{\{C_{s-}<\infty\}} ds)$. Again exchanging the integrals by Fubini's theorem and noting that since C_{s-} is predictable, $\mathbf{E}((\Pi_p X)_{C_{s-}} 1_{\{C_{s-}<\infty\}}) = \mathbf{E}(X_{C_{s-}} 1_{\{C_{s-}<\infty\}})$ for all $s \in \mathbf{T}$, we have shown that $\mu_A(\Pi_p X) = \mu_A(X)$. Conversely, suppose A is an integrable raw increasing process and $\mu_A(X) = \mu_A(\Pi_p X)$. To show that A is predictable we first show that A is accessible and then that A_T is $\mathscr{F}_{T-}$-measurable for all predictable T (cf. Theorem 3.5.9).

Thus let S be a totally inaccessible stopping time and suppose T is a predictable stopping time. Define $X = 1_{[\![S]\!]}$, then $X_T = 1_{\{S=T\}}$, so $\mathbf{E}((\Pi_p X)_T 1_{\{T<\infty\}}) = \mathbf{E}(X_T 1_{\{T<\infty\}}) = \mathbf{E}(1_{\{S=T<\infty\}}) = 0$ since $[\![S]\!] \cap [\![T]\!]$ is evanescent. Thus $(\Pi_p X)$ is evanescent by Corollary 3.5.5. Consequently $\mathbf{E}(A_S - A_{S-}) = \mu_A(1_{[\![S]\!]}) = \mu_A(\Pi_p 1_{[\![S]\!]}) = 0$, so that A charges no totally inaccessible stopping time, hence is accessible by Corollary 3.5.18.

Now to show that A is predictable we need only prove that $A_T 1_{\{T<\infty\}}$ is $\mathscr{F}_{T-}$-measurable for all predictable stopping times T. In fact, we have $A_T = A_T 1_{\{T<\infty\}} + A_\infty 1_{\{T<\infty\}}$ and $A_\infty 1_{\{T<\infty\}}$ is always $\mathscr{F}_{T-}$-measurable, so it will suffice to show that A_T is $\mathscr{F}_{T-}$-measurable. So let f in L^∞ have

$\mathbf{E}(f|\mathscr{F}_{T-})=0$. As in Theorem 3.6.11 we shall show that $\mathbf{E}(fA_T)=0$, which will complete the proof. First we show that the martingale $Y_t=\mathbf{E}(f|\mathscr{F}_t)$ is zero on $[\![0,T[\![$: by right-continuity it suffices to prove that $Y_t 1_{\{t<T\}}=0$ for all t. Hence let $B\in\mathscr{F}_t$ be given, then since $B\bigcap\{t<T\}\in\mathscr{F}_{T-}\bigcap\mathscr{F}_t$, we have $0=\int_{B\bigcap\{t<T\}} f\,dP=\int_{B\bigcap\{t<T\}} Y_t dP$. Since B was arbitrary, the result follows. Hence (Y_{t-}) is zero on $[\![0,T]\!]$, and as $1_{[\![0,T]\!]}$ is predictable, the predictable projection of $X=f\cdot 1_{[\![0,T]\!]}$ is $Y_-\cdot 1_{[\![0,T]\!]}=0$. Thus $0=\mu_A(\Pi_p X)=\mu_A(X)=\mathbf{E}(fA_T)$.

The proof of Theorem 3.6.8 is now complete.

Theorems 3.6.8, 3.6.11 and 3.6.14 allow us to define the 'duals' of the projections Π_o and Π_p.

Let A be an integrable raw increasing process and let $X\in B^\infty(\Sigma)$. Define $\mu_o(X)=\mu_A(\Pi_o X)$, $\mu_p(X)=\mu_A(\Pi_p X)$. Then $\mu_o(\mu_p)$ is determined by its values on $\Sigma_o(\Sigma_p)$, hence is generated by an optional (predictable) integrable increasing process $\Pi_o^* A(\Pi_p^* A)$, which we call the *dual optional projection* (*dual predictable projection*) of A. These processes are uniquely determined by the relations $\mu_{\Pi_o^* A}(X)=\mu_A(\Pi_o X)$ and $\mu_{\Pi_p^* A}(X)=\mu_A(\Pi_p X)$ for all $X\in B^\infty(\Sigma)$.

The importance of the dual predictable projection is evident:

3.6.15. ***Theorem:*** Let A be an integrable increasing process. The following statements are equivalent:

(i) A is a martingale.

(ii) μ_A is the zero measure on Σ_p.

(iii) $\Pi_p^* A$ is indistinguishable from the zero process.

Proof: (i)⟶(ii): Σ_p is generated by sets of the form $\{0\}\times F_0$ for $F_0\in\mathscr{F}_0$ and $]s,t]\in F_s$ for $F_s\in\mathscr{F}_s$, $s\leqslant t$ in $\mathbf{T}$. Now since $A_0=0$ a.s., $\mu_A(\{0\}\times F_0)=0$, and since A is a martingale and $F_s\in\mathscr{F}_s$, $\mu_A(]s,t]\times F_s)=\int_F(A_t-A_s)\,dP=0$. So $\mu_A\equiv 0$ on Σ_p.

(ii)⟶(iii): If $X\in B^\infty(\Sigma)$, $\mu_{\Pi_p{}^* A}(X)=\mu_A(\Pi_p X)=0$ since μ_A is zero on predictable sets. So $\Pi_p^* A=0$ up to evanescent sets.

(iii)⟶(i): If $s\leqslant t$ in $\mathbf{T}$ and $G\in\mathscr{F}_s$, then $\mu_A(]s,t]\times G)=\mu_A(]\!]s_G,t_G]\!])=\mu_{\Pi_p^* A}(]\!]s_G,t_G]\!])=0$ since $]\!]s_G,t_G]\!]\in\Sigma_p$. Hence $\int_G A_s dP=\int_G A_t dP$, i.e. A is a martingale.

3.6.16. ***Corollary:*** A predictable integrable increasing process A which is also a martingale is indistinguishable from the zero process.

Proof: Since A is predictable, $\Pi_p^* A=A$. Since A is a martingale, $\Pi_p^* A=0$ up to evanescent sets.

3.6.17. ***Proposition:*** Let A be an integrable increasing process and $X \in \mathscr{B}^\infty(\Sigma)$. If S, T are stopping times, $s \leqslant T$, then we have

$$\mathbf{E}\left(\int_S^T (\Pi_p X)_t \, \mathrm{d}A_t | \mathscr{F}_S\right) = \mathbf{E}\left(\int_S^T X_t \mathrm{d}(\Pi_p^* A)_t | \mathscr{F}_S\right).$$

Proof: Let $B \in \mathscr{F}_S$. We need to show that $\mathbf{E}(1_B \int_S^T (\Pi_p X)_t \mathrm{d}A_t) = \mathbf{E}(1_B \int_S^T X_t \mathrm{d}(\Pi_p^* A)_t)$. Recalling that

$$S_B = \begin{cases} S & \text{on } B \\ +\infty & \text{on } \Omega \backslash B \end{cases}$$

we see that the desired identity is $\mu_A(1_{]\!]S_B, T_B]\!]} \Pi_p X) = \mu_{\Pi_p^* A}(1_{]\!]S_B, T_B]\!]} X)$. Since $]\!]S_B, T_B]\!] \in \Sigma_p$, $1_{]\!]S_B, T_B]\!]}(\Pi_p X) = \Pi_p(1_{]\!]S_B T_B]\!]} X)$ so the result follows at once from the definition of $\Pi_p^* A$.

3.6.18. ***Corollary:*** Two integrable increasing processes A and B have the same dual predictable projection if and only if $M = A - B$ is a martingale.

Proof: Suppose that $\Pi_p^* A = \Pi_p^* B$. Apply Proposition 3.6.17 to the predictable process $X \equiv 1$ and constant stopping times $s \leqslant t$, then

$$\mathbf{E}(A_t - A_s | \mathscr{F}_s) = \mathbf{E}\left(\int_s^t 1 \, \mathrm{d}A_u | \mathscr{F}_s\right) = \mathbf{E}\left(\int_s^t 1 \, \mathrm{d}(\Pi_p * A)_u | \mathscr{F}_s\right)$$

$$= \mathbf{E}\left(\int_s^t 1 \, \mathrm{d}B_u | \mathscr{F}_s\right) = \mathbf{E}(B_t - B_s | \mathscr{F}_s),$$

so $A - B$ is a martingale.

Conversely, if $M = A - B$ is a martingale, we shall show that $\mu_A(X) = \mu_B(X)$ for all predictable X. For this it suffices to consider X of the form $1_{]\!]s_G, t_G]\!]}$, $G \in \mathscr{F}_s$, as in Theorem 3.6.15. We have

$$\begin{aligned} \mu_A(]\!]s_G, t_G]\!]) &= \mathbf{E}(1_G(A_t - A_s)) = \mathbf{E}(1_G \mathbf{E}(A_t - A_s | \mathscr{F}_s)) \\ &= \mathbf{E}(1_G(\mathbf{E}(B_t - B_s | \mathscr{F}_s)) = \mu_B(]\!]s_G, t_G]\!]), \end{aligned}$$

since $M = A - B$ is a martingale. The result follows.

3.6.19. ***Definition:*** Let X be a right-continuous supermartingale. We say that X has a *Doob–Meyer decomposition* $X = M - A$ if M is a right-continuous martingale and A is a predictable integrable increasing process.

This decomposition is *unique* when it exists: if we also have $X = M' - A'$ with the same properties, then $M - M' = A' - A$ is a martingale, so the integrable increasing processes A and A' have the same dual predictable projection, B, say. But A and A' are predictable, so $A' = B = A$.

Finally, predictability of increasing processes is characterised in terms of left-continuous martingales – this is Meyer's original definition of 'natural' increasing processes, see [63].

3.6.20. ***Proposition:*** An integrable increasing process A is predictable if and only if for each bounded right-continuous uniformly integrable martingale M, $\mathbf{E}(M_\infty A_\infty) = \mathbf{E}(\int_0^\infty M_{s-}\,\mathrm{d}A_s)$.

Proof: Write $Y = M_\infty$, then $M_t = \mathbf{E}(Y|\mathscr{F}_t)$ a.s. for all t. If A is predictable, $\mu_A(M) = \mu_A(\Pi_p M)$ by Theorem 3.6.14. But $\mu_A(M) = \mathbf{E}(\int_0^\infty \mathbf{E}(Y|\mathscr{F}_t)\mathrm{d}A_t) = \mathbf{E}(\int_0^\infty Y\mathrm{d}A_t) = \mathbf{E}(YA_\infty)$, while $\Pi_p M$ is the left-continuous modification of M, so $(\Pi_p M)_t = M_{t-}$.

Conversely, apply the identity $\mathbf{E}(M_\infty A_\infty) = \mathbf{E}(\int_0^\infty M_{s-}\mathrm{d}A_s)$ separately to $M_t = \mathbf{E}(Y|\mathscr{F}_t)$ and $M_{t\wedge u} = \mathbf{E}(M_u|\mathscr{F}_t)$. Subtracting we have $\mathbf{E}((Y - M_u)A_\infty) = \mathbf{E}(\int_u^\infty (M_{s-} - M_u)\mathrm{d}A_s) = \mathbf{E}(\int_0^\infty M_{s-}\mathrm{d}A_s) - \mathbf{E}(M_u(A_\infty - A_u))$. But $\mathbf{E}(YA_u) = \mathbf{E}(\mathbf{E}(Y|\mathscr{F}_u)A_u) = \mathbf{E}(M_u A_u)$ since A is adapted, so $\mathbf{E}(Y(A_\infty - A_u)) = \mathbf{E}(\int_u^\infty M_{s-}\mathrm{d}A_s)$ for all $u \in \mathbf{T}$. Combining this with $\mathbf{E}(YA_\infty) = \mathbf{E}(\int_0^\infty M_{s-}\mathrm{d}A_s)$ we see that $\mu_A(X) = \mu_A(\Pi_p X)$ whenever $X = Y \cdot 1_{[\![0,u]\!]}$, and a monotone class argument now shows that $\mu_A(X) = \mu_A(\Pi_p X)$ for all bounded measurable processes X, so that A is predictable by Theorem 3.6.14.

3.6.21. ***Exercise:*** Show that a stopping time T is predictable if and only if for each bounded right-continuous uniformly integrable martingale M, $\mathbf{E}(M_T 1_{\{T<\infty\}}) = \mathbf{E}(M_{T-} 1_{\{T<\infty\}})$. (Hint: apply Proposition 3.6.20 to the increasing process $A_t = 1_{\{T\leqslant t\}}$.)

3.7. Existence of the Doob–Meyer decomposition

We have shown that the Doob–Meyer decomposition of a right-continuous supermartingale $X = M - A$ must be unique if it exists. To prove the existence of the decomposition we first consider the case when X is a potential in class (D) (see Definition 3.6.3). In this case a more precise statement is possible.

3.7.1. ***Theorem:*** Every potential in class (D) is generated by a unique integrable predictable increasing process.

Proof: The idea of the proof is to apply the discrete-time Doob decomposition of the discrete potentials $X^{(n)} = (X_{i/2^n})_{i \geqslant 0}$ for each n. Each $X^{(n)}$ is generated by a predictable (*discrete-time*) integrable increasing process $A^{(n)}$, so that $X_{i/2^n} = \mathbf{E}(A_\infty^{(n)} | \mathscr{F}_{i/2^n}) - A_{i/2^n}^{(n)}$. We shall select our increasing process A by means of a weak cluster point of the sequence $(A_\infty^{(n)})_n$.

To do this we show first that $(A_\infty^{(n)})$ is uniformly integrable (it is here that the hypothesis that X lies in class (D) is used). Fix $\lambda > 0$ and $n \geqslant 1$ and define $T_{n\lambda} = \inf(i/2^n : A_{(i+1)/2^n}^{(n)} > \lambda)$. Since $A_{(i+1)/2^n}^{(n)}$ is $\mathscr{F}_{i/2^n}$-measurable (by definition of predictability in discrete time), the random variable $T_{n\lambda}$ is a stopping time. Moreover, we have $A_\infty^{(n)} > \lambda$ iff $T_{n\lambda} < \infty$. Applying optional sampling we obtain $X_{T_{n\lambda}} = \mathbf{E}(A_\infty^{(n)} | \mathscr{F}_{T_{n\lambda}}) - A_{T_{n\lambda}}^{(n)}$, so that

$$\int_{(A_\infty^{(n)} > \lambda)} A_\infty^{(n)} \, dP = \int_{(T_{n\lambda} < \infty)} (A_{T_{n\lambda}}^{(n)} + X_{T_{n\lambda}}) dP \leqslant \lambda P(A_\infty^{(n)} > \lambda) + \int_{(T_{n\lambda} < \infty)} X_{T_{n\lambda}} \, dP \quad (3.5)$$

since $A_{T_{n\lambda}}^{(n)} \leqslant \lambda$ by definition. Hence

$$\int_{(A_\infty^{(n)} > 2\lambda)} (A_\infty^{(n)} - \lambda) dP \leqslant \int_{(A_\infty^{(n)} > \lambda)} (A_\infty^{(n)} - \lambda) dP \leqslant \int_{(T_{n\lambda} < \infty)} X_{T_{n\lambda}} \, dP$$

and as $A_\infty^{(n)} > 2\lambda$ iff $A_\infty^{(n)} - \lambda > \lambda$, this implies that

$$\lambda P(A_\infty^{(n)} > 2\lambda) \leqslant \int_{(T_{n\lambda} < \infty)} X_{T_{n\lambda}} \, dP. \quad (3.6)$$

Now replace λ by 2λ in (3.5) to obtain, using (3.6),

$$\int_{(A_\infty^{(n)} > 2\lambda)} A_\infty^{(n)} \, dP \leqslant 2\lambda P(A_\infty^{(n)} > 2\lambda) + \int_{(T_{n,2\lambda} < \infty)} X_{T_{n,2\lambda}} \, dP$$

$$\leqslant 2 \int_{(T_{n\lambda} < \infty)} X_{T_{n\lambda}} \, dP + \int_{(T_{n,2\lambda} < \infty)} X_{T_{n,2\lambda}} \, dP. \quad (3.7)$$

By definition, for any $\mu > 0$,

$$\mu P(T_{n,\mu} < \infty) = \mu P(A_\infty^{(n)} > \mu) \leqslant \mathbf{E}(A_\infty^{(n)}) = \mathbf{E}(\mathbf{E}(A_\infty^{(n)} | \mathscr{F}_0)) = \mathbf{E}(X_0).$$

Since X is in class (D), the right-hand side of (3.7) can thus be made small, uniformly in n, by choosing λ large enough. So $(A_\infty^{(n)})_n$ is uniformly integrable.

By Theorem 1.2.11, we can therefore find a subsequence $(A_\infty^{(n_k)})_k$ converging weakly in L^1. We denote the limit by A_∞. Choosing a right-continuous modification (M_t) of the martingale $(\mathbf{E}(A_\infty|\mathscr{F}_t))_t$ we let $A_t = M_t - X_t$ for all $t \in \mathbf{T}$. Then $A = (A_t)$ is right-continuous, $\lim_{t\to\infty} A_t = \lim_{t\to\infty} M_t = A_\infty$ a.s. since X is a potential, and A_∞ and all A_t are integrable. Furthermore, by the weak continuity of the conditional expectation operator,

$$\mathbf{E}(A_\infty|\mathscr{F}_0) = w\text{-}\lim_{k\to\infty} \mathbf{E}(A_\infty^{(n_k)}|\mathscr{F}_0) = w\text{-}\lim_{k\to\infty}(X_0 + A_0^{(n_k)}) = X_0,$$

so $A_0 = 0$ a.s. Finally we show that A has increasing paths: given two dyadic rationals r, s with $r \leqslant s$ we can write both in the form $i/2^n$ for large enough n. Hence $A_r^{(n_k)} \leqslant A_s^{(n_k)}$ for the above (n_k), and $\mathbf{E}(A_\infty^{(n_k)}|\mathscr{F}_r) - X_r \leqslant \mathbf{E}(A_\infty^{(n_k)}|\mathscr{F}_s) - X_s$. Again using weak continuity of conditional expectations and letting $k \to \infty$, we conclude that $M_r - X_r \leqslant M_s - X_s$ a.s. So $t \to A_t$ is a.s. increasing on the dyadic rationals, hence on $\mathbf{T}$ by right-continuity.

It remains to show that the integrable increasing process A is predictable, which by our previous remarks also ensures that A is unique.

By Proposition 3.6.20, it suffices to show that $\mathbf{E}(\int_0^\infty Y_{s-}\,dA_s) = \mathbf{E}(Y_\infty A_\infty)$ for each bounded, right-continuous uniformly integrable martingale Y. Write $X_i, Y_i, A_i, A_i^{(n)}, \mathscr{F}_i$ instead of $X_{i/2^n}$, etc. for fixed $n \geqslant 0$. By bounded convergence

$$\left(\int_0^\infty Y_{s-}\,dA_s\right) = \lim_n \sum_{i=0}^{\infty} (Y_i(A_{i+1} - A_i)).$$

Now as $M = X + A$ is a martingale.

$$\begin{aligned}\mathbf{E}(A_{i+1} - A_i|\mathscr{F}_i) &= \mathbf{E}(X_i - X_{i+1}|\mathscr{F}_i) = \mathbf{E}(X_i^{(n)} - X_{i+1}^{(n)}|\mathscr{F}_i)\\ &= \mathbf{E}[(\mathbf{E}(A_\infty^{(n)}|\mathscr{F}_i) - A_i^{(n)}) - (\mathbf{E}(A_\infty^{(n)}|\mathscr{F}_{i+1}) - A_{i+1}^{(n)})|\mathscr{F}_i]\\ &= \mathbf{E}(A_{i+1}^{(n)} - A_i^{(n)}|\mathscr{F}_i).\end{aligned}$$

As Y is adapted, we also have $\mathbf{E}(Y_i(A_{i+1} - A_i)|\mathscr{F}_i) = \mathbf{E}(Y_i(A_{i+1}^{(n)} - A_i^{(n)})|\mathscr{F}_i)$. On the other hand, $A_{i+1}^{(n)}$ is $\mathscr{F}_i$-measurable and Y is a martingale, so $\mathbf{E}(Y_i A_{i+1}^{(n)}) = \mathbf{E}(A_{i+1}^{(n)}\mathbf{E}(Y_{i+1}|\mathscr{F}_i)) = \mathbf{E}(Y_{i+1}A_{i+1}^{(n)})$. Hence for all $n \geqslant 0$,

$$\sum_{i=0}^{\infty} \mathbf{E}(Y_i(A_{i+1} - A_i)) = \sum_{i=0}^{\infty} \mathbf{E}(Y_i(A_{i+1}^{(n)} - A_i^{(n)})) = \sum_{i=0}^{\infty} \mathbf{E}(Y_{i+1}A_{i+1}^{(n)} - Y_i A_i^{(n)}) = \mathbf{E}(Y_\infty A_\infty^{(n)}).$$

In particular, this holds for the sequence (n_k). Letting $k \to \infty$ we see that the left-hand side converges to $\mathbf{E}(\int_0^\infty Y_{s-}\,dA_s)$ while the right-hand side converges to $\mathbf{E}(Y_\infty A_\infty)$, since Y is bounded and $A_\infty \in L^1$. Hence A is predictable and thus unique. (Note that the uniqueness of A_∞ shows *a posteriori* that the whole sequence $(A_\infty^{(n)})$ converges weakly in L^1 to A_∞, not just $(A_\infty^{(n_k)})$.)

In the discrete-time case we used the general Doob decomposition to decompose a supermartingale into a martingale and a potential. Such a 'Riesz decomposition' is again possible in continuous time (with a very similar proof) and allows us to extend the Doob–Meyer decomposition theorem to more general supermartingales:

3.7.2. ***Theorem:*** Every right-continuous uniformly integrable supermartingale X can be decomposed uniquely in the form $X=Y+Z$, where Y is a right-continuous uniformly integrable martingale and Z is a potential.

Proof: Let Y be a right-continuous modification of $(\mathbf{E}(X_\infty|\mathscr{F}_t))_t$ and set $Z=X-Y$. Then Z is right-continuous, uniformly integrable, and $\mathbf{E}(Z_t|\mathscr{F}_s)=\mathbf{E}(X_t|\mathscr{F}_s)-Y_s\leqslant X_s-Y_s=Z_s$ for $s\leqslant t$. Finally, $\|Z_t\|_1=\|X_t-\mathbf{E}(X_\infty|\mathscr{F}_t)\|_1\to 0$ as $t\to\infty$, so Z is a potential. If $X=Y'+Z'$ is another such decomposition, $Y'_\infty=X_\infty=Y_\infty$ and Y'_t and Y_t are thus modifications of the same process $(\mathbf{E}(X_\infty|\mathscr{F}_t))_t$ and both are right-continuous by hypothesis. Hence they are indistinguishable.

3.7.3. ***Remark:*** The term 'Riesz decomposition' has a slightly wider use in the literature: in fact if X is a right-continuous supermartingale with $\lim_{t\to\infty}\mathbf{E}(X_t)>-\infty$, then X can be decomposed uniquely as the sum of a right-continuous martingale Y and a potential Z. The proof is slightly more complicated since X_∞ need not be defined. Instead one can consider $Y_s^t=\mathbf{E}(X_{t+s}|\mathscr{F}_s)$ for $s,t\in\mathbf{T}$. Then (Y_s^t) is decreasing as t increases and $\mathbf{E}(Y_s^t)$ is bounded below, so that $Y_s=\lim_{t\to\infty}Y_s^t$ and $Z_s=X_s-Y_s$ are well-defined. The reader may verify easily that Y is a martingale and Z is a potential (or see [63] for details).

Combining Theorems 3.7.1 and 3.7.2 we obtain the following result:

3.7.4. ***Theorem:*** Every class (D) supermartingale has a Doob–Meyer decomposition $X=M-A$, where the martingale M is also uniformly integrable. (In fact, $M_t=\mathbf{E}(X_\infty+A_\infty|\mathscr{F}_t)$ a.s. for all $t\in\mathbf{T}$.)

Proof: In particular, X is right-continuous and uniformly integrable by Definition 3.6.3, so has a Riesz decomposition $X=Y+Z$, where Y is a right-continuous uniformly integrable martingale and Z a potential. Thus $Z=X-Y$ is in class (D), and is generated by a predictable increasing process A, so that $Z_t = \mathbf{E}(A_\infty|\mathscr{F}_t) - A_t$ a.s. for all t. Moreover, Y is a right-continuous modification of $(\mathbf{E}(X_\infty|\mathscr{F}_t))_t$. So $X_t = Y_t + Z_t = \mathbf{E}((X_\infty + A_\infty)|\mathscr{F}_t) - A_t$ a.s. for all $t \in \mathbf{T}$.

3.7.5. ***Remarks:*** Let us now compare the Doob–Meyer decomposition with the Doob decomposition (Definition 2.8.1) of a discrete-time supermartingale X: this was obtained by letting $a_0 = 0$, $M_0 = X_0$ and defining $A_n = \sum_{k=0}^n a_k$, $M_n = \sum_{k=0}^n m_k$, where, for $k \geqslant 1$, $a_k = X_{k-1} - \mathbf{E}(X_k|\mathscr{F}_{k-1})$, and $m_k = X_k - \mathbf{E}(X_k|\mathscr{F}_{k-1})$. So $m_k - a_k = \Delta X_k = X_k - X_{k-1}$, and $a_k \geqslant 0$, so that A is increasing (and $\mathscr{F}_{k-1}$-measurable) and M is a martingale, since $\mathbf{E}(m_k|\mathscr{F}_{k-1}) = 0$.

If we 'translate' the role of these processes into the continuous-time framework, $(\mathrm{d}X)_t$ will be $(X_t - X_{t-})$ and sums are replaced by integrals. Since $a_k = X_{k-1} - \mathbf{E}(X_k|\mathscr{F}_{k-1}) = -\mathbf{E}(\Delta X_k|\mathscr{F}_{k-1})$ we are led to the symbolic 'formula' $A_t = \int_0^t -\mathbf{E}(\mathrm{d}X_s|\mathscr{F}_{s-})$.

This suggests a relation between A_t, X_{t-} and $\mathbf{E}(X_t|\mathscr{F}_{t-})$. In fact, the jump of A at $t \in \mathbf{T}$ is just the difference between X_{t-} and $(\Pi_p X)_t$, so that A is continuous iff the predictable projection of the supermartingale X equals the left-continuous process (X_{t-}). Finally we show that if these processes coincide, then also $X_{t-} = \mathbf{E}(X_t|\mathscr{F}_{t-})$.

3.7.6. ***Theorem:*** Let X be a class (D) supermartingale and let $X = M - A$ be its Doob–Meyer decomposition. Then the 'jump process' ΔA defined by $(\Delta A)_t = A_t - A_{t-}$ equals $X_- - \Pi_p X$ where $(X_-)_t = X_{t-}$ for $t > 0$ and $X_{0-} = X_0$, $X_{\infty -} = \lim_{t \to \infty} X_t$.

In particular, A is continuous iff $X_- = \Pi_p X$.

Proof: Clearly $X_- = M_- - A_-$ and $\Pi_p X = \Pi_p M - \Pi_p A$ by the linearity of Π_p. As A is predictable, $\Pi_p A = A$. Since M is a uniformly integrable martingale, $M_t = \mathbf{E}(M_\infty|\mathscr{F}_t)$ and, as in the proof of the projection theorem, $M_- = \Pi_p M$. Hence $X - \Pi_p X = \Delta A$.

3.7.7. ***Remark:*** The relations $X_- = \Pi_p X$ can be further extended for bounded X: for all predictable stopping times T we have that $\mathbf{E}(X_{T-}) = \mathbf{E}((\Pi_p X)_T) = \mathbf{E}(X_T)$. On the other hand, let $X = Y + Z$ be the Riesz decomposition of X, and let (T_n) be a sequence announcing T. Since Y is a uniformly integrable martingale, we can apply the (discrete-time) convergence theorem (Theorem 2.6.6) and the optional sampling theorem (Theorem 2.9.4) to conclude that $Y_{T-} = \lim_n Y_{T_n} = \mathbf{E}(Y_T|\sigma(\bigcup_n \mathscr{F}_{T_n})) = \mathbf{E}(Y_T|\mathscr{F}_{T-})$. In particular, $Y_{T-} \in L^1$. Moreover, $(\mathbf{E}(Z_{T_n}))_n$ is decreasing, so by Fatou's lemma $Z_{T-} \in L^1$. So X_{T-} is integrable, and applying optional sampling once more we have $X_{T_n} \geqslant \mathbf{E}(X_T|\mathscr{F}_{T_n})$ hence also $X_{T-} \geqslant \lim_n \mathbf{E}(X_T|\mathscr{F}_{T_n}) = \mathbf{E}(X_T|\mathscr{F}_{T-})$. Combining this with the above identity $\mathbf{E}(X_{T-}) = \mathbf{E}(X_T)$, it is clear that for all bounded X with $X_- = \Pi_p X$ we have $X_{T-} = \mathbf{E}(X_T|\mathscr{F}_{T-})$.

More generally, we call a uniformly integrable right-continuous supermartingale *regular* if $\mathbf{E}(X_{T-}) = \mathbf{E}(X_T)$ for all predictable stopping times T. The proof of Theorem 3.7.6 shows that a class (D) supermartingale is regular iff its associated increasing process is continuous.

In particular, any *continuous* class (D) supermartingale X is regular, since $X_{T-} = X_T$ a.s. Hence in its Doob–Meyer decomposition $X = M - A$, both A and M are continuous.

Note that we have proved that any uniformly integrable martingale is regular. Hence a supermartingale is regular iff the potential in its Riesz decomposition is a regular supermartingale.

It can be shown that for class (D) processes the above definition of regularity is equivalent to: for each increasing, uniformly bounded sequence (T_n) of stopping times with $T = \lim_n T_n$, $\mathbf{E}(X_T) = \lim \mathbf{E}(X_{T_n})$. (See [19, Ch. VI, p. 50]. This hypothesis is fundamental in the applications of the general theory of processes to optimal stopping problems (see [5]).)

3.7.8. ***Remark:*** If A is an integrable adapted process with a.s. increasing paths, then A is obviously a submartingale. Its Doob–Meyer decomposition is then a predictable increasing process B such that A-B is a martingale. So $A - (B + A_0)$ is also a martingale, hence by Corollary 3.6.18 $B + A_0 = \Pi_p^* A$. In particular, $(\Pi_p^* A)_0 = A_0$.

3.8. Local martingales and locally integrable increasing processes

We saw in section 3.6 how stochastic measures can be characterised among finite (positive) measures on $\mathbf{T} \times \Omega$ and how the extension of such measures from optional (predictable) processes to arbitrary bounded measurable processes led to the concept of the dual optional (dual predictable) projection of an increasing process. If the increasing process is a martingale then its dual predictable projection is zero. This led us to the uniqueness of the Doob–Meyer decomposition of a class (D) supermartingale.

3.8.1. ***Notation:*** We now undertake a similar analysis of *σ-finite* stochastic measures, in which we generalise the concept of dual predictable projections and obtain a generalised Doob–Meyer decomposition for all right-continuous supermartingales. To this end we introduce the idea of *localisation* of a process.

If $\mathscr{C}$ is any class of processes, denote by $\mathscr{C}_{\text{loc}}$ the class of processes X for which there exists a sequence (T_n) of stopping times, increasing a.s. to $+\infty$, such that each stopped process $X^{T_n}=(X_{t\wedge T_n})_t$ belongs to $\mathscr{C}$. We call (T_n) a *fundamental sequence* for X.

We shall concern ourselves primarily with two examples.

(i) Let $\mathscr{A}$ denote the family of all integrable increasing processes (see Definition 3.6.1). Then $A\in\mathscr{A}_{\text{loc}}$ (*A is locally integrable*) if $A_0=0$ and A is an adapted process with right-continuous increasing paths and there is a sequence of stopping times $T_n\uparrow+\infty$ a.s. such that $\mathbf{E}(A_{T_n})=\mathbf{E}(A^{T_n}_\infty)<\infty$ for all n.

We shall see that the dual predictable projections of the A^{T_n} can be combined to produce a predictable process in $\mathscr{A}_{\text{loc}}$. But first we investigate the properties of our other example of localised processes (cf. also section 2.4.10).

(ii) Let $\mathscr{M}$ be the family of all uniformly integrable right-continuous martingales. Then $\mathscr{M}_{\text{loc}}$, the space of *local martingales*, consists of processes M such that M^{T_n} is a uniformly integrable martingale for each n, where (T_n) is a sequence of stopping times increasing to $+\infty$ a.s.

Note that we need not demand uniform integrability explicitly: by using $T'_n=T_n\wedge n$ we can always ensure that $M^{T'_n}$ is uniformly integrable. In particular, this shows that each right-continuous martingale M is a local martingale: take $T_n=n$, then $M^{T_n}_t=M_{n\wedge t}=\mathbf{E}(M_n|\mathscr{F}_t)=\mathbf{E}(M_{T_n}|\mathscr{F}_t)$, so that M^{T_n} is a uniformly integrable martingale (and $M^{T_n}_t\to M_{T_n}$ a.s. and in L^1).

On the other hand there exist local martingales which are not martingales: an example is given by the uniformly integrable potential, defined by Johnson and Helms [44], which fails to belong to class (D). Such potentials can never belong to class (D) unless identically zero, in view of the following result.

3.8.2. ***Proposition:*** A local martingale is in class (D) iff it is a uniformly integrable martingale.

Proof: The sufficiency is clear by optional sampling. Conversely suppose that $M\in\mathscr{M}_{\text{loc}}$ belongs to class (D). Then M is uniformly integrable and we need only show that M is a martingale. Now there exists $T_n\uparrow+\infty$ such that $M^{T_n}\in\mathscr{M}$ for all n. So $M_{s\wedge T_n}=M^{T_n}_s=\mathbf{E}(M^{T_n}_t|\mathscr{F}_s)=\mathbf{E}(M_{t\wedge T_n}|\mathscr{F}_s)$ for $s\leqslant t$, $n\in\mathbf{N}$. Since $\{M_T: T \text{ is a stopping time}\}$ is uniformly integrable, and $M_{s\wedge T_n}\to M_s$ a.s. when $n\to\infty$, $M_{s\wedge T_n}\to M_s$ in L^1-norm, and similarly $\mathbf{E}(M_{t\wedge T_n}|\mathscr{F}_s)\to\mathbf{E}(M_t|\mathscr{F}_s)$ in L^1-norm. So in the limit $M_s=\mathbf{E}(M_t|\mathscr{F}_s)$ for $s\leqslant t$, as required.

3.8.3. ***Exercises:*** Suppose X is adapted to $(\mathscr{F}_t)$ and T is a stopping time for $(\mathscr{F}_t)$.

(i) Show that X^T is adapted to $(\mathscr{F}_t)$ iff it is adapted to $(\mathscr{F}_{t\wedge T})$.

(ii) Show that X^T is a martingale w.r.t. $(\mathscr{F}_t)$ iff it is a martingale w.r.t. $(\mathscr{F}_{t\wedge T})$.

(iii) Deduce that M is a local martingale with fundamental sequence (T_n) iff $(M_{t\wedge T_n}, \mathscr{F}_{t\wedge T_n})$ is a uniformly integrable martingale for all n.

3.8.4: We say that a stopping time T *reduces* a process X (assumed to be cadlag, adapted and with $X_0 \in L^1$) if X^T is a uniformly integrable martingale. Thus a local martingale X is reduced by each stopping time T_n in its fundamental sequence. This allows us to state succinctly some simple stability properties of local martingales:

3.8.5. ***Proposition:*** Let X be a local martingale, S, T stopping times.

(i) If T reduces X and $S \leqslant T$, then S reduces X.

(ii) If S and T reduce X, so does $S \vee T$.

(iii) $\mathscr{M}_{\text{loc}}$ is a vector space.

(iv) $\mathscr{M}_{\text{loc}}$ is closed under stopping (i.e. X^T is a local martingale for all T).

(v) If Y is a right-continuous adapted process such that for some sequence of stopping times $T_n \uparrow +\infty$ a.s., each Y^{T_n} is a local martingale, then Y is itself a local martingale.

Proof: (i) Since T reduces X, X^T is a uniformly integrable martingale, so by optional stopping $(X^T)^S = X^S$ is also a martingale, and is clearly uniformly integrable.

(ii) $X^{S\vee T} = X^S + X^T - X^{S\wedge T}$. By (i) $S \wedge T$ reduces X, so $X^{S\vee T}$ is a linear combination of uniformly integrable martingales.

(iii) If $X, Y \in \mathscr{M}_{\text{loc}}$ have fundamental sequences (S_n), (T_n) then $X + \alpha Y$ has fundamental sequence $(S_n \wedge T_n)$, by (i).

(iv) Let (T_n) be a fundamental sequence for X. Since for all t, $(X^T)_t^{T_n} = X_{T\wedge T_n \wedge t} = (X^{T_n})_t^T$, and each X^{T_n} is a uniformly integrable martingale, each $(X^T)^{T_n}$ is also a uniformly integrable martingale, by optional stopping.

(v) If S is the essential supremum of the set of stopping times reducing Y, we can find a sequence $S_n \uparrow S$ of stopping times reducing Y (we can choose (S_n) increasing by (ii)). But since $Y^{T_n} \in \mathscr{M}$ for all n, we can find sequences $U_m^{(n)}$ of stopping times with $U_m^{(n)} \uparrow +\infty$ a.s. as $m \to \infty$, $T_n \wedge U_m^{(n)}$ reducing Y for each $m, n \in \mathbb{N}$. So $T_n \wedge U_m^{(n)} \leqslant S$ and as $U_m^{(n)} \uparrow +\infty$ a.s., $T_n \leqslant S$ for all n. As $T_n \uparrow +\infty$, this means that $S = +\infty$ a.s., so (S_n) is a fundamental sequence for Y.

3.8.6. ***Remarks:*** Our definition of local martingale is less general than that of [19], where integrability of X_0 is not required. In our applications we can assume $X_0=0$ a.s. without loss, however. See [19; Ch. VI, pp. 27–36] for more results on local martingales, and for an interesting discussion of their discrete counterparts, which again highlights how concepts and results are extended to the context of σ-finite measures by localisation.

3.8.7: We turn to locally integrable increasing processes. We extend the notion of 'dual predictable projection' as follows: given a locally integrable increasing process A, we find a predictable increasing process B such that $A-B$ is a *local* martingale. (The converse also holds: given an increasing process A such that $A-B$ is a local martingale for some predictable increasing B, then A is locally integrable – see [19; Ch. VI, p. 50]).

The following lemma shows that the above statement is 'symmetrical' in relation to integrability properties:

3.8.8. ***Lemma:*** Any predictable increasing process is locally integrable.

Proof: By hypothesis $A_0=0$. Writing $S_n=\inf\{t:A_t\geqslant n\}$ it is clear that S_n is a stopping time. Moreover, if we set $H_n=\{(t,w):A_t\geqslant n\}$, S_n is the début of H_n and as A is right-continuous $[\![S_n]\!]\subset H_n$. Hence $[\![S_n]\!]=H_n\setminus]\!]S_n,\infty[\![$ is the difference of two sets in Σ_p, so S_n is predictable. Let $(S_{n,k})$ be a sequence announcing S_n. Then $A_{S_{n,k}}<n$ for all k, so $\mathbf{E}(A_{S_{n,k}})<\infty$, while, on the other hand, $\sup_{n,k}S_{n,k}=+\infty$. So take $T_m=\sup_{n,k\leqslant m}S_{n,k}$ for each m, then $\mathbf{E}(A_{T_m})<\infty$ for all m, and $T_m\uparrow+\infty$. Hence $A\in\mathscr{A}_{\text{loc}}$.

3.8.9. ***Theorem:*** Let $A\in\mathscr{A}_{\text{loc}}$. Then there is a unique predictable increasing process B such that $M=A-B$ is a local martingale with $M_0=0$.

We call B the *dual predictable projection* of A and write $B=\Pi_p^*A$.

Proof: The uniqueness follows from Corollary 3.6.18 by localisation: if $M_1=A-B_1$ and $M_2=A-B_2$ are local martingales and B_1, B_2 are predictable increasing processes, let (T_n) be a fundamental sequence for $M=M_2-M_1\in\mathscr{M}_{\text{loc}}$, and let (U_n) be an increasing sequence of stopping times with $\lim_n U_n=+\infty$ a.s. and $\mathbf{E}(A_{U_n})<\infty$ for all n. Let $S_n=T_n\wedge U_n\wedge n$, then S_n is a fundamental sequence for M by Proposition 3.8.5(i), so $M^{S_n}=B_1^{S_n}-B_2^{S_n}$ is a martingale and is the difference of two predictable integrable increasing processes, hence is evanescent by Corollary 3.6.18. Letting $n\to\infty$ we obtain the same result for M.

To construct $B=\Pi_p^* A$, let $T_n \uparrow +\infty$ such that $\mathbf{E}(A_{T_n})<\infty$ and define a measure μ_A on $\mathbf{T}\times\Omega$ by $\mu_A(X)=\mathbf{E}(\int_0^\infty X_s\,dA_s)$ for positive measurable X. Then μ is σ-finite, since $\mu([\![0,T_n]\!])<\infty$ for all n. Further, let $\mu_p(X)=\mu_A(\Pi_p X)$. Again $\mu_n^p=1_{[\![0,T_n]\!]}\cdot\mu_p$ is a finite positive measure, determined by its values on Σ_p, hence as in Theorem 3.6.14 it is generated by a predictable integrable increasing process B^n, so that $\mu_n^p(X)=\mu_{B^n}(X)=\mathbf{E}(\int_0^\infty X_S\,dB_s^n)$. But μ_{n+1}^p and μ_n^p coincide on $[\![0,T_n]\!]$, so $B_t^n=B_{t\wedge T_n}^{n+1}$ for all t. Hence we can 'paste together' the processes B^n to find a single predictable increasing process $B=\lim_n B^n$ such that for all n, $B_t^n=B_{t\wedge T_n}$ (each B_t^n is constant after T_n).

Clearly the σ-finite measures μ_p and μ_B agree on Σ_p, so that $\mu_B(X)=\mu_A(\Pi_p X)$ for all positive measurable X. Since $A^{T_n}-B^{T_n}$ is the difference of two integrable increasing processes whose difference is a martingale, $A-B$ is a local martingale with fundamental sequence (T_n).

We have now 'localised' all the concepts involved in the Doob–Meyer decomposition. It remains to extend the decomposition theorem itself:

3.8.10. ***Theorem:*** Let X be a right-continuous supermartingale. Then X has a unique decomposition $X=M-A$, where $M\in\mathcal{M}_{\text{loc}}$ and A is a predictable increasing process (and therefore in $\mathcal{A}_{\text{loc}}$).

Proof: The uniqueness follows from Theorem 3.8.9: if $X=M-A=M'-A'$, so that $A-A'=M'-M$ is a local martingale. But A' is predictable, hence by Theorem 3.8.9 $A'=\Pi_p^* A=A$, and so $M'=M$. Hence the decomposition is unique when it exists.

Now suppose that there are two stopping times S and T and increasing predictable processes A and B (with $A_0=B_0=0$) such that $X^S-A=M$ and $X^T-B=N$ are both local martingales. Then $X^{S\wedge T}=(X^S)^T=M^T+A^T$ and $X^{S\wedge T}=(X^T)^S=N^S+B^S$. Since M^T and N^S are local martingales, the uniqueness of the decomposition for $X^{S\wedge T}$ implies that $A^T=B^S$. Hence $A=B$ up to time $S\wedge T$. This shows that it will suffice to find, for each n in $\mathbf{N}$, a predictable increasing process $A^{(n)}$ such that $X_{t\wedge n}+A_t^{(n)}$ defines a local martingale. For then we have $A_t^{(n)}=A_{t\wedge n}^{(n+1)}$ by the above remarks and we can define the predictable increasing process A by $A_t=\lim_n A_t^{(n)}$, so that $A_{t\wedge n}=A_t^{(n)}$ and $X+A$ is a local martingale by Proposition 3.8.5(v).

Hence we need only prove the existence of the decomposition for the case when X is stopped at time $t=n$. The non-negative process Y defined by $Y_t=X_t-\mathbf{E}(X_n|\mathcal{F}_t)$ is then 0 for $t\geqslant n$, and is clearly a supermartingale. Hence Y is a potential.

Letting $S_n=\inf\{t:Y_t\geqslant n\}\wedge n$ yields a sequence (S_n) of stopping times increasing to $+\infty$, and Y^{S_n} is bounded above by n on $[\![0,S_n[\![$, by Y_{S_n} on

$[\![S_n, \infty[\![$, hence is bounded by $n \vee Y_{S_n}$ on $\mathbf{T} \times \Omega$. By optional stopping, $n \vee Y_{S_n} \in L^1$ hence Y^{S_n} is a potential of class (D). It therefore has a unique Doob–Meyer decomposition and we have found a unique predictable increasing process $A^{(n)}$ such that $Y_t^{S_n} + A_t^{(n)} = \mathbf{E}(A_\infty^{(n)} | \mathscr{F}_t)$. The proof is complete, since the local martingale $Y + A$ is reduced by each S_n.

3.8.11. ***Remark:*** The final part of the above proof is valid for *any* non-negative supermartingale Y. If we write $A = \lim_n A^{(n)}$ for the $A^{(n)}$ associated with Y, then $Y_0 - \mathbf{E}(Y_\infty | \mathscr{F}_0) \geqslant \mathbf{E}(A_\infty | \mathscr{F}_0)$, where $Y_\infty = \lim_{t \to \infty} Y_t$, $A_\infty = \lim_{t \to \infty} A_t$. (To see this, note that $Y_t^{S_n} - \mathbf{E}(Y_\infty^{S_n} | \mathscr{F}_t) = \mathbf{E}(A_\infty^{(n)} | \mathscr{F}_t) - A_t^{(n)}$ as $Y^{S_n} + A^{(n)}$ is a martingale, set $t = 0$ and let $n \to \infty$.) Applying this to $Y_{t \wedge S}$ and $A_{t \wedge S}$, we have $Y_0 \geqslant \mathbf{E}(Y_0 - Y_S | \mathscr{F}_0) \geqslant \mathbf{E}(A_s | \mathscr{F}_0)$. Setting $Y'_u = Y_{r+u}$, $\mathscr{F}'_u = \mathscr{F}_{r+u}$, which has the associated increasing process $B'_u = B_{r+u} - B_r$ (check!), we have finally $Y_r \geqslant \mathbf{E}(Y_r - Y_{r+s} | \mathscr{F}_r) \geqslant \mathbf{E}(A_{r+s} - A_r | \mathscr{F}_r)$ for all $r, s \in \mathbf{T}$. In particular, putting $U_t = \mathbf{E}(A_\infty | \mathscr{F}_t)$ and $N_t = Y_t + A_t - \mathbf{E}(A_\infty | \mathscr{F}_t)$, we have shown, upon letting $s \to \infty$ in the above inequalities, that $Y = U + N$ is a decomposition of Y into the sum of a potential U of class (D) and a non-negative local martingale N. Note the similarity of this *Riesz decomposition* and that obtained in the discrete-time case (Example 2.8.3).

3.8.12. ***Examples:*** We proved in Chapter 0 that the Wiener process W is a martingale relative to its natural filtration $(\mathscr{F}_t^W)$. Setting $A_t(w) = t$ for all $(t, w) \in \mathbf{T} \times \Omega$ we obtain a predictable increasing process, and since $\mathbf{E}((W_t - W_s)^2 | \mathscr{F}_s^W) = t - s = \mathbf{E}((A_t - A_s) | \mathscr{F}_s^W)$ for $0 \leqslant s \leqslant t < \infty$, $W^2 - A$ is a martingale. Since W^2 is a *sub*martingale, we conclude that A is the unique predictable increasing process such that $M = (-W^2 + A)$ is a local martingale, i.e. $-W^2 = M - A$ is the Doob–Meyer decomposition of the supermartingale $-W^2$.

For obvious reasons we call A the *quadratic variation* process of W. Note that the continuous process $-W^2$ has a continuous associated increasing process.

Continuity of the supermartingale is *sufficient* to ensure continuity of the associated increasing process (see Remark 3.7.7) but it is not *necessary*: for example, denote the Poisson process with parameter $\lambda > 0$ by (N_t). N has independent increments and $\mathbf{E}(N_t - N_s) = \lambda(t - s)$, hence $M_t = N_t - \lambda t$ is a martingale for the natural filtration $(\mathscr{F}_t^N)$. Thus $-N$ is a supermartingale with associated increasing process λt. But we also have $\mathbf{E}(M_t - M_s)^2 | \mathscr{F}_s^N) = \lambda(t - s)$ for $0 \leqslant s \leqslant t < \infty$. Hence the predictable increasing process λt is also the increasing process associated with the supermartingale $-M_t^2 = -(N_t - \lambda t)^2$. This quadratic variation process is therefore continuous while the associated supermartingale is not.

In neither of these examples is the supermartingale under consideration in class (D). On the other hand if $\mathbf{T}=[0,t]$ for some *finite* t, then the ordinary Doob–Meyer decomposition theorem applies.

Supplement: Capacitability and cross-sections

The cross-section theorems of section 3.5 are usually considered too abstract for proofs to be included in books on probability theory, although the theorems are valid for the general theory of processes as developed in Chapter 3. Several routes to the cross-section theorems exist. Here we simply reproduce with only minor changes the direct proof given by Dellacherie in [18]. For the theory of Choquet capacities, analytic sets and their application to measure theory we refer the reader to [19], [16], [76], where many of the definitions and results are given in greater generality.

S.1. ***Definition:*** A *paving* $\mathscr{E}$ on a set E is a collection of subsets of $\mathbf{E}$ which contains the empty set and is closed under *finite* unions and *countable* intersections. The pair $(E,\mathscr{E})$ is a *paved space.*

We shall primarily consider two pavings:

(a) if $(\Omega,\mathscr{F})$ is a measurable space, $\mathscr{F}$ is trivially a paving on Ω,

(b) if $E=\mathbf{R}^+\times\Omega$ let $\mathscr{R}$ be the collection of rectangles $K\times A$, where $K\subseteq\mathbf{R}^+$ is compact and $A\in\mathscr{F}$, and let $\mathscr{E}$ be the paving generated by $\mathscr{R}$.

If we let $\mathscr{U}$ denote the collection of all finite unions of sets in $\mathscr{R}$, then each element L of $\mathscr{E}$ can be written as $\bigcap_n L_n$, where (L_n) is a decreasing sequence of sets in $\mathscr{U}$. (Since $\mathscr{R}$ is closed under finite intersections, so is $\mathscr{U}$, hence the (L_n) can be chosen decreasing.)

Notation: (i) We shall use $(E,\mathscr{E})$ and $(F,\mathscr{F})$ for general paved spaces, although the symbols $(E,\mathscr{E})$ and $\mathscr{F}$ also have particular meanings in the above examples – however, our only applications will concern precisely these examples.

(ii) Write $A_n\uparrow A$ (respectively $A_n\downarrow A$) if (A_n) is an increasing (respectively decreasing) sequence with union (respectively intersection) A. Similarly, write $t_n\uparrow t$ and $t_n\downarrow t$ for monotone convergence of sequences of real numbers.

S.2. ***Definition:*** A (Choquet-) *capacity* C on a paved space $(E,\mathscr{E})$ is a mapping from the subsets of E into $[0,\infty]$ satisfying

(i) $A\subseteq B\Rightarrow C(A)\leqslant C(B)$,

(ii) $A_n\uparrow A\Rightarrow C(A_n)\uparrow C(A)$,

(iii) for (L_n) in $\mathscr{E}$ such that $L_n\uparrow L$, $C(L_n)\downarrow C(L)$.

We consider only two examples of capacities:

(a) outer measure (see Definition 0.1.1) $P^*(A)=\inf\{P(B):B\supseteq A,B\in\mathcal{F}\}$ associated with $(\Omega,\mathcal{F},P)$. This is obviously a capacity on $(\Omega,\mathcal{F})$.

(b) If $E=\mathbf{R}^+\times\Omega$ as before and $\pi:\mathbf{R}^+\times\Omega\to\Omega$ denotes the natural projection map, set $C(A)=P^*(\pi(A))$ for $A\subseteq E$. To see that this defines a capacity we need the following lemma:

S.3. ***Lemma:*** Considered as a map from $\mathcal{P}(E)$ to $\mathcal{P}(\Omega)$, the projection π has the following properties:

(i) $A\subseteq B\Rightarrow\pi(A)\subseteq\pi(B)$,

(ii) $A_n\uparrow A\Rightarrow\pi(A_n)\uparrow\pi(A)$,

(iii) for (L_n) *in* $\mathcal{E}$ such that $L_n\downarrow L$, $\pi(L_n)\in\mathcal{F}$ and $\pi(L_n)\downarrow\pi(L)$.

Proof: Only (iii) requires further proof. Since $L\subseteq L_n$ for all n, $\pi(L)\subseteq\bigcap_n\pi(L_n)$ by (i). To prove the reverse inclusion, first write $L_n=\bigcap_{m=1}^{\infty}S_{nm}$, where all S_{nm} belong to $\mathcal{U}$ and $S_{n(m+1)}\subseteq S_{nm}$ for all m,n. So $L=\bigcap_n L_n=\bigcap_N(\bigcap_{m,n=1}^{N}S_{nm})$. Now $\bigcap_{m,n=1}^{N}S_{nm}\supseteq\bigcap_{n=1}^{N}(\bigcap_{m=1}^{\infty}S_{nm})=\bigcap_{n=1}^{N}L_n$. Since the result is clearly true for sets in $\mathcal{R}$, and hence for sets in $\mathcal{U}$, we conclude that $\pi(L)\supseteq\bigcap_n\pi(L_n)$ also. In particular, $\pi(L)=\bigcap_n\pi(L_n)\in\mathcal{F}$, since each $\pi(L_n)\in\mathcal{F}$.

S.4. ***Definition:*** Given two paved spaces $(E,\mathcal{E})$ and $(F,\mathcal{F})$, any map π satisfying (i)–(iii) of Lemma S.3 will be called an $(\mathcal{E},\mathcal{F})$-*capacitary operation* from E to F.

S.5. ***Definition:*** Given a capacity C on a paved space $(E,\mathcal{E})$, we say that $A\subseteq E$ is C-*capacitable* if $C(A)=\sup\{C(L):L\in\mathcal{E},\ L\subseteq A\}$.

In our examples it is clear that π is an $(\mathcal{E},\mathcal{F})$-capacitary operation and that a set $A\subseteq\Omega$ is P^*-capacitable iff A belongs to the P-completion $\bar{\mathcal{F}}$ of $\mathcal{F}$ (see Definition 0.1.1).

Our main interest is in the example $E=\mathbf{R}^+\times\Omega$ with $\mathcal{E}$ generated by $\mathcal{R}$, and $C(\cdot)=P^*(\pi(\cdot))$. If $A\subseteq E$ is C-capacitable, then since $\pi(L)\in\mathcal{F}$ for all $L\in\mathcal{E}$ by Lemma S.3 we also have $\pi(A)\in\bar{\mathcal{F}}$ by the preceding remark. We can now characterise the C-capacitable sets:

S.6. ***Lemma:*** Let E, $\mathcal{E}$ and C be as in the above example. Then $A\subseteq E=\mathbf{R}^+\times\Omega$ is C-capacitable iff for each $\varepsilon>0$ there exists a random variable S with values in $[0,\infty]$ such that $[\![S]\!]\subseteq A$ and $P\{S<\infty\}\geqslant P^*(\pi(A))-\varepsilon$.

Proof: $\Rightarrow$: Fix $\varepsilon>0$. Since A is capacitable we can find $L\in\mathcal{E}$, $L\subseteq A$, such that $P(\pi(L))>P^*(\pi(A))-\varepsilon$. Let $S=D_L$ be the début of L, that is,

$S(\omega)=\inf\{t\geqslant 0:(t,\omega)\in L\}$, where $\inf\emptyset=+\infty$. Then S is a random variable: for if $L_n\in\mathscr{U}$ and $L_n\downarrow L$, then $D_{L_n}\uparrow D_L$ and it is clear that each D_{L_n} is a random variable. Moreover, the sections $L(\omega)=\{t:(t,\omega)\in L\}$ are compact, so the inf $S(\omega)$ is attained, hence $[\![S]\!]\subseteq A$.

$\Rightarrow$: Conversely, we can assume that $S\leqslant k$ on $\{S<\infty\}$, for some large enough k. Approximating S by stopping times of finite range, we can ensure that $[\![S]\!]\subseteq L$, and this suffices to prove that $C(A)<C(L)+\varepsilon$.

S.7: In order to deduce the first cross-section theorem, we now need only show that each element A of $\mathscr{B}(\mathbf{R}^+)\times\mathscr{F}$ is C-capacitable: for we find S as in the lemma, and since $\pi(A)\in\bar{\mathscr{F}}=\mathscr{F}$ (this is where the *completeness* assumption on $\mathscr{F}$ is important!) we also have $P^*(\pi(A))=P(\pi(A))$. Hence Lemma S.6 implies Theorem 3.5.3.

S.8: The C-capacitability of all sets in $\mathscr{B}(\mathbf{R}^+)\times\mathscr{F}$ will follow from Choquet's capacitability theorem. Let us call a collection $\mathscr{M}$ of subsets of a set E a *mosaic* if $\mathscr{M}$ contains $\emptyset$ and is closed under countable unions and countable intersections. The mosaic $\hat{\mathscr{E}}$ generated by a paving $\mathscr{E}$ on E is also the monotone class generated by $\mathscr{E}$: to see this, note first that $\hat{\mathscr{E}}$ is a monotone class, so $\hat{\mathscr{E}}$ contains the monotone class $\mathscr{M}(\mathscr{E})$ generated by $\mathscr{E}$. For the reverse inclusion it suffices to show that $\mathscr{M}(\mathscr{E})$ is a mosaic. Now for $F\subseteq E$ let $\mathscr{K}(F)=\{G\subseteq E: G\bigcap F, G\bigcup F\in\mathscr{E}\}$. Then $\mathscr{K}(F)$ is a monotone class and $G\in\mathscr{K}(F)$ iff $F\in\mathscr{K}(G)$. If $F,G\in\mathscr{E}$, then $G\in\mathscr{K}(F)$, so $\mathscr{E}\subseteq\mathscr{K}(F)$, hence $\mathscr{M}(\mathscr{E})\subseteq\mathscr{K}(\mathscr{F})$. Thus for $F\in\mathscr{E}$ and $G\in\mathscr{M}(\mathscr{E})$ we have $G\in\mathscr{K}(F)$, so $F\in\mathscr{K}(G)$, hence again $\mathscr{M}(\mathscr{E})\subseteq\mathscr{K}(G)$ for all G in $\mathscr{M}(\mathscr{E})$. But since any finite intersection in $\mathscr{E}$ can be replaced by a decreasing one, this means that $\mathscr{M}(\mathscr{E})$ is a mosaic. (For a different proof and variants of this Monotone Class Theorem, see [19; Th. 1.19].)

Consequently we see that $\hat{\mathscr{E}}$ is a σ-field iff the complement of each set in $\mathscr{E}$ belongs to $\hat{\mathscr{E}}$. In the example where $E=\mathbf{R}^+\times\Omega$ and $\mathscr{E}$ is generated by rectangles this holds, since the complement of a compact $A\subseteq\mathbf{R}^+$ is a countable union of compact sets. Therefore $\hat{\mathscr{E}}=\mathscr{B}(\mathbf{R}^+)\times\mathscr{F}$ in this example. Hence to prove the section theorem (Theorem 3.5.3) it suffices to show that $\hat{\mathscr{E}}$ is C-capacitable.

S.9. ***Definition:*** Let $(E,\mathscr{E})$ be a paved space. A set $A\subseteq E$ is *$\mathscr{E}$-capacitable* if it is C-capacitable for all $\mathscr{E}$-capacities C.

S.10. ***Theorem (Choquet):*** Let $(E,\mathscr{E})$ be a paved space. Then every element of $\hat{\mathscr{E}}$ is $\mathscr{E}$-capacitable.

The main effort in this supplement is devoted to a proof of this result. Before proceeding to the proof, we note that it also allows us to show that the début D_H of a progressive set is a stopping time.

We assume that a stochastic base $(\Omega,\mathscr{F},P,(\mathscr{F}_t),\mathbf{R}^+)$ is given. So we must check that $\{D_H<t\}\in\mathscr{F}_t$ for all t. But $\{D_H<t\}=\pi(H_t)$, where $H_t=H\bigcap([0,t[\times\Omega)\in\mathscr{B}_t\times\mathscr{F}_t$. This σ-field is the mosaic on $[0,t]\times\Omega$ generated by rectangles $K\times L$ with K compact in $[0,t]$ and $L\in\mathscr{F}_t$. Therefore H_t is $P^*(\pi(\cdot))$-capacitable by Choquet's Theorem (S.10) and, by the remark preceding Lemma S.6, $\pi(H_t)\in\bar{\mathscr{F}}_t=\mathscr{F}_t$, since $\mathscr{F}_t$ is complete. Hence D_H is a stopping time for $(\mathscr{F}_t)$.

S.11: Turning to the proof of Choquet's theorem, we begin by reducing the problem to that when E is a topological space with countable base and $\mathscr{E}$ consists of the closed subsets of E:

(i) Let $(E,\mathscr{E})$ be a paved space. Any $A\in\mathscr{E}$ belongs to the mosaic generated by some sequence (L_n) in $\mathscr{E}$: let $\mathscr{C}$ be the union of all the mosaics generated by countable subclasses of $\mathscr{E}$. Then $\mathscr{E}\subseteq\mathscr{C}\subseteq\hat{\mathscr{E}}$ and $\mathscr{C}$ is a monotone class, so $\mathscr{C}=\hat{\mathscr{E}}$ by S.8.

(ii) Fix (L_n) as in (i) and let $\mathscr{T}$ be the coarsest topology on $\mathscr{E}$ such that each L_n is $\mathscr{T}$-closed. Let $\mathscr{F}$ be the paving of all $\mathscr{T}$-closed sets on E. Then $\mathscr{F}\subseteq\mathscr{E}\bigcup\{E\}$ and $A\in\hat{\mathscr{F}}$. Further, each $\mathscr{E}$-capacity C is an $\mathscr{F}$-capacity: if $F_n\downarrow F$ in $\mathscr{F}$, then since (L_n) forms a basis for the $\mathscr{T}$-closed sets and $C(L_n)\downarrow C(\bigcap_n L_n)$ by hypothesis, we have $C(F_n)\downarrow C(F)$.

(iii) Thus each $\mathscr{F}$-capacitable set $A\subseteq E$ is also $\mathscr{E}$-capacitable, or equals E. Now if $E\in\hat{\mathscr{E}}$, then $E_n\uparrow E$ for some sequence (E_n) in $\mathscr{E}$, hence if C is an $\mathscr{E}$-capacity on E, $C(E_n)\uparrow C(E)$, so that E is also $\mathscr{E}$-capacitable.

Thus in the hypothesis of Theorem S.10 we may (and shall) take E to be a topological space and $\mathscr{E}$ the paving of its closed subsets. This hypothesis will apply until the end of the proof of Theorem S.10.

S.12. ***Definition:*** Let $A\subseteq E$ and let C be a capacity on E such that $C(A)>t$ for some fixed $t\in\mathbf{R}^+$. Associated with (A,C,t) we construct a two-person *Sierpinski-game* as follows: denote the players by I,II. I chooses $B_1\subseteq A$ such that $C(B_1)>t$, then II chooses $A_1\subseteq B_1$ such that $C(A_1)>t$, then I chooses $B_2\subseteq A_1$ with $C(B_2)>t$, and so on, *ad infinitum.* Thus I constructs a decreasing sequence (B_n) and II a decreasing sequence (A_n) of subsets of A. We say that II *wins the game* if the closed set $\bigcap_n\bar{A}_n(=\bigcap_n\bar{B}_n)$ is a subset of A. If II wins, $C(\bigcap_n\bar{A}_n)=\lim_n C(\bar{A}_n)\geqslant t$, and if this is true for all t, $C(A)=\sup\{C(L):L\subseteq A,L\in\mathscr{E}\}$, since $\mathscr{E}$ consists of the closed sets in E. Hence A is $\mathscr{E}$-capacitable if II wins the game for all choices of C and t.

A *winning strategy* for II, with given (A,C,t) consists of finding a function f which chooses the subsets $A_n = f(B_1,\ldots,B_n)$ of B_n such that $C(A_n) > t$ for all n and $\bigcap_n \bar{A}_n \subseteq A$. We say that A is *supercapacitable* if II always has a winning strategy in the Sierpinski game, for all capacities C and $t \in \mathbf{R}^+$ such that $t < C(A)$. Clearly supercapacitable sets are $\mathscr{E}$-capacitable. The proof of Theorem S.10 is then complete when we prove the following result:

S.13. ***Theorem:*** Each closed subset of E is supercapacitable. The class $\mathscr{S}$ of supercapacitable subsets of E is closed under countable intersections and increasing countable unions. In particular, each element of $\hat{\mathscr{E}}$ is supercapacitable.

Proof: If A is closed, $\bar{A}_n \subseteq A$ for any subset of A, hence any closed set is supercapacitable. Thus we need only show that $\mathscr{S}$ has the required stability properties.

Let $A^{(m)} \uparrow A$, where $A^{(m)} \in \mathscr{S}$ for all m. If C is a capacity on E and $C(A) > t$ for some $t \in \mathbf{R}^+$, then we can find m so large that $C(A^{(m)}) > t$. Now suppose that in the game (A,C,t) player I chooses $B_1 \subseteq A$ (with $C(B_1) > t$). We can then find $k \in \mathbf{N}$ such that $C(B_1 \bigcap A^{(k)}) > t$. But in the game $(A^{(k)},C,t)$ player II has a winning strategy $f^{(k)}$, since $A^{(k)} \in \mathscr{S}$. So II responds by selecting $A_1 = f^{(k)}(A^{(k)} \bigcap B_1)$. This forces I to play as if the game $(A^{(k)},C,t)$ were in progress, so $A_n = f^{(k)}(A^{(k)} \bigcap B_1, B_2, \ldots, B_n)$ is a winning strategy for II. Hence $A \in \mathscr{S}$.

Next, let A^1 and A^2 belong to $\mathscr{S}$ and set $A = A^1 \bigcap A^2$. If $C(A) > t$ for some C,t, and if f^1 and f^2 represent the winning strategies for II in games (A^1,C,t) and (A^2,C,t), then a winning strategy for II in (A,C,t) is constructed by alternating between f^1 and f^2. Given that I chooses $B_1 \subseteq A$ with $C(B_1) > t$, set $A_1 = f^1(B_1)$. Then I chooses $B_2 \subseteq A_1$ and II plays $A_2 = f^2(B_2)$, as if I had played B_2 as his first choice. Continuing in this fashion, we obtain the winning strategy

$$A_n = \begin{cases} f^1(B_1,B_3,\ldots,B_n) & \text{if } n \text{ is odd} \\ f^2(B_2,B_4,\ldots,B_n) & \text{if } n \text{ is even.} \end{cases}$$

Hence $A \in \mathscr{S}$.

For a general intersection $A = \bigcap_m A^m$, where (A^m) is a finite or infinite sequence in $\mathscr{S}$, first partition $\mathbf{N}$ into finite subsets (N^m) according to the elements of (A^m). Now if $C(A) > t$ and f^m is the winning strategy for II in the game (A^m,C,t), II can respond to choices $B_1,\ldots,B_n$ by I as follows: if $n \in N^m$ set $A_n = f^m(B_{n^1}, B_{n^2}, \ldots, B_n)$, where $n^1, n^2, \ldots, n$ are the elements of $\{1,2,\ldots,n\} \bigcap N^m$ arranged in increasing order. It is not difficult to check that this produces a winning strategy, so $A \in \mathscr{S}$.

Finally we proceed to the proof of Theorem 3.5.4 (the second cross-section theorem). Let B be an optional set ($B \in \Sigma_o$). Since $\Sigma_o \subseteq \Sigma_\pi$, B is progressive, so its début D_B is a stopping time, and thus $\pi(B) = \{D_B < \infty\} \in \mathscr{F}$. So $P(\pi(B))$ is well-defined. Also, $\mathscr{B}(\mathbf{R}^+) \times \Omega$ is the mosaic generated by $\mathscr{R}$, hence B is $P(\pi(\cdot))$-capacitable and, by Lemma S.6, given $\varepsilon > 0$ there is a random variable S such that $[\![S]\!] \subseteq B$ and $P\{S < \infty\} > P(\pi(B)) - \varepsilon/2$. To replace S by a stopping time, define a measure on Σ_o by $m(A) = P(\pi(A) \cap [\![S]\!])$ for $A \in \Sigma_o$. The outer measure m^* is an $\mathscr{E}$-capacity for the paving $\mathscr{E}$ generated by stochastic intervals $[\![U,V[\![$, U,V stopping times. By Theorem S.10, the set $B \in \hat{\mathscr{E}}$ is m^*-capacitable, so we can find $L \in \mathscr{E}$ with $m(L) > m(B) - \varepsilon/2$. Now $L = \bigcap_n L_n$ where each L_n is a finite union of left-closed stochastic intervals (cf. Definition S.1) and hence each non-empty section $L(\omega) = \{t : (t,\omega) \in L\}$ contains its infimum. Hence $[\![D_L]\!] \subseteq L$. Setting $T = D_L$ we can easily verify that T has the required properties. (See Fig. 4.)

Fig. 4

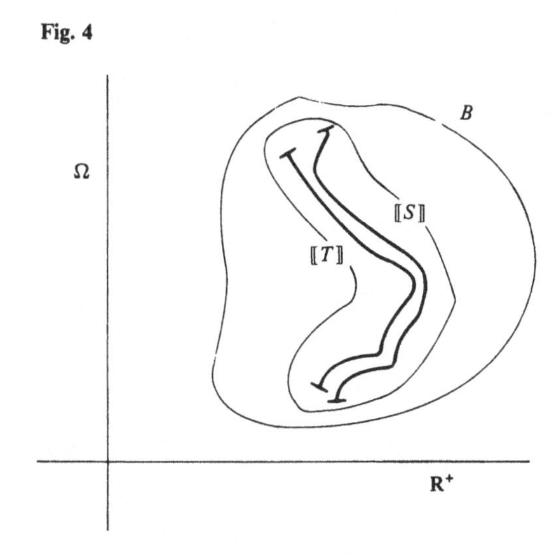

If $B \in \Sigma_p$ we can find S as above and define m on Σ_p instead. The paving $\mathscr{E}$ is now that generated by intervals $[\![U,V]\!]$, where U,V are predictable and finite. Since any stopping time is approximated from above by predictable stopping times, it is clear that $\Sigma_p = \hat{\mathscr{E}}$. So B contains a set $L \in \mathscr{E}$ with $m(L) > m(B) - \varepsilon/2$ and, writing $L = \bigcap_n L_n$, the stopping times D_{L_n} and $D_L = \sup D_{L_n}$ are predictable, since each section $L_n(\omega)$ is compact, consisting of a finite union of closed bounded intervals (see Proposition 3.4.9). Hence $T = D_L$ again has the required properties.

We leave the case of accessible sets (which finds no significant applications in this book) to the reader. (Compare also the proof of Theorem

3.5.4 given in [49], which makes more essential use of the results of section 3.4.)

This supplement contains only the essentials needed for the proof of the cross-section theorems. For further applications and extensions, the reader should consult [18], [19], [76] – probably in that order!

4

Stochastic integrals

In this chapter we discuss our principal application of martingale theory: the formulation of a stochastic calculus using martingales as integrators. We can only indicate briefly some of the many applications of this extremely useful theory; detailed discussions of stochastic differential equations can be found elsewhere [42], [55], [81]. As we saw in the case of Brownian motion, we cannot hope to define $\int_0^t H_s dM_s$ (e.g. for $H \in \mathscr{B}^\infty(\Sigma)$ and a martingale M) as a Stieltjes integral for each $\omega \in \Omega$, as M need not be of bounded variation in $[0,t]$. In fact, we saw that the set of paths of Brownian motion which are locally of bounded variation has measure zero (see 0.2.7).

However, we can exploit the *isometry* $Y \to Y \cdot X$ described in Theorem 2.8.5 for a certain discrete-time martingale transform to define our stochastic integral, first for 'elementary' predictable processes, and then extending by continuity. This procedure, which is essentially that followed by Itô in defining the integral relative to Brownian motion, yields a stochastic integral which is again a martingale and which coincides with the Stieltjes integral whenever the latter exists.

As in the case of martingale transforms, the *integrands* will be predictable. The role of this restriction is obscured when the integrator (e.g. Brownian motion) has continuous paths, but we shall indicate why this restriction is essential in the general case. (It is also indispensable in certain applications, such as filtering theory, where the filtration is not right-continuous.)

4.1. The stochastic integral operator

4.1.1: As always, we fix a stochastic base $(\Omega, \mathscr{F}, P, (\mathscr{F}_t), \mathbf{T})$. Recall that, for $1 \leqslant p \leqslant \infty$, a process X is L^p-bounded if $\sup_{t \in \mathbf{T}} \|X_t\|_p < \infty$. We denote the vector space of L^p-bounded cadlag martingales by $\mathscr{M}^p$. If $1 < p < \infty$, $M \in \mathscr{M}^p$ is uniformly integrable (see Corollary 1.1.8) so that $\lim_{t \to \infty} M_t = M_\infty$ exists as an a.s. and L^p-limit and closes M. Moreover, by

Doob's inequality (Proposition 3.3.2(ii), $M^* := \sup_{t\in\mathbf{T}}|M_t| \in L^p$ since $|M_t|^p$ is a submartingale. We can therefore write Doob's inequality as: $\|M^*\|_p \leqslant (p/(p-1))\|M_\infty\|_p$, so that $\mathcal{M}^p$ possesses the *equivalent* norms $M \xrightarrow{\lambda} \|M\|_p$, $M \xrightarrow{\nu} \|M_\infty\|_p$. Since the map $M \to M_\infty$ identifies $\mathcal{M}^p$ with the Banach space $L^p(\Omega, \mathcal{F}_\infty, P)$, $\mathcal{M}^p$ is a Banach space under either of these norms (as always, we identify indistinguishable processes.)

4.1.2. ***Proposition:*** If $M^{(n)}, M \in \mathcal{M}^p$, $1 < p < \infty$, and $\lim_{n\to\infty} \lambda(M^{(n)} - M) = 0$ (equivalently, $\lim_{n\to\infty} \nu(M^{(n)} - M) = 0$), then $\sup_{t\in\mathbf{T}}|M_t^{(n)} - M_t| \to 0$ in probability. Hence there is then a subsequence $M^{(n_k)}$ such that $M_t^{(n_k)}(\omega)$ converges to $M_t(\omega)$ uniformly on $\mathbf{T}$ for almost all $\omega \in \Omega$.

Proof: As $\lambda(M^{(n)} - M) = \|(M^{(n)} - M)^*\|_p = \|\sup_{t\in\mathbf{T}}|M_t^{(n)} - M_t|\|_p \to 0$, the result follows from elementary measure theory (*Exercise*!).

4.1.3. ***Corollary:*** If $M^{(n)} \to M$ in $\mathcal{M}^p$, then $\Delta M_t^{(n)} \to \Delta M_t$ for all $t \in \mathbf{T}$. In particular if $M^{(n)}$ is continuous for each n, then M is continuous.

4.1.4. ***Definition:*** The space $\mathcal{M}^2$ of square-integrable martingales is thus a Hilbert space under the inner product $(M,N) \to \mathbf{E}(M_\infty N_\infty)$ and the space $\mathcal{M}_c^2 = \{M \in \mathcal{M}^2 : t \to M_t(\omega)$ is continuous for almost all $\omega\}$ is a *closed* subspace of $\mathcal{M}^2$. We call $\mathcal{M}_c^2$ the space of continuous (square-integrable) martingales, and denote its orthogonal complement by $\mathcal{M}_d^2$, the space of (purely) *discontinuous* martingales. We shall examine the structure of $\mathcal{M}_d^2$ in detail later.

As always, there is a slight difficulty in maintaining a consistent notation at $t=0$: we shall adopt the convention that $X_{0-}=0$ a.s. for all processes. This convention was already used in the definition of the process X_- (see Theorem 3.7.6), and has the effect that $\Delta X_0 = X_0$.

We often consider processes with $X_0 = 0$. In particular, $\mathcal{M}_0^2$ denotes the subspace of $\mathcal{M}^2$ of martingales M with $M_0 = 0$. Our convention then forces $\mathcal{M}_c^2 \subset \mathcal{M}_0^2$, since $M \in \mathcal{N}_c^2$ is continuous at $t=0$ by definition.

Now fix $M \in \mathcal{M}^2$. Since $M^* \in L^2$, M^2 is a class (D) submartingale since $M_T^2 \leqslant M^{*2} \in L^1$ for all T. Applying the Doob–Meyer decomposition theorem 3.7.4 to $(-M^2)$ we thus obtain a unique predictable increasing process A such that $X = M^2 - A$ is a uniformly integrable cadlag martingale with $X_0 = M_0^2$. We call the process $(A + M_0^2)$ the *predictable quadratic variation* of M and write it as $\langle M,M \rangle$ or $\langle M \rangle$. The reason for this terminology is clear: X and M are martingales, hence if $s \leqslant t$ in $\mathbf{T}$,

$$\begin{aligned} &\mathbf{E}((M_t - M_s)^2 | \mathcal{F}_s) = \mathbf{E}((M_t^2 - M_s^2)|\mathcal{F}_s) = \mathbf{E}((A_t - A_s)|\mathcal{F}_s) \\ &= \mathbf{E}((\langle M \rangle_t - \langle M \rangle_s)|\mathcal{F}_s). \end{aligned} \tag{4.1}$$

Hence $\langle M\rangle = A + M_0^2$ is the unique predictable increasing process such that $\langle M\rangle - M^2 \in \mathscr{M}_0^1$. The process $\langle M\rangle$ also characterised by the potential it generates: let Z be the positive supermartingale $Z_t = \mathbf{E}(\mathscr{M}_\infty^2 | \mathscr{F}_t) = M_t^2$. Then $Z_t \to 0$ in L^1-norm as $t \to \infty$, hence Z is a potential.

Since M^2 is dominated in L^1 by M^{*2}, Z is in class (D). By Theorem 3.7.1, Z is generated by a predictable increasing process. But this process is $\langle M\rangle$, since by (4.1) we have $Z_t = \mathbf{E}(M_\infty^2 = M_t^2 | \mathscr{F}_t) = \mathbf{E}(\langle M\rangle_\infty - \langle M\rangle_t | \mathscr{F}_t)$.

4.1.5. ***Example:*** In Examples 3.8.12 we encountered the predictable quadratic variation processes of Brownian motion and of the Poisson process. Note that in the first case predictability followed from the path-continuity of the original process and that in both cases the variation process was actually *deterministic.*

We now use the martingales $M_t = N_t - t$ of the standard (rate 1) Poisson process (N_t) to indicate why predictability of the integrands is essential: it is an easy exercise to verify that the 'jump times' $T_i(\omega) := \inf\{t \in \mathbf{T} : N_t(\omega) = i\}$ of N are stopping times, and that the Stieltjes integral $(H \cdot M)_t := \int_0^t H_s \mathrm{d}M_s = \sum_{i=1}^N H_{T_i} - \int_0^t H_s \mathrm{d}s$ is well-defined for each bounded measurable process H. (See section 0.3.4.) However, when H is not predictable $H \cdot M$ need *not* be a martingale.

To see this, note that N is also an increasing process (in the sense of Definition 3.6.1) and that $M_t = N_t - t$ defines a martingale. Since $(t,\omega) \mapsto t$ is a continuous, hence predictable, increasing process, it is the dual predictable projection $\Pi_p^* N$, of N, by Definition 3.6.19. This means that the jump times T_i are totally inaccessible: let S be a predictable stopping time, then as $(\Pi_p^* N)_t = t$ a.s., we have, by Theorem 3.6.15, $\mathbf{E}(\Delta N_S) = \mu_N(1_{[\![S]\!]}) = \mu_{\Pi_p^* N}(1_{[\![S]\!]}) = 0$ since $1_{[\![S]\!]}$ is predictable. Thus $[\![T_i]\!] \cap [\![S]\!]$ is evanescent for all predictable stopping times S. Hence the optional process $H = 1_{[\![T_1]\!]}$ is not predictable. We show that $H \cdot M$ is not a martingale: clearly $H \cdot M = 1_{[\![T_1,\infty[\![}$. This process is not a martingale, since the expectation $\mathbf{E}((H \cdot M)_t) = P(N_t \geq 1) = 1 - e^t$ is not constant. (See also [45; Remark 3.3.1] for a further example.)

Restricting to predictable integrands also allows one to apply the theory of situations where the filtration $(\mathscr{F}_t)$ is not right-continuous. For if we let $\mathscr{G}_t = \mathscr{F}_{t+}$, then any process X which is predictable relative to $(\mathscr{G}_t)$ is also $(\mathscr{F}_t)$-adapted. This follows easily from the fact that the predictable σ-field is generated by sets of the form $]s,t] \times B_s$, $B_s \in \mathscr{G}_s$, $\{0\} \times B_0$, $B_0 \in \mathscr{G}_0$. (See [45] for details and for applications to the 'innovations process' in filtering theory.)

We now construct stochastic integrals for martingales in $\mathscr{M}^2$:

4.1.6. ***Definition:*** Denote by $\mathscr{E}$ the vector space of elementary predictable processes on $[0,\infty]$: $H \in \mathscr{E}$ if there is a partition

$0=t_0<t_1<t_2<\dots t_n<t_{n+1}=\infty$ of $[0,\infty]$ and $H_t=H_{t_i}$ for t in $]t_i,t_{i+1}]$ for $i=0,\dots,n$, where H_{t_i} is $\mathscr{F}_{t_i}$-measurable and bounded, while H also takes the value H_0 at $t=0$.

Fix $M\in\mathscr{M}^2$. Define the *stochastic integral* $H\cdot M:=\int H\,\mathrm{d}M$ by

$$(H\cdot M)_t=\int_0^t H_s\mathrm{d}M_s=H_0M_0+\sum_{i=0}^n H_{t_i}(M_{t_{i+1}\wedge t}-M_{t_i\wedge t}) \tag{4.2}$$

for $t\in[0,\infty]$. Since H is left-continuous and M is cadlag, it is also clear that $H\cdot M$ is cadlag, adapted, and that $\Delta(H\cdot M)_t=H_t\Delta M_t$: if $t\in]t_k,t_{k+1}]$ then $\Delta(H\cdot M)_t=H_{t_k}(M_t-M_{t_k})-H_{t_k}-(M_{t-}-M_{t_k})=H_{t_k}(M_t-M_{t-})=H_t\Delta M_t$ (since $H_{t_k}=H_{t_k-}$). In fact we also have the following basic result:

4.1.7. ***Proposition:*** If $M\in\mathscr{M}^2$, $H\in\mathscr{E}$, then $H\cdot M\in\mathscr{M}^2$ and

$$\mathbf{E}((H\cdot M)^2_\infty)=\mathbf{E}\left(\int_0^\infty H_s^2\mathrm{d}\langle M\rangle_s\right). \tag{4.3}$$

Proof: The map $H\to H\cdot M$ is a (finite) martingale transform, applied to the discrete-time stopped martingale $(M^t_{t_i})_{i=0,\dots,n}$. It is thus a martingale, so that for $0\leqslant i<j\leqslant n$, $\mathbf{E}((H\cdot M^t)_{t_j}|\mathscr{F}_{t_i})=(H\cdot M^t)_{t_i}$. Since $(H\cdot M^t)=(H\cdot M)^t$, this implies that for $s\leqslant t$ in $\mathbf{T}$, we have $\mathbf{E}((H\cdot M)_t|\mathscr{F}_s)=(H\cdot M)_s$. As $\|(H\cdot M)^*\|_2$ is bounded by a multiple of $\|M^*\|_2$, we have proved that $H\cdot M\in\mathscr{M}^2$. Now

$$\mathbf{E}((H\cdot M)^2_\infty)=\mathbf{E}\left(\sum_{i,j}H_{t_i}H_{t_j}(M_{t_{i+1}}-M_{t_i})(M_{t_{j+1}}-M_{t_j})\right).$$

For $j>i$, $H_{t_i}H_{t_j}(M_{t_{i+1}}-M_{t_i})$ is $\mathscr{F}_{t_j}$-measurable (cf. Theorem 2.8.5) so all cross-terms vanish and (with appropriate conventions at $t=0$)

$$\begin{aligned}\mathbf{E}((H\cdot M)^2_\infty)&=\mathbf{E}\left(\sum_k H^2_{t_k}(M_{t_{k+1}}-M_{t_k})^2\right)\\&=\mathbf{E}\left(\sum_k H^2_{t_k}\mathbf{E}((M_{t_{k+1}}-M_{t_k})^2|\mathscr{F}_{t_k})\right)\\&=\mathbf{E}\left(\sum_k H^2_{t_k}(\langle M\rangle_{t_{k+1}}-\langle M\rangle_{t_k})\right)\quad\text{by (4.1)}\\&=\mathbf{E}\left(\int_0^\infty H_s^2\mathrm{d}\langle M\rangle_s\right).\end{aligned}$$

4.1.8. ***Definition:*** Let $\mathscr{L}^2(M)$ denote the space $L^2(\mathbf{T}\times\Omega,\Sigma_p,\mathrm{d}\langle M\rangle)$ of predictable processes for which the seminorm $\|H\|_M := \mathbf{E}(\int_0^\infty H_s^2 \mathrm{d}\langle M\rangle_s)^{\frac{1}{2}}$ is finite. (Here $\mathrm{d}\langle M\rangle$ is the Stieltjes measure generated by $\langle M\rangle$.) Since the sets of the forms $\{0\}\times F_0$, $F\in\mathscr{F}_0$, $]s,t]\times F_s$, $F_s\in\mathscr{F}_s$, generate Σ_p, it follows that $\mathscr{E}$ is dense in $\mathscr{L}^2(M)$ in the topology defined by this seminorm. By proposition 4.1.7, the map $H\to H\cdot M$ is an *isometry* from $(\mathscr{E},\|\ \|_M)$ into $(\mathscr{M}^2,\nu)$. Hence it extends uniquely to an isometry from $(\mathscr{L}^2(M),\|\ \|_M)$ into $(\mathscr{M}^2,\nu)$. We state this fundamental result as follows:

4.1.9. ***Theorem:*** The map $H\to H\cdot M$ from $\mathscr{E}$ into $\mathscr{M}^2$ has a unique extension as a linear isometry, again denoted by $H\to H\cdot M$, from $(\mathscr{L}^2(M),\|\ \|_M)$ into $(\mathscr{M}^2,\nu)$. Hence (4.3) holds for all $H\in\mathscr{L}^2(M)$. Moreover, the processes $(\Delta(H\cdot M)_t)$ and $(H_t\Delta M_t)$ are indistinguishable.

Proof: Only the final sentence still requires proof. For $H\in\mathscr{E}$, the result was verified in the remarks following Definition 4.1.6. For general H, let $(H^{(n)})$ be a sequence in $\mathscr{E}$ approximating H in the sense of $\|\ \|_M$. By (4.3), $\mathbf{E}((H^{(n)}\cdot M - H\cdot M)_\infty^2) = \mathbf{E}(\int_0^\infty (H_s^{(n)} - H_s)^2 \mathrm{d}\langle M\rangle_s)$ tends to 0 as $n\to\infty$, so that $\|H^{(n)}\cdot M - H\cdot M)^*\|_2 \to 0$ also. By Proposition 4.1.2, we can find a subsequence $H^{(n_k)}$ such that $(H^{(n_k)}M)_t$ converges uniformly on $\mathbf{T}$ to $(H\cdot M)_t$ for almost all $\omega\in\Omega$. So $\Delta(H^{(n_k)}M)_t = H_t^{(n_k)}\Delta M_t$ converges to $\Delta(H\cdot M)_t$ a.s. for all t, and $H_t^{(n_k)}$ converges to H_t a.s. (P) since $(\int_0^\infty (H_t^{(n_k)} - H_t)^2 \mathrm{d}\langle M\rangle_t) \to 0$. Hence for almost all $\omega\in\Omega$, the paths $t\to\Delta(H\cdot M)_t(\omega)$ and $t\to H_t(\omega)\Delta M_t(\omega)$ are identical.

4.2. The integral as a stochastic process

So far we have described the *operator* $H\to H\cdot M$ as an isometry from $\mathscr{L}^2(M)$ into $\mathscr{M}^2$. We now wish to characterise the *stochastic process* $H\cdot M$: first recall that $H\cdot M\in\mathscr{M}^2$, so that $(H\cdot M)_t = \mathbf{E}((H\cdot M)_\infty|\mathscr{F}_t)$ for all t. We also write $\int_0^t H_s \mathrm{d}M_s$ for $(H\cdot M)_t$. We have already shown that if $M\in\mathscr{M}^2$ and $H\in\mathscr{L}^2(M)$, then $H\cdot M$ is again in $\mathscr{M}^2$ and has the same jump times as M.

4.2.1. ***Definition:*** Let $M,N\in\mathscr{M}^2$, and define a predictable process of bounded variation by $\langle M,N\rangle = \frac{1}{2}(\langle M+N\rangle - (\langle M\rangle + \langle N\rangle))$. This 'bilinear form' can be characterised as follows:

4.2.2. ***Lemma:*** Given $M,N\in\mathscr{M}^2$, $\langle M,N\rangle$ is the unique predictable process of bounded variation for which $\langle M,N\rangle - MN\in\mathscr{M}_0^1$.

Proof: From the definition of $\langle M\rangle$ (Definition 4.1.4) we have: $\langle M\rangle$ is the unique predictable increasing process such that $\langle M\rangle - M^2 \in \mathcal{M}_0^1$, and $M_0^2 = \langle M\rangle_0$. Similarly for $\langle N\rangle$ and $\langle M+N\rangle$. So $\langle M,N\rangle$ is the difference of increasing processes, hence of bounded variation, and $\langle M,N\rangle - MN = \frac{1}{2}((\langle M+N\rangle - (M+N)^2) - (\langle M\rangle - M^2) - (\langle N\rangle - N^2))$ is a martingale in $\mathcal{M}_0^1$, while $M_0N_0 = \langle M,N\rangle_0$. The uniqueness of $\langle M,N\rangle$ follows from Corollary 3.6.18.

4.2.3. *Remarks:*

(1) The map $(M,N) \rightarrow \langle M,N\rangle$ has properties close to those of an inner product it is obviously symmetric, and it is easy to prove (*Exercise!*) that it is additive in both variables. Moreover, $\langle \alpha M,M\rangle = \alpha\langle M\rangle = \langle M,\alpha M\rangle$ for $\alpha \in R$.

(2) The relation $\langle M,N\rangle = 0$ expresses *orthogonality* of martingales in $\mathcal{M}^2$ in the following sense: $M,N \in \mathcal{M}^2$ are (*strongly*) *orthogonal* iff MN is a martingale with $(MN)_0 = 0$. We write $M \perp\!\!\!\perp N$. Now since $MN - \langle M,N\rangle$ is a martingale and $(MN)_0 = \langle M,N\rangle_0$, we see that $M \perp\!\!\!\perp N$ iff $\langle M,N\rangle = 0$.

This concept of orthogonality may seem unnatural at first sight since the Hilbert space $(\mathcal{M}^2,\nu)$ already has the 'natural' orthogonality $M \perp N$ iff $\mathbf{E}(M_\infty N_\infty) = 0$. However, strong orthogonality will turn out to be an extremely useful tool in analysing the structure of $\mathcal{M}_d^2$. For the present, we shall only show why the concept is stronger than the usual orthogonality in $\mathcal{M}^2$:

4.2.4. *Proposition:* $M \perp\!\!\!\perp N$ iff $M_0N_0 = 0$ and $\mathbf{E}(M_TN_T) = 0$ for all stopping times T. In particular $M \perp\!\!\!\perp N$ implies $M \perp N$.

Proof: Since $(MN)_t$ is dominated by $M^*N^* \in L^1$, $MN \in \mathcal{M}^1$. Now MN is a martingale with $\mathbf{E}(M_0N_0) = 0$, so if T is a stopping time, optional stopping implies $\mathbf{E}(M_TN_T) = \mathbf{E}(M_0N_0) = 0$. The converse follows upon applying the following extremely useful result to MN:

4.2.5. *Lemma:* If X is a right-continuous adapted process with $\mathbf{E}(|X_T|) < \infty$ and $\mathbf{E}(X_T) = 0$ for all stopping times, then X is a uniformly integrable martingale.

Proof: Let $t \in \mathbf{T}$, $A \in \mathcal{F}_t$ and let

$$T = \begin{cases} t \text{ on } A \\ +\infty \text{ on } A^c. \end{cases}$$

Then $0 = \mathbf{E}(X_T) = \int_A X_t \,dP + \int_{A^c} X_\infty \,dP$, while $\mathbf{E}(X_\infty) = \int_A X_\infty \,dP + \int_{A^c} X_\infty \,dP$. Hence $\int_A X_t \,dP = \int_A X_\infty \,dP$, as required.

4.2.6. ***Lemma:*** If T is a stopping time and M,N are in $\mathcal{M}^2$, then $\langle M,N\rangle^T = \langle M,N^T\rangle = \langle M^T,N^T\rangle$.

Proof: $MN - \langle M,N\rangle$ is a martingale in $\mathcal{M}_0^1$, hence by optional stopping so is $(MN)^T - \langle M,N\rangle^T$. On the other hand the process $MN^T - M^TN^T = (M - N^T)N^T$ is the martingale $(\mathbf{E}((M_\infty - M_T)N_T|\mathcal{F}_t))_t$, since

$$(M_t - M_t^T)N_t^T = (M_t - M_{t\wedge T})N_{t\wedge T} = \begin{cases} 0 \text{ if } t \leqslant T \\ (M_t - M_T)N_T \text{ if } t > T. \end{cases}$$

So $MN^T - \langle M,N\rangle^T$ is also a martingale in $\mathcal{M}_0^1$. Since $MN^T - \langle M,N^T\rangle \in \mathcal{M}_0^1$ by definition of $\langle M,N^T\rangle$, the process $\langle M,N^T\rangle - \langle M,N\rangle^T$ is a martingale which is also the difference of two predictable increasing processes, and zero at $t=0$. By Definition 3.6.19 it must be identically zero. The final identity is now clear.

The key to the characterisation of the integral as a process in $\mathcal{M}^2$ lies in the following 'Schwarz inequality' due to Kunita and Watanabe:

4.2.7. ***Theorem:*** Let H and K be measurable processes and let $M,N \in \mathcal{M}^2$. Then we have, a.s. (P)

$$\int_0^\infty |H_s|\,|K_s|\,|d\langle M,N\rangle_s| \leqslant \left(\int_0^\infty H_s^2 d\langle M\rangle_s\right)^{1/2} \left(\int_0^\infty K_s^2 d\langle N\rangle_s\right)^{1/2}. \quad (4.4)$$

Hence if $1 < p < \infty$ and $1/p + 1/q = 1$, then

$$\mathbf{E}\left(\int_0^\infty |H_s|\,|K_s|\,|d\langle M,N\rangle_s|\right) \leqslant \left\|\left(\int_0^\infty H_s^2 d\langle M\rangle_s\right)^{1/2}\right\|_p \left\|\left(\int_0^\infty K_s^2 d\langle N\rangle_s\right)^{1/2}\right\|_q. \quad (4.5)$$

Proof: Let s,t be non-negative rationals with $s < t$. For any process X write $\Delta_s^t X = X_t - X_s$. If λ is rational, then $\Delta_s^t\langle M + \lambda N\rangle \geqslant 0$ a.s. since $\langle M + \lambda N\rangle$ is increasing. Hence $\Delta_s^t\langle M,M\rangle + 2\lambda\Delta_s^t\langle M,N\rangle + \lambda^2\Delta_s^t\langle N,N\rangle \geqslant 0$ holds for all rational λ outside a fixed P-null set Ω_0. But then it also holds for all real λ by continuity of the polynomial in λ. So if $\omega \notin \Omega$, $|\Delta_s^t\langle M,N\rangle(\omega)| \leqslant (\Delta_s^t\langle M\rangle(\omega))^{1/2}(\Delta_s^t\langle N\rangle(\omega))^{1/2}$ whenever $s,t \in [0,\infty]$, $s < t$. Also $|\langle M,N\rangle_0| = |M_0N_0| = (\langle M\rangle_0)^{1/2}(\langle N\rangle_0)^{1/2}$. Let $0 = t_0 < t_1 < \ldots < t_{n+1} = \infty$ be a finite partition of $[0,\infty]$ and let H and K be bounded step processes relative to this partition.

The above inequalities with $s = t_i$, $t = t_{i+1}$ then give

$$\left|\int_0^\infty H_s K_s \mathrm{d}\langle M,N\rangle_s\right| \leqslant |H_0 K_0 \langle M,N\rangle_0| + \sum_{i=0}^n |H_{t_i} K_{t_i}|\,|\Delta_{t_i}^{t_{i+1}}\langle M,N\rangle|$$

$$\leqslant \left(H_0^2\langle M\rangle_0 + \sum_{i=0}^n H_{t_i}^2 \Delta_{t_i}^{t_{i+1}}\langle M\rangle\right)^{1/2} \left(K_0^2\langle N\rangle_0 + \sum_{i=0}^n K_{t_i}^2 \Delta_{t_i}^{t_{i+1}}\langle N\rangle\right)^{1/2}$$

by the Schwarz inequality on sums (i.e. in $\mathbf{R}^{n+2}$). But the right-hand side equals $(\int_0^\infty H_s^2 \mathrm{d}\langle M\rangle_s)^{1/2}(\int_0^\infty K_s^2 \mathrm{d}\langle N\rangle_s)^{1/2}$. Since the algebra of these step processes generates the product σ-field $B(\mathbf{T}) \times \mathscr{F}$ on $\mathbf{T} \times \Omega$, a monotone class argument yields the inequality

$$\left|\int_0^\infty H_s K_s \mathrm{d}\langle M,N\rangle_s\right| \leqslant \left(\int_0^\infty H_s^2 \mathrm{d}\langle M\rangle_s\right)^{1/2} \left(\int_0^\infty K_s^2 \mathrm{d}\langle N\rangle_s\right)^{1/2}$$

for all bounded measurable H,K. To prove (4.4) it now suffices to apply the inequality to H_s and $K_s J_s$, where J_s is the measurable process with range $\{-1,1\}$ such that $|\mathrm{d}\langle M,N\rangle_s| = J_s \mathrm{d}\langle M,N\rangle_s$ (where $|\mathrm{d}\langle M,N\rangle|$ denotes the total variation of the Stieltjes measure $\mathrm{d}\langle M,N\rangle$.)

4.2.8. ***Corollary:*** Let $M \in \mathscr{M}^2$, $H \in \mathscr{L}^2(M)$. Then for all $N \in \mathscr{M}^2$, $\mathbf{E}(\int_0^\infty |H_s|\,|\mathrm{d}\langle M,N\rangle_s|) < \infty$.

Proof: Set $K \equiv 1$, $p = q = 2$ in (4.5).

The characterisation of $H \cdot M$ among elements of $\mathscr{M}^2$ follows:

4.2.9. ***Theorem:*** Let $M \in \mathscr{M}^2, H \in \mathscr{L}^2(M)$. Then $H \cdot M$ is the unique process L in $\mathscr{M}^2$ such that for all $N \in \mathscr{M}^2$,

$$\mathbf{E}(L_\infty N_\infty) = \mathbf{E}\left(\int_0^\infty H_s \mathrm{d}\langle M,N\rangle_s\right). \tag{4.6}$$

Denoting the Stieltjes integral $\int_0^t H_s \mathrm{d}\langle M,N\rangle_s$ by $H \cdot \langle M,N\rangle_t$, the stochastic integral $H \cdot M$ is characterised as the unique element of $\mathscr{M}^2$ satisfying

$$\langle H \cdot M, N\rangle = H \cdot \langle M,N\rangle \quad \text{for all } N \in \mathscr{M}^2. \tag{4.7}$$

Proof: The linear form ϕ on $\mathscr{L}^2(M)$ defined by $\phi(H) = \mathbf{E}(H \cdot M)_\infty N_\infty - H \cdot \langle M,N\rangle_\infty)$ is bounded (Corollary 4.2.8) and we verify that it vanishes on elementary predictable processes: to do this it suffices to consider $H \in \mathscr{E}$ with $H_s = h \cdot 1_{]u,v]}(s)$, where h is $\mathscr{F}_u$-measurable. Then $(H\cdot M)_\infty = h(M_v - M_u)$ while $H\cdot\langle M,N\rangle_\infty = h\cdot(\langle M,N\rangle_v - \langle M,N\rangle_u)$, hence using the notation of Theorem 4.2.7, $\phi(H) = \mathbf{E}(h(N_\infty \Delta_u^v M - \Delta_u^v\langle M,N\rangle)) = \mathbf{E}(h\mathbf{E}(N_\infty \Delta_u^v M - \Delta_u^v\langle M,N\rangle | \mathscr{F}_u)) = 0$ since $\mathbf{E}(N_\infty M_u | \mathscr{F}_u) = M_u N_u$ and $\mathbf{E}(N_\infty M_v | \mathscr{F}_v) = M_v N_v$, so that the final expectation reduces to $\mathbf{E}\Delta_u^v(MN - \langle M,N\rangle)$, which is zero, since $MN - \langle M,N\rangle$ is in $\mathscr{M}_0^1$.

Since $\mathscr{E}$ is dense in $\mathscr{L}^2(M)$, ϕ vanishes on $\mathscr{L}^2(M)$, proving (4.6). To see that $L = H\cdot M$ is uniquely defined by (4.6), note that, given H and M, $\psi(N) = \mathbf{E}(\int_0^\infty H_s \mathrm{d}\langle M,N\rangle_s)$ defines a bounded linear form on the Hilbert space $\mathscr{M}^2$ (by Corollary 4.2.8) and that L is the unique element of $\mathscr{M}^2$ such that $\psi(N) = (L,N)$, where $(L,N) = \mathbf{E}(L_\infty N_\infty)$ denotes the inner product in $\mathscr{M}^2$.

To prove (4.7) we again write $L = H\cdot M$ and consider the process $\phi_t = L_t N_t - H\cdot\langle M,N\rangle_t$, which is dominated by the integrable function $L^* N^* + \int_0^\infty |H_s|\,|\mathrm{d}\langle M,N\rangle_s|$. Now apply (4.6) to $N^T \in \mathscr{M}^2$, where T is any stopping time. Since $\langle M,N^T\rangle = \langle M,N\rangle^T$ (by Lemma 4.2.6) we obtain

$$\mathbf{E}(\phi_T) = \mathbf{E}(L_T N_T - H\cdot\langle M,N\rangle_T) = \mathbf{E}(N_T \mathbf{E}(L_\infty | \mathscr{F}_T) - \int_0^T H_s \mathrm{d}\langle M,N\rangle_s)$$

$$= \mathbf{E}(L_\infty N_\infty^T - \int_0^\infty H_s \mathrm{d}\langle M,N^T\rangle_s) = 0.$$

By Lemma 4.2.5 this means that (ϕ_t) is a martingale. But then $\langle L,N\rangle_t - H\cdot\langle M,N\rangle_t = \phi_t - (L_t N - \langle L,N\rangle_t)$ also defines a martingale, zero at $t=0$, while the left-hand side is a predictable process of bounded variation. Hence both sides are zero. The uniqueness assertion follows as in the first part of the proof.

4.2.10. ***Corollary:*** If $H \in \mathscr{L}^2(M)$ and K is a bounded predictable process, then $(KH)\cdot M = K\cdot(H\cdot M)$.

Proof: Given N in $\mathscr{M}^2$ we have

$$\mathbf{E}(((KH)\cdot M)_\infty N_\infty) = \mathbf{E}\left(\int_0^\infty K_s H_s \mathrm{d}\langle M,N\rangle_s\right) \qquad \text{by 4.6}$$

$$= \mathbf{E}\left(\int_0^\infty K_s \mathrm{d}\langle H\cdot M,N\rangle_s\right) \qquad \text{by 4.7}$$

$$= \mathbf{E}\Big((K\cdot(H\cdot M))_\infty N_\infty\Big) \qquad \text{by 4.6}$$

and as $N_\infty \in L^2$ was arbitrary, $((KH)\cdot M)_\infty = (K\cdot(H\cdot M))_\infty$. The result now follows.

4.2.11. *Remarks:*

(1) The martingale $H\cdot M \in \mathcal{M}^2$ has predictable quadratic variation $\langle H\cdot M\rangle_t = \int_0^t H_s^2 \mathrm{d}\langle M\rangle_s$. To see this, let $N = H\cdot M$ in (4.7) so $\langle H\cdot M\rangle_t = \int_0^t H_s \mathrm{d}\langle M, H\cdot M\rangle_s = \int_0^t H_s^2 \mathrm{d}\langle M\rangle_s$ by Corollary 4.2.10. In particular, if M is standard Brownian motion on a finite interval $[0,t_0]$ and $f \in \mathscr{L}^2(B)$, then $\langle f\cdot B\rangle_t = \int_0^t f_s^2 \mathrm{d}s$ for all t. (Cf. 0.2.10)

(2) If T is a stopping time, then $(H\cdot M)^T = (H 1_{[\![0,T]\!]})\cdot M$. In fact, by Corollary 4.2.10 this reduces to showing that $1_{[\![0,T]\!]}\cdot M = M^T$. Although $1_{[\![0,T]\!]}$ need not be predictable, we can approximate T from above by predictable stopping times T_n with range $\{k/2^n : k \leqslant 2^{2n}\} \cup \{+\infty\}$. Then it is clear that $1_{[\![0,T_n]\!]}\cdot M = M^{T_n}$ for all n and the result follows on letting $n\to\infty$. Consequently we write $\int_0^t H_s 1_{[\![0,T]\!]}(s)\mathrm{d}M_s$ also as $\int_0^{t\wedge T} H_s \mathrm{d}M_s$.

(3) We saw in Definition 4.1.4 that $\mathcal{M}^2$ decomposes into the direct sum of $\mathcal{M}_c^2$ and $\mathcal{M}_d^2$, the spaces of continuous and purely discontinuous martingales. By Theorem 4.1.9, if $M \in \mathcal{M}_c^2$ and $H \in \mathscr{L}^2(M)$, then $H\cdot M \in \mathcal{M}_c^2$. But the question remains whether the stochastic integral respects the decomposition of M into a continuous and a purely discontinuous part, i.e. whether $(H\cdot M)^c = H\cdot M^c$ and $(H\cdot M)^d = H\cdot M^d$. To see that this is indeed the case we make the following definition:

4.2.12. *Definition:* A closed subspace $K \subseteq \mathcal{M}^2$ is *stable* if it is also closed under stochastic integration, i.e. if $M \in K$ implies $H\cdot M \in K$ for all $H \in \mathscr{L}^2(M)$.

Hence $\mathcal{M}_c^2$ is a stable subspace of $\mathcal{M}^2$. The next result shows how stable subspaces form the link between ordinary (or 'weak') orthogonality ($M \perp N \Leftrightarrow \mathrm{E}(M_\infty N_\infty) = 0$) and strong orthogonality in $\mathcal{M}^2$ as defined in Remark 4.2.3(ii) ($M \perp\!\!\!\perp N \Leftrightarrow MN \in \mathcal{M}_0^1 \Leftrightarrow \langle M,N\rangle = 0 \Leftrightarrow M_0 N_0 = 0$ and $\mathrm{E}(M_T N_T) = 0$ for all stopping times T).

Let K be a subspace of $\mathcal{M}^2$. Write $K^\perp$ for the usual orthogonal complement of K, i.e. $N \in K^\perp$ iff $\mathrm{E}(M_\infty N_\infty) = 0$ for all M in K. We also say that a subspace L of $\mathcal{M}^2$ is *stable under stopping and restriction* if $1_A N^T \in L$ whenever $N \in L$, $A \in \mathscr{F}_0$ and T is a stopping time. We have the following result:

4.2.13. *Theorem:* Suppose K is a stable subspace of $\mathcal{M}^2$. Then

(i) K and $K^\perp$ are both stable under stopping and restriction.

(ii) If $M \in K$ and $N \in K^\perp$ then $M \perp\!\!\!\perp N$, hence $\langle M,N\rangle = 0$.

Proof: Since $1_A M^T = 1_{[\![0_A,T]\!]} \cdot M$ by Remark 4.2.11(ii), $L = 1_A M^T \in K$ for all $M \in K$, $A \in \mathscr{F}_0$ and stopping times T. So if $N \in K^\perp$, $0 = \mathbf{E}(L_\infty N_\infty) = \mathbf{E}(1_A M_T N_\infty) = \mathbf{E}(1_A M_T \mathbf{E}(N_\infty | \mathscr{F}_T)) = \mathbf{E}(1_A M_T N_T)$. In particular, with $A = \Omega$, this shows that MN is a martingale. With $T = 0$ we obtain $\int_A M_0 N_0 dP = 0$ for all $A \in \mathscr{F}_0$, so $M_0 N_0 = 0$. Hence $M \perp\!\!\!\perp N$. On the other hand, we can write $0 = \mathbf{E}(1_A M_T N_T) = \mathbf{E}(M_\infty (1_A N^T)_\infty)$, hence $K^\perp$ is stable under stopping and restriction.

4.2.14. ***Remarks:***

(1) It is now easy to see that $K \subseteq \mathscr{M}^2$ is stable iff it is a closed subspace of $\mathscr{M}^2$ which is stable under stopping and restriction: the necessity was proved above. For the sufficiency we show that if $M \in K$ and $H \in \mathscr{L}^2(M)$, then $H \cdot M \in K^{\perp\perp} = K$ (as K is closed in $\mathscr{M}^2$). If $N \in K^\perp \langle M,N \rangle = 0$ by Theorem 4.2.13, so $0 = H \cdot \langle M,N \rangle = \langle H \cdot M, N \rangle$ by (4.7). Hence $H \cdot M \in K^{\perp\perp}$.

(2) It is now clear that if $M \in \mathscr{M}^2$, its purely discontinuous part M^d is strongly orthogonal to any continuous martingale, since $\mathscr{M}_c^2$ (and hence $\mathscr{M}_d^2$) is a stable subspace. So if M in $\mathscr{M}^2$ is given, with decomposition $M = M^c + M^d$, then $\langle H \cdot M^d, L \rangle = H \cdot \langle M^d, L \rangle = 0$ for all $L \in \mathscr{M}_c^2$. In particular, $H \cdot M^c \perp\!\!\!\perp H \cdot M^d$, so that these martingales are respectively the continuous and pure discontinuous parts of $H \cdot M$ (since $\mathscr{M}^2 = \mathscr{M}_c^2 \oplus \mathscr{M}_d^2$ the decomposition is unique).

(3) If $M \in \mathscr{M}^2$, the stable subspace generated by M is obviously the set $\{H \cdot M : H \in \mathscr{L}^2(M)\}$, i.e. the range in $\mathscr{M}^2$ of the isometry $H \to H \cdot M$. Denote this closed subspace of $\mathscr{M}^2$ by $\mathscr{I}_M$. Then $\mathscr{I}_M$ and $\mathscr{I}_N$ are orthogonal subspaces in $\mathscr{M}^2$ iff $\langle M,N \rangle = 0$: this follows immediately from (4.6). (The analogies between the above ideas and those outlined for discrete-time martingales should now be obvious).

(4) Given $M \in \mathscr{M}^2$, the projection of $N \in \mathscr{M}^2$ onto $\mathscr{I}_M$ is given as the stochastic integral $D \cdot M$, where D is a predictable density of the measure $d\langle M,N \rangle$ with respect to $d\langle M \rangle$. To see this let $N \in \mathscr{M}^2$ and write $N = N_1 + N_2$, where $N_1 \in \mathscr{I}_M$ and $N_2 \perp \mathscr{I}_M$. Then $\langle N,M \rangle = \langle N_1, M \rangle$ by Theorem 4.2.13, as $\mathscr{I}_M$ is stable. But $N_1 = D \cdot M$ for some $D \in \mathscr{L}^2(M)$, since $N_1 \in \mathscr{I}_M$. Hence $\langle N,M \rangle = \langle D \cdot M, M \rangle = D \cdot \langle M,M \rangle$, which shows that D is a density, as claimed. Also D is uniquely defined up to $\mathrm{d}\langle M \rangle$-null sets, so that $\mathbf{E}(\int_0^\infty (D_s - D_s')^2 \mathrm{d}\langle M \rangle_s) = 0$ if D' is a second such density. In particular, $D' \in \mathscr{L}^2(M)$ and $D \cdot M = D' \cdot M$, by Remark 4.2.11(1).

(5) Our definition of stable subspaces is not the one usually found in the literature (e.g. [66]), where the basic notion is what we have called 'stable under stopping and restriction' for closed subspaces of $\mathscr{M}^2$. We have shown, however, that this definition is equivalent to our definition of stable subspaces.

4.3. The structure of $\mathcal{M}_d^2$ and the optional quadratic variation

We must still prove that our stochastic integral $H \cdot M$ coincides with the Stieltjes integral $\int_0^t H_s dM_s$, which is well-defined for $H \in \mathcal{L}^2(M)$ if $\mathbf{E}(\int_0^\infty |H_s|\,|dM_s|) < \infty$ and $M \in \mathcal{M}^2$ is of integrable variation. Before we turn to this problem we shall first investigate martingales of integrable variation. It is clear that the only continuous (therefore predictable) martingales with this property are constant (see Definition 3.6.19). Remarkably, the structure theory of martingales of integrable variation can be carried over in large measure to the whole of $\mathcal{M}_d^2$. This allows us to consider an 'optional quadratic variation' process $[M]$ for general $M \in \mathcal{M}^2$. The dual predictable projection of this process turns out to be $\langle M \rangle$. Hence the characterisations of the stochastic integral developed in sections 4.1 and 4.2 have counterparts with $[M]$ instead of $\langle M \rangle$, and in this form are suitable for extension to local martingales and finally to semimartingales.

4.3.1. ***Definition:*** Recall that an adapted process V has *integrable variation* if it is the difference of two integrable increasing processes. We extend this slightly to allow $V_0 \neq 0$; this causes no difficulties. Denote the vector space of such processes by $\mathcal{V}$. The associated signed measure $\mu_V(X) = \mathbf{E}(\int_0^\infty X_s dV_s)$ for $X \in \mathcal{B}^\infty(\Sigma)$ gives rise to the dual predictable projection $\Pi_p^* V$ of V if we put $\mu_{\Pi_p^* V}(X) = \mu_V(\Pi_p X)$. Then $\Pi_p^* V$ is a predictable process of integrable variation, as an easy extension of the results of section 3.6 shows. $\Pi_p^* V$ is called the *compensator* of V. The *compensated process* $\bar{V} = V - \Pi_p^* V$ is then clearly a martingale, since $\mu_{\bar{V}}$ vanishes on Σ_p (see Corollary 3.6.16) and also $\bar{V}_0 = \mu_{\bar{V}}(\{0\} \times \Omega) = 0$. Similar considerations lead to the following:

4.3.2. ***Proposition:***

(i) A process $M \in \mathcal{V}$ is a martingale iff it has the form $M_t = M_0 + \bar{V}_t$, where $V \in \mathcal{V}$. In fact, V can be chosen in the form $V_0 = 0$, $V_t = \sum_{0 < s \leqslant t} \Delta M_s$, so that $\Pi_p^* V$ is continuous.

(ii) With M as above, and any bounded martingale N we have

$$\mathbf{E}(M_\infty N_\infty) = \mathbf{E}\left(\sum_{s \geqslant 0} \Delta M_s \Delta N_s\right), \quad \text{where} \quad \Delta M_0 \Delta N_0 = M_0 N_0 \qquad (4.8)$$

and $X_t = M_t N_t = \sum_{0 \leqslant s \leqslant t} \Delta M_s \Delta N_s$ defines a uniformly integrable martingale X with $X_0 = 0$.

Proof: (i) Let $N = M - M_0$, then μ_N vanishes on Σ_p. Hence $\Pi_p^* N = 0$ and $N = \bar{N} = \bar{M}$. So assume without loss that $M_0 = 0$ and let $V_t = \sum_{0 \leqslant s \leqslant t} \Delta M_s$.

Then $W = M - V \in \mathscr{V}_0$ and W is continuous, hence predictable. So $W = \Pi_p^* W = \Pi_p^* M - \Pi_p^* V = -\Pi_p^* V$ and hence $M = V - \Pi_p^* V = \bar{V}$.

(ii) Since the martingale $N = (\mathbf{E}(N_\infty | \mathscr{F}_t))_t$ is the optional projection of the constant process N_∞, $\mu_M(N_\infty) = \mu_M(N)$, as M is optional. This means that the Stieltjes integral $\int_0^\infty N_s dM_s$ satisfies $\mathbf{E}(\int_0^\infty N_s dM_s) = \mu_M(N) = \mathbf{E}(M_\infty N_\infty)$. On the other hand, the left-continuous process $N_- = (\mathbf{E}(N_\infty | \mathscr{F}_{t-}))_t$ is predictable, while μ_M vanishes on Σ_p. Hence $\mathbf{E}(\int_0^\infty N_{s-} dM_s) = 0$. Therefore $\mathbf{E}(M_\infty N_\infty) = \mathbf{E}(\int_0^\infty \Delta N_s dM_s) = \mathbf{E}(\sum_{s \leqslant t} \Delta M_s \Delta N_s)$. Finally, apply (4.8) with N^T instead of N, where T is any stopping time. Then $\mathbf{E}(M_T N_T) = \mathbf{E}(M_\infty N_T) = \mathbf{E}(M_\infty (N^T)_\infty) = \mathbf{E}(\sum_{s \leqslant T} \Delta M_s \Delta N_s)$. Since T is arbitrary, Lemma 4.2.5 shows that $M_t N_t - \sum_{s \leqslant t} \Delta M_s \Delta N_s$ defines a martingale.

4.3.3. ***Remarks:***

(1) Although we refer to $N \in \mathcal{M}_d^2$ as a 'purely discontinuous process' the *paths* $t \to N_t(\omega)$ need not be purely discontinuous. The centred Poisson process $N_t = N'_t - \lambda t$, where N' is a standard Poisson process of rate λ, illustrates this when restricted to a bounded time interval: the paths are constant between jump times, yet $N \in \mathcal{M}_d^2$. In fact, any square-integrable martingale N of integrable variation is orthogonal to every bounded continuous martingale M: for if $M \in \mathcal{M}_c^2$ then $\Delta M_s = 0$ for all s, so that if $N \in \mathscr{V}$, MN is a uniformly integrable martingale with $M_0 N_0 = 0$, by Proposition 4.3.2(ii). Hence $N \perp\!\!\!\perp M$ and $N \in \mathcal{M}_d^2$, as may easily be seen by approximation, using Proposition 4.1.2.

(2) The same argument shows that a martingale M in $\mathscr{V}$ is (strongly) orthogonal to any martingale N which has no jumps in common with M.

(3) Proposition 4.3.2 gives a representation of any M in $\mathscr{V}$ as a (trivially convergent) sum of 'compensated jump martingales'. With this in mind we now examine the structure of $\mathcal{M}_d^2$. The extension of the representation theorem to *arbitrary* M in $\mathcal{M}_d^2$ presents the difficulty that the sum of the jumps $\sum_{s \leqslant t} \Delta M_s$ need not converge. A similar problem was solved by Paul Levy in the study of processes with independent increments by noticing that the jumps ΔX_s with $\varepsilon < |\Delta X_s| < 1$, where $\varepsilon > 0$, form a process $A_t^\varepsilon = \sum_{s \leqslant t} \Delta X_s 1_{\{\varepsilon < |\Delta X_s| < 1\}}$ in $\mathscr{V}$. Then, although neither A^ε nor $\Pi_p^* A^\varepsilon$ converge as $\varepsilon \to 0$, Levy showed that the compensated processes $\bar{A}^\varepsilon$ converge in L^2-norm as $\varepsilon \to 0$.

(4) The jump times of the processes Levy considers are totally inaccessible. Now if T is a totally inaccessible stopping time and $V = f 1_{[\![T,\infty[\![}$, where $f \in L^1(\mathscr{F}_T)$, then $\Pi_p^* V$ is *continuous*: since $\Pi_p^* V$ is predictable, it charges no totally inaccessible stopping time, by Corollary 3.5.18. But μ_V and $\mu_{\Pi_p^* V}$ coincide on Σ_p and μ_V is supported on $[\![T]\!]$. Hence $\Pi_p^* V$ charges no predictable stopping time ($[\![S]\!] \cap [\![T]\!]$ is evanescent for any predictable

stopping time S). So the increasing process $\Pi_p^* V$ has a.s. no jumps, in other words, it has continuous paths. This simple observation provides the starting point for our analysis of $\mathcal{M}_d^2$:

4.3.4. ***Definition:*** If T is a stopping time, let $\mathcal{M}(T)$ denote the subspace of martingales in $\mathcal{M}_d^2$ which are continuous outside the graph of T, i.e.

$$\mathcal{M}(T) = \{N \in \mathcal{M}_d^2 : \{(t,\omega) : N_t(\omega) \neq N_{t-}(\omega)\} \subset [\![T]\!]\}.$$

4.3.5. ***Remarks:***

(1) $\mathcal{M}(T)$ is a stable subspace (for H in $\mathcal{L}^2(N)$, $\Delta(H \cdot N)_t = H_t \Delta N_t$).

(2) If $T=0$ each N in $\mathcal{M}(T)$ is constant: since N is then continuous on $]0,\infty[$ and hence $N - N_0 \in \mathcal{M}_c^2 \bigcap \mathcal{M}_d^2 = \{0\}$. So $N_t = N_0$ for all t. Note that continuity *at* 0 of a process X demands that $X_0 = 0$ a.s. because of the convention that $X_{0-} = 0$ (Definition 4.1.4).

(3) Hence we shall henceforth only consider the case $T>0$. Then N in $\mathcal{M}(T)$ is continuous at 0 and $N=0$ a.s. Therefore $\mathcal{M}(T) \subset (\mathcal{M}_d^2)_0$.

(4) Note that since $(\mathcal{M}^2, v)$ is isometrically isomorphic to $L^2(\Omega, \mathcal{F}, P)$ under the map $M \to M_\infty$, the spaces $\mathcal{M}_c^2$ and $\mathcal{M}_d^2$, as well as $\mathcal{M}(T)$, can be considered as subspaces of L^2. This leads to an alternative interpretation of the orthogonal projections onto these spaces – see [49; Th. 12.8].

We now describe the basic martingales in $\mathcal{M}(T)$:

4.3.6. ***Theorem:*** Suppose T is totally inaccessible and $f \in L^2(\mathcal{F}_T)$. The integrable increasing process $V = f \cdot 1_{[\![T,\infty[\![}$ has a continuous dual predictable projection and the compensated martingale $\bar{V}$ belongs to $\mathcal{M}(T)$.

Proof: Since T is totally inaccessible, $T>0$ a.s. We can also take $f \geq 0$. The continuity of $W = \Pi_p^* V$ was proved in Remark 4.3.3(iv). Thus $\bar{V}$ is continuous outside $[\![T]\!]$ and it remains to show $\bar{V} \in \mathcal{M}^2$. Now $V_\infty = f \in L^2$ and $\mathbf{E}(W_\infty) = \mathbf{E}(V_\infty)$, so $W_\infty \in L^1$. Since $\bar{V}_\infty = V_\infty - W_\infty$ it suffices to prove that $W_\infty \in L^2$. To do this, note that since $(W - V)$ is a martingale, $Z_t = \mathbf{E}(V_\infty | \mathcal{F}_t) - V_t = \mathbf{E}(W_\infty | \mathcal{F}_t) - W_t$. Let $M_t = \mathbf{E}(W_\infty | \mathcal{F}_t)$, then $\Pi_p Z = \Pi_p M - \Pi_p W = M_- - W_- = Z_-$. Now by Corollary 3.6.13, $\mu_W(M) = \mathbf{E}(\int_0^\infty M_t \mathrm{d}W_t) = \mathbf{E}(M_\infty W_\infty) = \mathbf{E}(W_\infty^2)$, and since $\int_0^\infty W_t \mathrm{d}W_t = \frac{1}{2} W_\infty^2$, we have $2\mathbf{E}(W_\infty^2) = 2\mathbf{E}(\int_0^\infty M_t \mathrm{d}W_t)$ and $\mathbf{E}(W_\infty^2) = 2\mathbf{E}(\int_0^\infty W_t \mathrm{d}W_t)$. Subtracting, we obtain $\mathbf{E}(W_\infty^2) = 2\mathbf{E}(\int_0^\infty (M_t - W_t) \mathrm{d}W_t) = 2\mu_W(Z) = 2\mu_V(Z_-)$. On the other hand, $Z = N - V$, where $N_t = \mathbf{E}(V_\infty | \mathcal{F}_t)$. So since $\mu_V(V_-) = 0$ and $\mu_V(N_-) \leq \mathbf{E}(\sup_t |N_t| \cdot V_\infty) \leq \|\sup |N_t|\|_2 \|V_\infty\|_2$ is finite we have shown that $\mathbf{E}(W_\infty^2) = 2\mu_V(Z_-)$ is finite.

4.3.7. ***Lemma:*** Let T be a predictable stopping time and let $f \in L^1(\mathcal{F}_T)$. The dual predictable projection of the process $V = f \cdot 1_{[\![T,\infty[\![}$ is of the form $W = \mathbf{E}(f \mid \mathcal{F}_{T-}) 1_{[\![T,\infty[\![}$.

Proof: In fact W is accessible by Corollary 3.5.18 and $W_T = \mathbf{E}(f \mid \mathcal{F}_{T-})$ is $\mathcal{F}_{T-}$-measurable. By Theorem 3.5.9, W is predictable. But the measures μ_V and μ_W are carried by $[\![T]\!]$ and the predictable subsets of $[\![T]\!]$ have the form $[\![T_A]\!]$, where $A \in \mathcal{F}_{T-}$, by Theorem 3.4.11 and Corollary 3.5.6. Hence the identities, for $A \in \mathcal{F}_{T-}$,

$$\mu_V([\![T_A]\!]) = \mathbf{E}(\textstyle\int_0^\infty 1_A 1_{[\![T]\!]} dV_t) = \mathbf{E}(1_A V_T) = \mathbf{E}(1_A f) = \mathbf{E}(1_A \mathbf{E}(f \mid \mathcal{F}_{T-})) = \mu_W([\![T_A]\!])$$

show that μ_V and μ_W coincide on Σ_p. Hence $W = \Pi_p^* V$.

This proves the second structure theorem for $\mathcal{M}(T)$:

4.3.8. ***Theorem:*** If $T > 0$ is a predictable stopping time and $f \in L^2(\mathcal{F}_T)$ satisfies $\mathbf{E}(f \mid \mathcal{F}_{T-}) = 0$ a.s., then $V = f \cdot 1_{[\![T,\infty[\![}$ belongs to $\mathcal{M}(T)$ and its dual predictable projection is zero.

4.3.9. ***Definition:*** If V is as in Theorem 4.3.6 or Theorem 4.3.8 we call $M = \bar{V}$ *a compensated jump martingale.*

In both cases $\Pi_p^* V$ is continuous, so $\Delta M_T = \Delta V_T = f$. Moreover, M is (strongly) orthogonal to any martingale in $\mathcal{M}^2$ which has no jump at T. To see this, we prove

4.3.10. ***Corollary:*** Let M be a compensated jump martingale. For any $N \in \mathcal{M}^2$ $L_t = M_t N_t - \Delta M_T \Delta N_T 1_{\{T \leq t\}}$ defines a uniformly integrable martingale, and $L_0 = 0$. In particular, $\mathbf{E}(M_\infty^2) = \mathbf{E}(\Delta M_T^2)$. Hence M is (strongly) orthogonal to any martingale in $\mathcal{M}^2$ which is continuous at T.

Proof: $M \in \mathcal{V}$, so by (4.8) and optional sampling $\mathbf{E}(M_\infty N_\infty) = \mathbf{E}(\Delta M_T \Delta N_T)$ holds for all bounded $N \in \mathcal{M}^2$, as M is continuous outside $[\![T]\!]$. But by Proposition 4.1.2 this extends to arbitrary $N \in \mathcal{M}^2$, since the jumps of the limit of a sequence of $\mathcal{M}^2$-martingales are the limits of jumps of that sequence. Again, using a stopped N^S instead of N the identity becomes $\mathbf{E}(M_S N_S) = \mathbf{E}(\Delta M_T \Delta N_T 1_{\{T \leq S\}})$, so $\mathbf{E}(L_S) = 0$ for all S. So the result follows by Lemma 4.2.5.

Thus if $N \in \mathcal{M}^2$ and T is as above, set $f = \Delta N_T$. We show that the associated increasing process $V = \Delta N_T 1_{[\![T,\infty[\![}$ has $M = \bar{V}$ as the orthogonal projection of N onto $\mathcal{M}(T)$: first note that $\mathbf{E}(\Delta N_T \mid \mathcal{F}_{T-}) = 0$ since any uniformly integrable martingale is regular (Remark 3.7.7). Thus Theorem 4.3.8 applies and $\Pi_p^* V=0$. Now since $\Delta M_T=\Delta N_T$, $N-M$ is continuous at T, and thus orthogonal to $\mathcal{M}(T)$.

We saw that martingales in $\mathcal{V}$ can be represented as sums of such compensated jump martingales. For general $N\in\mathcal{M}_d^2$, we now prove a similar result by considering its orthogonal projection onto $\mathcal{M}(T)$ for each T as a compensated jump martingale and showing that these converge in $(\mathcal{M}^2,\nu)$. Note that this implies that $\mathcal{V}$ is dense in $\mathcal{M}_d^2$.

4.3.11. ***Theorem:*** If $N\in\mathcal{M}_d^2$ and $N_0=0$, then N is the sum of a series of compensated jump martingales, and is orthogonal to each martingale in $\mathcal{M}^2$ which does not charge a common jump time with N.

Proof: By Theorem 3.5.17 the set $\{(t,\omega):\Delta N_t\neq 0\}$ is contained, up to evanescent sets, in the union of the graphs of a sequence (T_n) of stopping times. As in Corollary 3.5.8 these graphs can be chosen disjoint and, as $N_0=N_{0-}=0$, each $T_n>0$ a.s. By Theorem 3.4.8 we can split each T_n into its accessible and totally inaccessible parts, T_n^a, T_n^i, and represent each $[\![T_n^a]\!]$ as a disjoint union of graphs of predictable stopping times, again using Theorem 3.4.8. Hence we may assume that the $[\![T_n]\!]$ are disjoint and that each T_n is either totally inaccessible or predictable.

Now let $V^{(n)}=\Delta N_{T_n} 1_{[\![T_n,\infty[\![}$ and $M^{(n)}=\bar{V}^{(n)}$ for all $n\in\mathbf{N}$. Since any uniformly integrable martingale is regular (Remark 3.7.7) we have $\mathbf{E}(\Delta N_{T_n}|\mathcal{F}_{T_n-})=0$ if T_n is predictable, so by Theorems 4.3.6 and 4.3.8 we have $M^{(n)}\in\mathcal{M}(T_n)$ for each $n\in\mathbf{N}$.

Let $Z^{(k)}=\sum_{n=1}^k M^{(n)}$. Then $N-Z^{(k)}$ is continuous at $T_1,\dots,T_k$, hence orthogonal to $M^{(1)},\dots,M^{(k)}$ by Corollary 4.3.10, and hence also to $Z^{(k)}$. As the $[\![T_n]\!]$ are disjoint, the $M^{(n)}$ are pairwise orthogonal, hence

$$\mathbf{E}(N_\infty^2)=\mathbf{E}(Z_\infty^{(k)2})+\mathbf{E}((N-Z^{(k)})_\infty^2)=\sum_{n=1}^k \mathbf{E}(M_\infty^{(n)2})+\mathbf{E}((N-Z^{(k)})_\infty^2)$$

$$=\sum_{n=1}^k \mathbf{E}(\Delta N_{T_n}^2)+\mathbf{E}((N-Z^{(k)})_\infty^2),$$

where the final identity follows from Corollary 4.3.10 (with $M=N=M^{(n)}$). Thus $\sum_{n=1}^k \mathbf{E}(M_\infty^{(n)2})$ converges and so $(Z^{(k)})_k$ converges in $\mathcal{M}^2$ to a martingale Z, and $(N-Z)$ is orthogonal to Z in $\mathcal{M}^2$. By Proposition 4.1.2 a subsequence of the $Z^{(k)}$ converges to Z uniformly on $\mathbf{T}$ for almost all $\omega\in\Omega$, so $(N-Z)$ has no jumps, being continuous at each T_n. So $(N-Z)\in\mathcal{M}_c^2$, and thus $N\in\mathcal{M}_d^2$ is orthogonal to $(N-Z)$. Consequently $(N-Z)$ is the orthogonal projection of $N\in\mathcal{M}_0^2$ onto $\mathcal{M}_c^2$ and Z its projection onto $\mathcal{M}_d^2$. But $N\in\mathcal{M}_d^2$, hence $N=Z$. Since therefore $\mathbf{E}(N_\infty^2)=\sum_{k=1}^\infty \mathbf{E}(\Delta N_{T_k}^2)$, it is clear that N is orthogonal to all martingales having no jumps at any of the T_n.

4.3.12. *Remarks:*

(1) By adding $T_0=0$ to (T_n) and using $\Delta N_0 = N_0$ we can extend this result to general $N\in\mathcal{M}_d^2$. Then $(N-Z)$ is the projection onto $\mathcal{M}_c^2$ (which by our convention is contained in $\mathcal{M}_0^2$!) and Z that onto $\mathcal{M}_d^2$.

(2) For any $M\in\mathcal{M}^2$ we can repeat the argument in the proof of Theorem 4.3.11 to obtain $\mathbf{E}(M_\infty^2)=\mathbf{E}(\sum_n \Delta M_{T_n}^2)+\mathbf{E}((M-Z)_\infty^2)$. Hence

$$\mathbf{E}\left(\sum_s \Delta M_s^2\right) \leqslant \mathbf{E}(M_\infty^2) \tag{4.9}$$

and equality holds iff $M\in\mathcal{M}_d^2$ (since then $M-Z=0$). In particular $\sum_{s\leqslant t}\Delta M_s^2<\infty$ a.s. for all $t\in\mathbf{T}$.

(3) By the Schwarz inequality we have for $M,N\in\mathcal{M}^2$ that $\sum_s|\Delta M_s \Delta N_s| \leqslant (\sum_s \Delta M_s^2)^{1/2}(\sum_s \Delta N_s^2)^{1/2}$. Hence

$$\left(\sum_s |\Delta M_s \Delta N_s|\right) \leqslant ||M||_2 \cdot ||N||_2. \tag{4.10}$$

Theorem 4.3.11 shows that $\mathscr{V}$ is *dense* in $\mathcal{M}_d^2$, since each $Z^{(k)}$ is of integrable variation. This fact now allows us to extend (4.8) and its consequences from $\mathscr{V}$ to $\mathcal{M}_d^2$:

4.3.13. *Theorem:* Let $M\in\mathcal{M}^2$, $N\in\mathcal{M}_d^2$. Then

(i) $\mathbf{E}(M_\infty N_\infty)=\mathbf{E}(\sum_s \Delta M_s \Delta N_s)$ (4.11)

(ii) $L_t = M_t N_t - \sum_{s\leqslant t}\Delta M_s \Delta N_s$ defines a martingale in $\mathcal{M}_0^1$.

Proof: Assume first that M and N are *both* in $\mathcal{M}_d^2$; then so is $(M+N)$ and (4.9) is an identity for all three cases. Hence, by polarisation,

$$\mathbf{E}(M_\infty N_\infty) = \tfrac{1}{2}(\mathbf{E}((M+N)_\infty^2 - M_\infty^2 - N_\infty^2))$$

$$= \tfrac{1}{2}\mathbf{E}\left(\sum_s ((\Delta M_s + \Delta N_s)^2 - \Delta M_s^2 - \Delta N_s^2)\right) = \mathbf{E}\left(\sum_s \Delta M_s \Delta N_s\right)$$

by (4.9). This proves (i). Applying (4.11) to M^T, N^T instead, we obtain $\mathbf{E}(L_T)=0$ for all stopping times T, hence L is a martingale with $|L_t| \leqslant M^*N^* + \sum_s |\Delta M_s \Delta N_s|$ and by Remark 4.3.12(ii) this shows that $L\in\mathcal{M}_0^1$. For general $M\in\mathcal{M}^2$, write $M=M^c+M^d$ with $M^c\in\mathcal{M}_c^2$, $M^d\in\mathcal{M}_d^2$. Then $N \perp\!\!\!\perp M^c$ so (M^cN) is in $\mathcal{M}_0^1$, so since $MN=M^cN+M^dN$ we need only deal with M^d and N, which are both in $\mathcal{M}_d^2$.

This result allows us to introduce a second 'quadratic variation' process associated with $M\in\mathcal{M}^2$:

4.3.14. ***Definition:*** Let $M \in \mathcal{M}^2$ and let M^c be its continuous part (orthogonal projection onto $\mathcal{M}_c^2$). Define the *optional quadratic variation* $[M]=[M,M]$ of M as the increasing process $[M]_t = \langle M^c \rangle_t + \sum_{s \leq t} \Delta M_s^2$.

This process is integrable by (4.9) and $[M]_t < \infty$ a.s. for all $t \in \mathbf{T}$. Writing $M = M^c + M^d$, we note that $M^2 = (M^c)^2 + 2M^c M^d + (M^d)^2$. Now $M^c \perp\!\!\!\perp M^d$, and $(M^d)_t^2 - \sum_{s \leq t} \Delta M_s^2$ is a martingale in $\mathcal{M}_0^1$ by Remarks 4.3.12, while $(M^c)^2 - \langle M^c \rangle$ is a martingale in $\mathcal{M}_0^1$ by Definition 4.1.4. Consequently $M_t^2 - [M]_t = ((M^c)_t^2 - \langle M^c \rangle)_t + ((M^d)_t^2 - \sum_{s \leq t} \Delta M_s^2) + 2M_t^c M_t^d$ defines a martingale in $\mathcal{M}_0^1$.

Moreover, $[M] - \langle M \rangle = (M^2 - \langle M \rangle) - (M^2 - [M])$ is in $\mathcal{M}_0^1$. Hence by Corollary 3.6.18 $\langle M \rangle$ is the dual predictable projection of $[M]$.

4.3.15. ***Exercises:***

(1) Show that $\langle M^c \rangle$ is continuous for each $M \in \mathcal{M}^2$.

(2) Prove that $[M] \equiv 0$ implies $M \equiv 0$ for $M \in \mathcal{M}_0^2$.

4.3.16. ***Definition:*** We can again 'polarise' the quadratic variation process. If $M,N \in \mathcal{M}^2$ let $[M,N] = \frac{1}{2}([M+N] - [M] - [N])$. Clearly $[M,N]_t = \langle M^c, N^c \rangle_t + \sum_{s \leq t} \Delta M_s \Delta N_s$, and the processes $[M,N] - \langle M,N \rangle$ and $MN - [M,N]$ are martingales in $\mathcal{M}_0^1$. Lemma 4.2.6 implies that $[M,N]^T = [M,N^T] = [M^T,N^T]$ for any stopping time T.

4.3.17. ***Note:*** Let $M \in \mathcal{M}^2$ be given. Since $\langle M \rangle = \Pi_p^*[M]$ we have $\mu_{[M]}(X) = \mu_{\langle M \rangle}(X)$ for any *predictable* X. In particular, (4.3) can be extended to $\mathbf{E}((H \cdot M)_\infty^2) = \mathbf{E}(\int_0^\infty H_s^2 \mathrm{d}\langle M \rangle_s) = \mathbf{E}(\int_0^\infty H_s^2 \mathrm{d}[M]_s)$ for $\mathrm{H} \in \mathcal{L}^2(M)$. Similarly, the Kunita–Watanabe inequalities (4.4) and (4.5) can be written using the optional instead of the predictable quadratic variation process, and it follows that for $H \in \mathcal{L}^2(M)$ the stochastic integral $H \cdot M$ can be characterised as the unique element $L \in \mathcal{M}^2$ such that $[L,N] = H \cdot [M,N]$ for all $N \in \mathcal{M}^2$: indeed, by (4.7) we have $\langle L,N \rangle = H \cdot \langle M,N \rangle$ characterising $L = H \cdot M$, and $(H \cdot M)^c = H \cdot M^c$, $(H \cdot M)^d = H \cdot M^d$ by Remark 4.2.14(ii). Hence $[L,N]_t = \langle H \cdot M^c, N^c \rangle_t + \sum_{s \leq t} H_s (\Delta M_s \Delta N_s)$ by Theorem 4.1.9, so that $[H \cdot M, N]_t = H \cdot \langle M^c, N^c \rangle_t + \sum_{s \leq t} H_s \Delta M_s \Delta N_s = H \cdot [M,N]_t$ as required. It follows that $H \cdot M$ can again be characterised as the unique L in $\mathcal{M}^2$ such that $\mathbf{E}(L_\infty N_\infty) = \mathbf{E}(\int_0^\infty H_s \mathrm{d}[M,N]_s)$ for all N in $\mathcal{M}^2$. It is now easy to see that this 'reformulated' stochastic integral coincides with the Stieltjes integral when $M \in \mathcal{M}^2 \cap \mathcal{V}$ – the reason for using $[M]$ instead of $\langle M \rangle$ lies in the characterisation of martingales in $\mathcal{V}$ as sums of compensated jump martingales, which have no continuous part, which makes $[M,N]$ easy to calculate: it is just $\sum_{s \geq t} \Delta M_s \Delta N_s$.

4.3.18. ***Theorem:*** If $M \in \mathscr{M}^2 \cap \mathscr{V}$ and $H \in \mathscr{L}^2(M)$ satisfy $\mu_M(|H|) < \infty$ then the stochastic integral $H \cdot M$ coincides with the Stieltjes integral $J = \int H \, dM$.

Proof: Assume $M_0 = 0$ without loss of generality. From the above discussion we must show that $\mathbf{E}(J_\infty N_\infty) = \mathbf{E}(\int_0^\infty H_s d[M,N]_s)$ for all N in $\mathscr{M}^2$. M is a martingale, so μ_M vanishes on Σ_p, hence so does the absolutely continuous measure $H\mu_M = \mu_J$, so J is a martingale in $\mathscr{V}$. By Proposition 4.3.2 this means that $\mathbf{E}(N_\infty J_\infty) = \mathbf{E}(\sum_s \Delta N_s \Delta J_s) = \mathbf{E}(\sum_s H_s \Delta N_s \Delta M_s)$ for all bounded martingales N. On the other hand, since $M \in \mathscr{V} \subseteq \mathscr{M}_d^2$, $M^c = 0$ and $[M,N]_t = \sum_{s \leq t} \Delta M_s \Delta N_s$, so that $\mathbf{E}(\int_0^\infty H_s d[M,N]_s) = \mathbf{E}(\sum_s H_s \Delta N_s \Delta M_s) = \mathbf{E}(N_\infty J_\infty)$. This proves the desired identity for bounded N. But $N_\infty \in L^\infty(P)$ is arbitrary and $L^\infty(P)$ is dense in $L^2(P)$, so $J_\infty = (H \cdot M)_\infty$. Conditioning with respect to $\mathscr{F}_T$ shows that $J_T = (H \cdot M)_T$ for all stopping times T.

4.3.19. ***Remark:*** We have defined the process $[M]$ in terms of $\langle M \rangle$, since our analysis of $\mathscr{M}^2$ rests heavily upon the Doob–Meyer decomposition. Since $\langle M \rangle = \Pi_p^*[M]$ it should be clear that $[M]$ is the 'simpler' object, despite appearances. This is borne out by a deeper analysis of these processes, which reveals that if $M \in \mathscr{M}_0^2$, and $t \in \mathbf{T}$, $[M]_t$ and $\langle M \rangle_t$ can both be expressed as limits of 'discrete skeletons': in fact, $[M]_t$ is the strong L^1-limit of sums of the form $\sum_{i=0}^{[2^n t]} (M_{(i+1)/2^n} - M_{i/2^n})^2$, while $\langle M \rangle_t$ is the weak L^1-limit of the sums $\sum_{i=0}^{[2^n t]} \mathbf{E}((M_{(i+1)/2^n} - M_{i/2^n}) | \mathscr{F}_{(i/2^n)-})^2$. (See Remark 4.5.7 and [19; Ch. VII, p. 43].)

These results further justify the terminology and provide an alternative proof that $\langle M \rangle = \Pi_p^*[M]$.

Finally, we shall prove in section 4.5, using the Itô formula, that for semimartingales X, Y the process $[X,Y]$ can be defined via an 'integration by parts' formula: $[X,Y] = XY - \int Y_- dX - \int X_- dY$. In characterising semimartingales as the most general 'integrators' yielding a sensible integration theory, Dellacherie [17] uses this definition to indicate how stochastic calculus may be viewed as a genuine extension of classical (deterministic) calculus.

4.4. Extension of the integral

The optional quadratic variation process is the appropriate tool for defining stochastic integrals relative to more general processes. We do this in two stages: first the results of section 4.2 are extended by 'localisation' to *local martingales*. Then the final extension of the integral to *semimartingales* results from the compatibility of stochastic and Stieltjes integrals.

The definition and basic properties of local martingales were discussed in section 3.8. Our aim is now to show that a local martingale can be decomposed *locally* into the sum of a square-integrable martingale and a process of integrable variation. We shall work primarily with $(\mathcal{M}_{\text{loc}})_0$ and *denote this space by* $\mathcal{L}$ for convenience.

4.4.1. ***Definition:*** If $M \in \mathcal{L}$ the stopping time T *strongly reduces* M if T reduces M and the martingale $(\mathbf{E}(|M_T| | \mathcal{F}_t))_t$ is bounded on $[\![0,T[\![$.

4.4.2: (i) If $S \leqslant T$ and T strongly reduces M, then M^T is uniformly integrable and by optional sampling on the submartingale $|M^T|$ we have $|M^S| \leqslant \mathbf{E}(|M_T| | \mathcal{F}_S)$, so that, as $\mathcal{F}_S \bigcap \mathcal{F}_t = \mathcal{F}_{S \wedge t}$, $\mathbf{E}(|M_S| | \mathcal{F}_t) \leqslant \mathbf{E}(|M_T| | \mathcal{F}_{S \wedge t})$. Since S obviously reduces M we have proved that S strongly reduces M.

(ii) If two stopping times S and T strongly reduce $M \in \mathcal{L}$, so does $S \vee T$: clearly (see Proposition 3.8.5) $S \vee T$ reduces M. We must also show that $\mathbf{E}(|M_{S \vee T}| | \mathcal{F}_t)$ is bounded on $[\![0,S \vee T[\![$. To see this, let $(t,\omega) \in [\![0,S \vee T[\![$. Since $|M_{S \vee T}| \leqslant |M_S| + |M_T|$ and

$$\mathbf{E}(|M_{S \vee T}| | \mathcal{F}_t)(\omega) = \begin{cases} \mathbf{E}(|M_{S(\omega)}| | \mathcal{F}_t)(\omega) & \text{if } S(\omega) \geqslant T(\omega) \\ \mathbf{E}(|M_{T(\omega)}| | \mathcal{F}_t)(\omega) & \text{if } T(\omega) \geqslant S(\omega) \end{cases}$$

the result follows as $\mathbf{E}(|M_S| | \mathcal{F}_t)$ is bounded on $[\![0,S[\![$ and $\mathbf{E}(|M_T| | \mathcal{F}_t)$ is bounded on $[\![0,T[\![$.

4.4.3. ***Proposition:*** If $M \in \mathcal{L}$ there is a sequence $T_n \uparrow \infty$ of stopping times such that each T_n strongly reduces M.

Proof: Let $m \geqslant 1$. If (R_n) reduces M, set $S_{nm} = R_n \wedge \inf\{t : \mathbf{E}(|M_{R_n}| | \mathcal{F}_t) \geqslant m\}$, arrange (S_{nm}) into a sequence (S_k), and set $T_n = \vee_{k=1}^n S_k$. To show that T_n strongly reduces M it is enough to show (by (ii) above) that each S_k strongly reduces M. Since each S_k reduces M, it remains to show that $\mathbf{E}(|M_S| | \mathcal{F}_t)$ is bounded on $[\![0,S[\![$ if $S = S_{nm}$. Also write $R = R_n$. Now if $Y_t = \mathbf{E}(|M_R| | \mathcal{F}_t)$, Y is bounded by m on $[\![0,S[\![$. By optional sampling we also have $M_S 1_{\{t<S\}} = \mathbf{E}(M_R 1_{\{t<S\}} | \mathcal{F}_S)$. It is easily shown that $\mathbf{E}(|M_S| 1_{\{t<S\}} | \mathcal{F}_t) \leqslant Y_t 1_{\{t<S\}} \leqslant m$ by using Jensen's inequality (*Exercise*!). The proposition follows.

4.4.4. ***Theorem:***

(i) Suppose $M \in \mathcal{L}$ and T strongly reduces M. Then there exist U in $\mathcal{M}_0^2$, V in $\mathcal{V}_0$, such that $M^T = U + V$.

(ii) If M is a local martingale there is a sequence $T_n \uparrow \infty$ of stopping times such that $M^{T_n} = M_0 + U^n + V^n$, where U^n and V^n are stopped at T_n and $U^n \in \mathcal{M}_0^2$, $V^n \in \mathscr{V}_0$. The decompositions are *not* in general unique.

Proof: By Proposition 4.4.3 only (i) needs proof. We may assume that $M = M^T$ since we only deal with M^T. Let $M_T = \lim M_t$ on $\{T = \infty\}$. As T strongly reduces M, we have M_T in L^1 and the martingale $(\mathbf{E}(|M_T| \,|\, \mathscr{F}_t))_t$ is bounded on $[\![0,T[\![$ by some constant K. First we prove the following result:

Lemma: Let Y in $L^1_+(\mathscr{F}_T)$ be given, with $Y = 0$ on $\{T = 0\}$. Set $C_t = Y1_{\{t \geqslant T\}}$ and $X_t = \mathbf{E}(Y|\mathscr{F}_t)1_{\{t < T\}}$, and assume that $|X_t| \leqslant K$ for some K. Then $U = X + \Pi_p^* C \in \mathcal{M}_0^2$.

Proof: X is a bounded positive supermartingale, since if $s \leqslant t$ and $A \in \mathscr{F}_s$, then $\int_A X_t \mathrm{d}P = \int_{A \cap \{s < T\}} Y \mathrm{d}P = \int_A X_s \mathrm{d}P$. On the other hand, U is a martingale: for all t, $U_t = X_t + (\Pi_p^* C)_t = \mathbf{E}(Y|\mathscr{F}_t) - (Y1_{[\![T,\infty[\![} - \Pi_p^*(Y1_{[\![T,\infty[\![}))(t,\cdot)$, and the term in brackets is a compensated process hence also a martingale.

Thus the predictable increasing process $A = \Pi_p^* C$ generates the potential part of X (by Theorem 3.7.1). But X is bounded by K, so we may assume that X is a potential. Hence $X_S = \mathbf{E}(A_\infty - A_S|\mathscr{F}_S)$ for any stopping time S and if S is predictable, we also have $X_{S-} = \mathbf{E}(A_\infty - A_{S-}|\mathscr{F}_{S-})$. (To see this, let (S_n) announce S and use Exercise 3.4.6(ii).) So X is the optional projection of $(A_\infty - A_s)_s$ and X_- is the predictable projection of $(A_\infty - A_{s-})_s$.

Integration by parts now yields $A_\infty^2 = \int_0^\infty (A_s + A_{s-}) \mathrm{d}A_s$ and clearly $\int_0^\infty A_\infty \mathrm{d}A_s = A_\infty^2$ also. Hence we obtain by subtraction that $\mathbf{E}(A_\infty^2) = \mathbf{E}(\int_0^\infty ((A_\infty - A_s) + (A_\infty - A_{s-}))\mathrm{d}A_s) = \mathbf{E}(\int_0^\infty (X_s + X_{s-})\mathrm{d}A_s)$ by the above remarks. But X is bounded by K, so the final term is bounded by $2K\mathbf{E}(\int_0^\infty \mathrm{d}A_s) = 2K\mathbf{E}(\mathbf{E}(A_\infty - A_0|\mathscr{F}_0)) = 2K\mathbf{E}(X_0) \leqslant 2K^2$. Hence $A_\infty = (\Pi_p^* C)_\infty \in L^2$ and so $U \in \mathcal{M}_0^2$. The lemma is proved.

We apply the lemma twice, with Y equal to M_T^+ and M_T^- respectively. Thus $M_t = M_T 1_{\{t \geqslant T\}} + \mathbf{E}(M_T|\mathscr{F}_t)1_{\{t < T\}}$ can be rewritten, in terms of the positive and negative parts of M_T, in the form $M = (C^1 + X^1) + (C^2 + X^2)$, where the C^i and X^i $(i = 1,2)$ are as in the lemma. But we have $C^i + X^i = (C^i - \Pi_p^*(C^i)) + (X^i + \Pi_p^*(C^i))$, and here the first term is in $\mathscr{V}_0$ and the second belongs to $\mathcal{M}_0^2$ by the lemma. This proves the theorem.

4.4.5. ***Definitions:*** We can now extend the quadratic variation processes to local martingales: first let $X = M + V$, where $M \in \mathcal{M}_0^2$ and $V \in \mathscr{V}_0$. We define the *continuous part* X^c of X as $X^c = M^c$, where $M = M^c + M^d$ is the decomposition of M into its projections onto $\mathcal{M}_c^2$ and $\mathcal{M}_d^2$ respectively.

Then X^c is well-defined: if $X=\bar{M}+\bar{V}$ with $\bar{M}\in\mathcal{M}_0^2$ and $\bar{V}\in\mathcal{V}_0$, then $M-\bar{M}=\bar{V}-V$ is a square-integrable martingale of integrable variation, hence belongs to $\mathcal{M}_d^2$ by Remark 4.3.3(i). So $M^c=\bar{M}^c$. Define the *optional quadratic variation* $[X]$ of X by $[X]_t=\langle X^c\rangle_t+\sum_{s\leqslant t}\Delta X_s^2$. The process $[X,Y]$ is again defined by polarisation, and clearly $[X,Y]_t=\langle X^c,Y^c\rangle_t+\sum_{s\leqslant t}\Delta X_s\Delta Y_s$. Now suppose $X\in\mathcal{L}$. If T strongly reduces X, Theorem 4.4.4 shows that $X^T=M+V$ with $M\in\mathcal{M}_0^2$, $V\in\mathcal{V}_0$, so that $(X^T)_c$ and $[X^T]_t=\langle(X^T)^c\rangle_t+\sum_{s\leqslant t\wedge T}\Delta X_s^2$ are well-defined as above. If S and T both strongly reduce X, it is clear that $(X^S)_t^c=(X^T)_t^c$ and $[X^S]_t=[X^T]_t$ for $t\leqslant S\wedge T$. Choosing a sequence $(T_n)\uparrow\infty$ strongly reducing X, we can thus define the *continuous part* X^c of X by setting $(X^c)^T=(X^T)^c$ whenever T strongly reduces X. Similarly $\langle X^c\rangle$ is the unique continuous increasing process (with $\langle X^c\rangle_0=0$) such that $\langle X^c\rangle_{T\wedge t}=\langle(X^T)^c\rangle_t$ for all such T, and since $\sum_{s\leqslant t}\Delta X_s^2<\infty$ a.s. (by Remark 4.3.12 $\sum_{s\leqslant t}\Delta M_s^2<\infty$, and $V\in\mathcal{V}_0$) we can finally define the *optional quadratic variation* of X by setting, for $t\in\mathbf{T}$, $[X]_t=\langle X^c\rangle_t+\sum_{s\leqslant t}\Delta X_s^2$. If X is a local martingale (not necessarily with $X_0=0$) we consider $\hat{X}=X-X_0\in\mathcal{L}$: set $X^c=\hat{X}_c$ and $[X]_t=[\hat{X}]_t+X_0^2$. $[X,Y]$ is again defined by polarisation in both cases.

4.4.6. ***Remarks:***

(1) The concept of orthogonality in $\mathcal{M}_0^2$ exploited in sections 4.2 and 4.3 is easily extended to $\mathcal{L}$: $X,Y\in\mathcal{L}$ are (strongly) *orthogonal* if $XY\in\mathcal{L}$. The continuous part X^c of $X\in\mathcal{L}$ then yields a unique decomposition of $X\in\mathcal{L}$ into X^c+X^d, where $X^d\in\mathcal{L}$ is orthogonal to each continuous local martingale.

The uniqueness of such a decomposition is easy to prove and left to the reader. As for the existence, we note that it suffices to consider $X^T=U+V$ as in Theorem 4.4.4, where T strongly reduces X. Since $U\in\mathcal{M}_0^2$, $U=U^c+U^d=(X^T)^c+U^d=(X^c)^T+U^d$ by definition. So we need only show that U^d+V is orthogonal to each continuous local martingale N. If $N\in\mathcal{M}_c^2$, this is simple: $V=X^T-U$ is a martingale of integrable variation, hence belongs to $\mathcal{M}_d^2$, and $U^d\perp\!\!\!\perp U^c$ by definition. So $(U^d+V)\perp\!\!\!\perp N$. The general case can be reduced to this be considering N^{S_n} instead, where the stopping time $S_n=\inf\{t:|N_t|\geqslant n\}$. Since by Theorem 4.4.4 the processes U^d and V are stopped at T and each $N^{S_n}\in\mathcal{M}_c^2$, the result follows upon letting $n\to\infty$.

(2) If $X\in\mathcal{L}$ is orthogonal to each continuous local martingale, we say that X is *purely discontinuous*, and write $X\in\mathcal{L}^d$ (write $\mathcal{L}^c$ for continuous elements of $\mathcal{L}$). $X\in\mathcal{L}^d$ is determined by its jumps: set $(\Delta X)_t=\sum_{s\leqslant t}\Delta X_s$, and suppose $\Delta X=\Delta Y$, so $\Delta M=0$, where $M=X-Y$. Hence $M\in\mathcal{L}^c\bigcap\mathcal{L}^d$, and by

the uniqueness of the decomposition of processes in $\mathscr{L}$ we have $M=0$ (recall that $M_0=0$). Hence $\Delta X=\Delta Y$ implies $X=Y$ for $X,Y\in\mathscr{L}^d$.

(3) Now suppose $X\in\mathscr{L}$ satisfies $[X]=0$. We shall show that $X=0$: first note that $\langle X^c\rangle=0$ and $\Delta X=0$, so in particular $X\in\mathscr{L}^c$. By considering stopping times which strongly reduce X we may assume that $X=X^c+M+V$ with $M\in\mathscr{M}_d^2$, $V\in\mathscr{V}_0$. In fact, one can choose V predictable, since $M'=M-(V-\Pi_p^*V)$ is also a uniformly integrable martingale (even if not necessarily in $\mathscr{M}_0^2$). By Remark 3.7.7 M is then regular, so $\mathbf{E}(\Delta X_T|\mathscr{F}_{T-})=0$ for all predictable T, and so $\Pi_p^*(\Delta M)=0$ by Theorem 4.3.8. Hence $0=\Pi_p^*(\Delta X)-\Delta V$, and so $\Delta M=\Delta X-\Delta V=0$. By remark (2) above, $M=0$. Moreover, $\langle X^c\rangle=0$ implies that $(X^c)^2\in\mathscr{L}$, since the relation $\langle X^c\rangle-(X^c)^2\in\mathscr{L}$ characterises $\langle X^c\rangle$. Thus $\mathbf{E}((X^c)_s^2)=0$ for all s, so that $X^c=0$.

(4) The above proof shows that the only predictable elements of $\mathscr{M}_{\mathrm{loc}}\bigcap\mathscr{V}$ are constant.

(5) We saw in Lemma 3.8.8 that all predictable increasing processes (and hence all predictable processes of bounded variation) are locally integrable. Theorem 4.4.4 allows a similar conclusion for local martingales V of bounded variation.

We may assume $V\in\mathscr{L}$. If V is strongly reduced by T_n for each n, where $T_n\uparrow\infty$, let $S_n=\inf\{t:\int_0^t|dV_s|\geqslant n\}$ and define $R_n=S_n\wedge T_n$. Then $\mathbf{E}(\int_{[\![0,R_n[\![}|dV_s|)\leqslant n$ by construction and it suffices to show that $\mathbf{E}(|\Delta V_{R_n}|)<\infty$. Now since R_n strongly reduces V, we may write $V^{R_n}=M+A$ as in Theorem 4.4.4, where $M\in\mathscr{M}_0^2$, $A\in\mathscr{V}_0$. In particular $\Delta M_{R_n}\in L^2$, $\Delta A_{R_n}\in L^1$, so $\Delta V_{R_n}\in L^1$ and $\mathbf{E}(\int_{[\![0,R_n[\![}|dV_s|)<\infty$ as required.

We are now in a position to define the stochastic integral for a wide class of processes relative to local martingales:

4.4.7. ***Definition:*** An optional process H is *locally bounded* if there exist stopping times $T_n\uparrow\infty$ and constants K_n such that $|H_t|1_{\{0<t\leqslant T_n\}}\leqslant K_n$ a.s. (Since H_0M_0 is treated separately in defining $H\cdot M$, we need impose no conditions on H_0.)

The stochastic integral $H\cdot M$, for H predictable, locally bounded, and M a local martingale, is now characterised as a stochastic process exactly as in section 4.2 for $M\in\mathscr{M}^2$.

4.4.8. ***Theorem:***

(i) Let M be a local martingale, and let H be a predictable locally bounded process. There is a unique local martingale $H\cdot M$ such that for each bounded martingale N we have

$[H\cdot M,N]=H\cdot[M,N]$, where the right-hand side denotes the Stieltjes integral relative to the bounded variation process $[M,N]$.

(ii) $(H\cdot M)_0=H_0M_0$, $(H\cdot M)^c=H\cdot M^c$ and $\Delta(H\cdot M)=H\cdot\Delta M$.

(iii) If $M\in\mathscr{V}_{\text{loc}}$, $H\cdot M$ coincides with the Stieltjes integral.

Proof: We can clearly take $M\in\mathscr{L}$. Choose a sequence $T_n\uparrow\infty$ of stopping times strongly reducing M and such that each H^{T_n} is bounded. We can write $M^n=M^{T_n}=U^n+V^n$ where, as in Theorem 4.4.4, $U^n\in\mathscr{M}_0^2$, $V^n\in\mathscr{V}_0$, and both are stopped at T_n. We can therefore define the stochastic integral $H\cdot M^n:=H\cdot V^n$. Since V^n is a martingale, so is the Stieltjes integral $H\cdot V^n$. Moreover $H\cdot U^n\in\mathscr{M}_0^2$ by Theorem 4.1.9. Hence $H\cdot M^n$ is a uniformly integrable martingale for each n. Since $(M^{T_n})^{T_m}=M^{T_m\wedge T_n}=M^{T_m}$ if $m\leqslant n$, we also have $(H\cdot M^m)=(H\cdot M^{T_n\wedge T_m})=(H\cdot M^n)^{T_m}$ by Remarks 4.2.11, and so we can define $H\cdot M$ by setting $(H\cdot M)^{T_n}=H\cdot M^n$ for all n. Then $H\cdot M\in\mathscr{L}$ and $\Delta(H\cdot M)_t^{T_n}=H_t\Delta M_t^n$ for all $t\in\mathbf{T}$, $n\in\mathbf{N}$. Again these identities are compatible for $m,n\in\mathbf{N}$, so $\Delta(H\cdot M)_t=H_t\Delta M_t$ for all $t\in\mathbf{T}$.

Since the continuous and discontinuous parts of $H\cdot U^n$ are $H\cdot(U^n)^c$ and $H\cdot(U^n)^d$ respectively, it follows similarly that $H\cdot M^c=(H\cdot M)^c$ (recall that $(M^n)^c=(U^n)^c$). Adding, we obtain for any bounded martingale N, $t\in\mathbf{T}$, $[H\cdot M,N]_t=[H\cdot M^c,N^c]_t+\sum_{s\leqslant t}H_s\Delta M_s\Delta N_s=(H\cdot[M,N])_t$.

Finally, if $M\in\mathscr{V}_{\text{loc}}\cap\mathscr{L}$, choose the (T_n) above so that $M^{T_n}\in\mathscr{V}$. But then $H\cdot M^n$ coincides with the Stieltjes integral $J_n=\int H\,\mathrm{d}M^n$ by Theorem 4.3.18. Now (iii) follows.

To see that $H\cdot M$ is unique, suppose A,B in $\mathscr{M}_{\text{loc}}$ both satisfy the theorem, so $[A,N]=[B,N]$ for all bounded N. Then $[A-B,N]=0$ for all bounded martingales N. This relation then also holds for all $N\in\mathscr{M}_{\text{loc}}$: choosing a sequence $T_n\uparrow\infty$ of stopping times strongly reducing N ensures that N^{T_n} is a bounded martingale for each n, and so $[A-B,N]^{T_n}=[A-B,N^{T_n}]=0$ for all n. Thus in particular, $[A-B,A-B]=0$, hence $A=B$ by Remark 4.4.6(3).

4.4.9. ***Remark:*** Let $L\in\mathscr{L}$ and let N be a bounded martingale. If T strongly reduces L, $L^T=U+V$ as in Theorem 4.4.4. Since U and N are in $\mathscr{M}^2$, the Kunita–Watanabe inequality, Theorem 4.2.7, ensures that $\mathbf{E}(\int_0^\infty|\mathrm{d}[U,N]_s|)$ is finite, and as $V\in\mathscr{M}_d^2$, $[V,N]_t=\sum_{s\leqslant t}\Delta V_s\Delta N_s$ by definition and so $\mathbf{E}(\int_0^\infty|\mathrm{d}[V,N]_s|)$ is also finite. Hence the bounded variation process $[L,N]$ is locally integrable, and we can define $\langle L,N\rangle$ as its dual predictable projection.

The final extension of the integral is now almost trivial:

4.4.10. ***Definition:*** A *semimartingale* is an adapted process X which admits a decomposition of the form $X=M+A$ where $M\in\mathscr{L}$ and A is a cadlag process of bounded variation. (Note that this implies that X is also cadlag.) Let $\mathcal{S}$ be the vector space of semimartingales.

4.4.11. ***Examples:***

(1) Local martingales and processes of bounded variation belong to $\mathcal{S}$.

(2) The Doob–Meyer decomposition theorem 3.8.10 shows that each right-continuous supermartingale has the form $X=M-A$, with $M\in\mathcal{M}_{\text{loc}}$, $A\in\mathcal{A}_{\text{loc}}$. Hence $X\in\mathcal{S}$.

(3) A less obvious example is provided by stationary processes with independent increments – see [42].

4.4.12. ***Remark:*** The decomposition $X=M+A$ is not, in general, unique: there are local martingales of bounded variation, e.g. the compensated jump martingales of Proposition 4.3.2. On the other hand, if there is a decomposition $X=M'+A'$ with A' locally integrable, then we can write $X=M+A$ where A is also predictable: since $A'-\Pi_p^*(A')\in\mathcal{L}$ we take $M=M'+(A'-\Pi_p^*(A'))$ and $A=\Pi_p^*(A')$. That this decomposition is unique follows from Remark 4.4.6(4). If X has such a *canonical decomposition*, X is called a *special semimartingale*. It is not hard to show that $X\in\mathcal{S}$ is special iff $X^*=\sup_t|X_t|$ is a locally integrable increasing process (see [42; 2.14]). Moreover, it follows from the Doob–Meyer decomposition theorem (Theorem 3.8.10) that each special semimartingale is the difference of two local supermartingales. The converse is also true (see [42; 2.18]). For more results on special semimartingales we refer the reader to [66], [42], [30]. We remark only that our definition is slightly more restrictive than that of Meyer [66]: he does not demand that X_0 be integrable.

4.4.13. ***Definitions:*** Let $X\in\mathcal{S}$. Writing $X_t=X_0+M_t+A_t$, we can also ensure that $A_0=0$. If $X_t=X_0+\bar{M}_t+\bar{A}_t$ is a second such decomposition, $N=M-\bar{M}=\bar{A}-A$ is a local martingale of bounded variation, hence by Remark 4.4.6(5) N is locally integrable. If T strongly reduces N and N^T has integrable variation, $N^T=U+V$ with $U\in\mathcal{M}_0^2$, $V\in\mathcal{V}_0$, by Theorem 4.4.4. Thus $U=N^T-V\in\mathcal{V}\bigcap\mathcal{M}^2\subseteq\mathcal{M}_d^2$ by Remark 4.3.3(1), so that N^T has no continuous part. Consequently the *continuous martingale part* $X^c:=M^c$ of X is independent of the decomposition $X=M+A$.

This again allows us to define the *optional quadratic variation* of X: define $[X]$ by $[X]_t=\langle X^c\rangle_t+\sum_{s\leqslant t}\Delta X_s^2$. (The sum is a.s. finite because A has bounded variation and $\sum_{s\leqslant t}\Delta M_s^2$ is finite, as we saw in Definition 4.4.5.) Finally, if $X,Y\in\mathcal{S}$, the process $[X,Y]$ is defined by polarisation: $[X,Y]=\frac{1}{2}([X+Y]-[X]-[Y])$. If $[X,Y]$ has locally integrable variation, we can also define $\langle X,Y\rangle$ as its dual predictable projection. Both these processes are bilinear and $[X],\langle X\rangle$ are positive, hence the proof of the Kunita–Watanabe inequalities (Theorem 4.2.7) applies, so the inequalities hold for all semimartingales in the case of $[X,Y]$ and for all special semimartingales with $\langle X,Y\rangle$.

4.4.14. ***Theorem:*** Suppose $X\in\mathscr{S}$ has decomposition $X_t=X_0+M_t+A_t$ with $M\in\mathscr{L}$, and A of bounded variation. If H is a locally bounded predictable process, the semimartingale $H\cdot X$ is well-defined by $H\cdot X=H_0X_0+H\cdot M+H\cdot A$. Moreover (up to indistinguishability), $(H\cdot X)^c=H\cdot X^c$, $\Delta(H\cdot X)=H\cdot\Delta X$, and if T is any stopping time, $(H\cdot X)^T=H\cdot X^T$.

Proof: Let $X_t=X_0+M_t+A_t=X_0+\bar{M}_t+\bar{A}_t$, so that $M-\bar{M}=\bar{A}-A$ is a local martingale of bounded variation, hence has locally integrable variation (Remark 4.4.6(v)). Thus $H\cdot(M-\bar{M})$ coincides with the Stieltjes integral $\int H\,\mathrm{d}(\bar{A}-A)$. Therefore $H\cdot M+H\cdot A=H\cdot\bar{M}+H\cdot\bar{A}$ as required. Moreover, $X^c=M^c$ and $H_t\Delta X_t=H_t(\Delta M_t+\Delta A_t)=\Delta(H\cdot M)_t+\Delta(H\cdot A)_t$ by Theorem 4.4.8. Finally, $H\cdot X^T=H\cdot(1_{[\![0,T]\!]}X)=(H\cdot X)^T$.

4.5. The Itô formula and stochastic calculus

4.5.1: The stochastic integral $H\cdot X$ was originally constructed as a generalisation of the Itô integral (0.2.10) for the purpose of formulating and solving stochastic differential equations. This also requires a 'change of variable' formula if we are to manipulate such equations. In ordinary calculus we have $\mathrm{d}f(X)=f'(X)\mathrm{d}X$ or more precisely $f(X_t)-f(X_0)=\int_0^t f'(X_s)\mathrm{d}X_s$ under certain hypotheses, when $X:\mathbf{T}\to\mathbf{R}$ and $f:\mathbf{R}\to\mathbf{R}$ are given. (f' denotes the derivative of f.) On the other hand, if X is a *continuous semimartingale*, the optional quadratic variation $[X]$ intervenes: we obtain $\mathrm{d}f(X)=f'(X)\mathrm{d}X+\frac{1}{2}f''(X)\mathrm{d}[X]$ or, again,

$$f(X_t)-f(X_0)=\int_0^t f'(X_s)\mathrm{d}X_s+\tfrac{1}{2}\int_0^t f''(X_s)\mathrm{d}[X]_s. \tag{4.12}$$

The reason for the presence of the final term arises from the possibility that X has unbounded variation: let (t_k) be a partition of $[0,t]$ and expand in terms of a first-order Taylor polynomial,

$$f(X_t)-f(X_0)=\sum_{k\geq 1}(f(X_{t_{k+1}})-f(X_{t_k}))=\sum_{k\geq 1}f'(X_{t_k})(X_{t_{k+1}}-X_{t_k})+\sum_{k\geq 1}R_k^*$$

$$\to\int_0^t f'(X_{s-})\mathrm{d}X_s+\lim_{\sup_t\Delta t_n\to 0}\Big(\sum_{k\geq 1}R_k^*\Big).$$

Here the final term need *not* converge to zero. On the other hand, taking one further term in the expansion leaves us with a remainder which is small compared to second-order differences, hence small compared to $[X]$. In the expression

$$f(X_t)-f(X_0)=\sum_{k\geq 1} f'(X_{t_k})(X_{t_{k+1}}-X_{t_k})+\tfrac{1}{2}\sum_{k\geq 1} f''(X_{t_k})(X_{t_{k+1}}-X_{t_k})^2+\sum_{k\geq 1} R_k,$$

the first term on the right will then converge to $\int_0^t f'(X_{s-})dX_s$, the second to $\frac{1}{2}\int_0^t f''(X_{s-})d[X]_s$, and the final sum will go to 0 in probability. When X is continuous this will yield (4.12).

When X is a general semimartingale, the jumps of $s\to X_s$ introduce a somewhat unsightly compensating term into the formula. This is already evident in the normal change of variable formula for Stieltjes integrals when the integrator is merely cadlag and not necessarily continuous: the reader may verify that if $X:\mathbf{T}\to\mathbf{R}$ has bounded variation and is cadlag then for any $f\in\mathscr{C}^1(\mathbf{R})$ we obtain the formula

$$f(X_t)-f(X_0)=\int_0^t f'(X_{s-})dX_s+\sum_{0<s\leq t}[f(X_s)-f(X_{s-})-f'(X_{s-})\Delta X_s].$$

When X is a *cadlag semimartingale* we need $f\in\mathscr{C}^2$ to make sense of the formula (4.12), and this formula is then augmented by a 'jump term' on the right of the form $\sum_{0<s\leq t}U_s$, where $U_s=f(X_s)-f(X_{s-})-f'(X_{s-})\Delta X_s-\frac{1}{2}f''(X_{s-})(\Delta X_s)^2$. It is clear that this term simply 'corrects' the right-hand side of (4.12) in the presence of jumps in $s\to X_s$.

We shall not prove the formula in this generality, but restrict attention to (4.12) for *continuous* semimartingales. (A detailed 'elementary' proof of the general formula appears in [30] and a much more sophisticated approach is taken in [17], [19], [62].)

4.5.2: Let us first consider a special case. We assume that $X_t=X_0+M_t+A_t$, where $M_0=A_0=0$, and M is a continuous martingale bounded by $K>0$ and A is a continuous process of integrable variation with total variation bounded by K. Let $f\in\mathscr{C}^2(\mathbf{R})$ then $\sup_{x\in[-3k,3k]}(|f'(x)|+|f''(x)|)\leq c$ for some $c>0$. Note that $|X_t|\leq 3K$ for all $t\in\mathbf{T}$. Now expand f into a second order Taylor polynomial: let $a,b\in[-3k,3k]$, then $f(b)-f(a)=(b-a)f'(a)+\frac{1}{2}(b-a)^2f''(a)+R(a,b)$, where $R(a,b)$ denotes the remainder term. Now f'' is uniformly continuous on $[-3k,3k]$, so $|R(a,b)/(b-a)^2|\to 0$ as $b\to a$.

Now partition $[0,t]$ as follows for each $\omega\in\Omega$ (note that the partition points depend on ω!): fix $a>0$, $t_0=0$ and set $t_{i+1}=t_i\wedge(t_i+a)\wedge\inf(s>t_i:|M_s-M_{t_i}|>a$ or $|A_s-A_{t_i}|>a)$. Then $|t_{i+1}-t_i|$, $|M_{t_{i+1}}-M_{t_i}|$ and $|A_{t_{i+1}}-A_{t_i}|$ are bounded by a for all i. Expanding $f(X_t)$ we obtain

$$f(X_t)-f(X_0)=\sum_i f'(X_{t_i})(X_{t_{i+1}}-X_{t_i})+\tfrac{1}{2}\sum_i f''(X_{t_i})(X_{t_{i+1}}-X_{t_i})^2+\sum_i r(X_{t_i},X_{t_{i+1}}),$$

where r denotes the remainder term as above. Take each sum on the right in turn:

(i)

$$S_1=\sum_i f'(X_{t_i})(X_{t_{i+1}}-X_{t_i})=\sum_i[f'(X_{t_i})(M_{t_{i+1}}-M_{t_i})+f'(X_{t_i})(A_{t_{i+1}}-A_{t_i})].$$

The 'martingale part' converges in L^2-norm to $\int_0^t f'(X_s)\mathrm{d}M_s$ when $a\to 0$, since the square of the L^2-distance between them is $\mathbf{E}(\sum_i\int_{t_i}^{t_{i+1}}(f'(X_{t_i})-f'(X_s)\mathrm{d}\langle M\rangle_s)$ and each integrand is bounded by $\sup_i(\sup_{s\in[t_i,t_{i+1}]}|f'(X_{t_i})-f'(X_s)|)$. Thus the integrands converge to 0 and since $\mathbf{E}(\langle M\rangle_t)<\infty$ the L^2-convergence is proved. An analogous argument shows that the other part of S_1 converges to $\int_0^t f'(X_s)\mathrm{d}A_s$ in L^1-norm as $a\to 0$. Hence $S_1\to\int_0^t f'(X_s)\mathrm{d}X_s$ in probability as $a\to 0$.

(ii) $S_2=\sum_i f''(X_{t_i})(X_{t_{i+1}}-X_{t_i})^2$ can be written as $K_1+K_2+K_3$, where $K_1=\sum_i f''(X_{t_i})(M_{t_{i+1}}-M_{t_i})^2$, $K_2=\sum_i f''(X_{t_i})(A_{t_{i+1}}-A_{t_i})(M_{t_{i+1}}-M_{t_i})$ and $K_3=\sum_i f''(X_{t_i})(A_{t_{i+1}}-A_{t_i})^2$. It is easy to see that K_2 and K_3 go to 0 in L^1, since $|K_3|\leqslant C\sup_i(A_{t_{i+1}}-A_{t_i})\int_0^t|\mathrm{d}A_s|\leqslant aCK$ and $|K_2|\leqslant 2C\sup_i(M_{t_{i+1}}-M_{t_i})\int_0^t|\mathrm{d}A_s|\leqslant 2aCK$.

To deal with K_1, we first note that since M is bounded by K, $\mathbf{E}(\langle M\rangle_\infty-\langle M\rangle_t|\mathscr{F}_t)=\mathbf{E}(M_\infty^2-M_t^2|\mathscr{F}_t)\leqslant K^2$, so that $\mathbf{E}(\langle M\rangle_\infty^2)=2\mathbf{E}(\int_0^\infty(\langle M\rangle_\infty-\langle M\rangle_t)\mathrm{d}\langle M\rangle_t)\leqslant 2K^2\mathbf{E}(\langle M\rangle_\infty)\leqslant 2K^4$ (cf. Theorem 4.3.6). So $\langle M\rangle\in L^2$ and $M^2-\langle M\rangle$ is a martingale in $\mathscr{M}_0^2$.

We shall now approximate K_1 in L^2-norm by $S_2^*=\sum_i f''(X_{t_i})(\langle M\rangle_{t_{i+1}}-\langle M_{t_i}\rangle)$, which clearly converges to $\int_0^t f''(X_s)\mathrm{d}\langle M\rangle_s$ in L^1-norm (by the same argument as in (i)). Now $\mathbf{E}(S_2^*-K_1)^2=\mathbf{E}(\sum_i f''(X_{t_i})[(\langle M\rangle_{t_{i+1}}-\langle M\rangle_{t_i})-(M_{t_{i+1}}-M_{t_i})^2])^2$. But distinct terms in the sum are orthogonal since $\langle M\rangle-M^2$ is a martingale, and $|f''(X_{t_i})|\leqslant C$. This, together with the obvious inequality $(\alpha-\beta)^2\leqslant 2(\alpha^2+\beta^2)$ applied to each term, yields $\mathbf{E}(S_2^*-K_1)^2\leqslant 2C^2(\sum_i\mathbf{E}(\langle M\rangle_{t_{i+1}}-\langle M\rangle_{t_i})^2+\sum_i\mathbf{E}(M_{t_{i+1}}-M_{t_i})^4)$. The first sum is dominated by $\mathbf{E}(\sup_i(\langle M\rangle_{t_{i+1}}-\langle M_{t_i}\rangle)\cdot\langle M\rangle_t)$ and so goes to 0 exactly like K_3. The second sum is majorised by $\mathbf{E}(\sup_i(M_{t_{i+1}}-M_{t_i})^2\sum_i(M_{t_{i+1}}-M_{t_i})^2)\leqslant a^2\mathbf{E}(M_t^2)\to 0$ as $a\to 0$. Hence we have shown that $S_2\to\int_0^t f''(X_s)\mathrm{d}\langle M\rangle_s$ in probability when $a\to 0$.

(iii) $S_3=\sum_i r(X_{t_i},X_{t_{i+1}})\leqslant\sum_i(X_{t_{i+1}}-X_{t_i})^2\varepsilon(|X_{t_{i+1}}-X_{t_i}|)$ by our remark on the Taylor remainder, where $\varepsilon(h)\to 0$ as $h\to 0$. But $|X_{t_{i+1}}-X_{t_i}|\leqslant 2a$ by choice of (t_i) so that $|S_3| \leqslant 2\varepsilon(2a)\sum_i((A_{t_{i+1}} - A_{t_i})^2 + (M_{t_{i+1}} - M_{t_i})^2)$. However, $\mathbf{E}(\sum_i(M_{t_{i+1}} - M_{t_i})^2) = \mathbf{E}(\sum_i(M^2_{t_{i+1}} - M^2_{t_i})) = \mathbf{E}(M^2_t) < \infty$ as M is a martingale, and as before $\mathbf{E}(\sum_i(A_{t_{i+1}} - A_{t_i})^2) \leqslant a\mathbf{E}(\int_0^t |\mathrm{d}A_s|)$. Hence $\mathbf{E}(|S_3|) \rightarrow 0$.

Putting all three terms together we have shown that

$$f(X_t) - f(X_0) = S_1 + \tfrac{1}{2}S_2 + S_3 \xrightarrow{P} \int_0^t f'(X_s)\mathrm{d}X_s + \tfrac{1}{2}\int_0^t f''(X_s)\mathrm{d}\langle M \rangle_s.$$

On the other hand, since X is continuous, $X^c = M$ and $[X] = \langle M \rangle$. So the limit coincides with the right-hand side of (4.12). Both sides are cadlag, hence the processes defined by the two sides of (4.12) are indistinguishable.

4.5.3: We now relax the conditions imposed on X. Assume that $X_t = X_0 + M_t + A_t$ is a continuous semimartingale with $M \in \mathscr{M}^2_0$ and $A \in \mathscr{V}_0$ and let f be as before. Fix stopping times $T_n = \inf\{t : |M_t| \geqslant N$ *or* $\int_0^t |\mathrm{d}A_t| \geqslant N\}$ and let $X_0^N = X_0 1_{\{|X_0| \leqslant N\}}$, $M^N = M^{T_N}$, $A^N = A^{T_N}$, $X_t^N = X_0^N + M_t^N + A_t^N$. Then it is easy to see that $X_0^N \xrightarrow{L^1} X_0$, $M_t^N \xrightarrow{L^2} M_t^N$ and $\mathbf{E}(\int_0^t |\mathrm{d}(A_t^N - A_t)|) \rightarrow 0$. The semimartingale X^N satisfies the conditions of the special case proved in section 4.5.2, so the Itô formula holds for X^N. To deduce its validity for X we note that $f(X_t^N) \xrightarrow{P} f(X_t)$ and show that the various differences of corresponding terms on the right will converge to 0 in probability as $N \rightarrow \infty$. This involves no new ideas, and the details are left to the reader.

The extension to general continuous semimartingales (and arbitrary $f \in \mathscr{C}^2$) is not entirely simple: if $X = M + A$ with $M \in \mathscr{M}_{\text{loc}}$ we first need to 'localise' X by finding stopping times (T_n) which strongly reduce M and such that $|X| \leqslant K$ on $[\![0, T_n[\![$. We refer the reader to [45], [30] for detailed proofs of the continuous and the general case, and merely state the general form of the formula:

4.5.4. ***Theorem:*** If X is a semimartingale and $f \in \mathscr{C}^2$, then $f(X)$ is a semimartingale and (up to indistinguishability)

$$f(X_t) = f(X_0) + \int_0^t f'(X_{s-})\mathrm{d}X_s + \tfrac{1}{2}\int_0^t f''(X_{s-})\mathrm{d}\langle X^c \rangle_s$$

$$+ \sum_{0 < s \leqslant t} (f(X_s) - f(X_{s-}) - f'(X_{s-})\Delta X_s). \tag{4.13}$$

4.5.5. ***Remarks:***

(1) The formula (4.13) can be re-written in the form (with f, X as in Section 4.5.2)

$$f(X_t)=f(X_0)+\int_0^t f'(X_{s-})\mathrm{d}X_s+\tfrac{1}{2}\int_0^t f''(X_{s-})\mathrm{d}[X]_s$$

$$+\sum_{0<s\leqslant t}(f(X_s)-f(X_{s-})-f'(X_{s-})\Delta X_s-\tfrac{1}{2}f''(X_{s-})(\Delta X_s)^2)$$

This formulation does not depend on the underlying probability measure: we shall see below that $[X]$ can be defined *directly* as $X^2-2\int X_-\mathrm{d}X$ for any semimartingale, and that the semimartingale property is invariant under changes of equivalent probability measures. On the other hand, $\langle X^c\rangle$ is derived from the continuous martingale part of X and this depends crucially on P. See [19].

(2) The Itô formula can be extended to $f\in\mathscr{C}^2(\mathbf{R}^n)$ and vector semimartingales $X=(X^1,\ldots,X^n)$ – this means that each X^i is a real semimartingale. Write D^i for the ith partial derivative and D^{ij} for the mixed second partials. Then $f(X)$ is a real semimartingale and the formula is written as

$$f\circ X_t=f\circ X_0+\sum_i\int_{]0,t]}D^if\circ X_{s-}\mathrm{d}X^i_s+\tfrac{1}{2}\sum_{ij}\int_0^t D^{ij}f\circ X_{s-}\mathrm{d}\langle X^{ic},X^{jc}\rangle_s$$

$$+\sum_{0<s\leqslant t}\left[f\circ X_s-f\circ X_{s-}-\sum_i D^if\circ X_{s-}(\Delta X^i_s)\right] \tag{4.14}$$

The proof is essentially the same as in the real case.

The extension of (4.14) to Hilbert space-valued semimartingales was undertaken in [62] and, independently, in [50], using an alternative formulation of stochastic integration which we outline in section 4.8.

As an immediate consequence of (4.14) consider the case $f(x,y)=xy$. We obtain an 'integration by parts' formula:

4.5.6. ***Proposition:*** If X and Y are semimartingales, so is their product XY. Moreover,

$$X_tY_t=\int_{]0,t]}X_{s-}\mathrm{d}Y_s+\int_{]0,t]}Y_{s-}\mathrm{d}X_s+[X,Y]_t \tag{4.15}$$

Proof: Using (4.14) with $n=2, f(x,y)=xy$ we obtain

$$X_tY_t = X_0Y_0 + \int_{]0,t]} Y_{s-}\mathrm{d}X_s + \int_{]0,t]} X_{s-}\mathrm{d}Y_s + 2\cdot\tfrac{1}{2}\int_{]0,t]} \mathrm{d}\langle X^c, Y^c\rangle$$
$$+ \sum_{0<s\leqslant t} (X_sY_s - X_{s-}Y_{s-} - Y_{s-}\Delta X_s - X_{s}\Delta Y_s).$$

Now since $X_0Y_0 = \Delta X_0 \Delta Y_0$ (recall that $X_{0-} = 0 = Y_{0-}$) we can incorporate it in the final sum, which then becomes $\sum_{s\leqslant t}\Delta X_s\Delta Y_s$.

If X and Y are both of bounded variation, the stochastic integrals reduce to Stieltjes integrals and a comparison with the corresponding integration by parts formula for Stieltjes integrals (cf. 4.5.1) yields $[X,Y]_t = \sum_{s\leqslant t}\Delta X_s\Delta Y_s$. Of course, this is also clear from the orthogonality results of section 4.3.

In fact, the reader may prove easily that if *one* of X and Y, say Y, has bounded variation, then $\langle X^c, Y^c\rangle = 0$, so that $[X,Y]_t = \sum_{s\leqslant t}\Delta X_s\Delta Y_s$.

4.5.7. *Remarks:*

(1) Equation (4.15) can also be used as the *definition* of the 'square brackets' process $[X,Y]$: in particular, this yields the definition $[X]_t = X_t^2 - 2\int_{]0,t]} X_{s-}\mathrm{d}X_s$ for the optional quadratic variation process of a semimartingale. From this, the nature of $[X]$ becomes evident: $[X]_t$ is the limit in probability as $k\to\infty$, of the sums $X_0^2 + \sum_{i=0}^n (X_{t_{i+1}} - X_{t_i})^2$, where we put $n = [2^k t]$, $t_i = i/2^k$ for $i \leqslant n$ and $t_{n+1} = t$.

To see this, note first that the stochastic integral has a simple *continuity property*: (*) if (H^n) is a sequence of elementary predictable processes (see Definition 4.1.6) which are uniformly bounded by some constant, supported on some fixed interval $[0,N]$ and such that $\sup_{t,\omega}|H^n(t,\omega)| \to 0$ as $n\to\infty$, then for any semimartingale X, $\int_0^t H_s^n \mathrm{d}X_s \xrightarrow{P} 0$ for all t. This is clear from the construction of the integral if $X \in \mathcal{M}^2$ or if X has bounded variation. The general case can be reduced to this, using the invariance of the convergence under changes of equivalent measures. (See [80].)

In particular $[X]_t = X_t^2 - 2\int_{]0,t]} X_{s-}\mathrm{d}X_s$ can be approximated in probability by $X_t^2 - 2\int_{]0,t]} X_{s-}^n \mathrm{d}X_s$, where X^n denotes the elementary predictable process defined by $X_0^n = X_0$, $X_t^n = X_{(n-1)/2^k}$ for $(n-1)/2^k < t < n/2^k < N$ and $X_t^n = X_t$ for $t \geqslant N$. But the integral $2\int_{]0,t]} X_{s-}^n \mathrm{d}X_s$ then reduces to a sum of the form $2X_{t_i}(X_{t_{i+1}} - X_{t_i})$ for a partition (t_i) as above, and since $X_t^2 = X_0^2 + \sum_{i=0}^n (X_{t_{i+1}}^2 - X_{t_i}^2)$, the identity $(x_{k+1} - x_k)^2 = x_{k+1}^2 - x_k^2 - 2x_k(x_{k+1} - x_k)$ suffices to prove that

$$[X]_t = X_t^2 - 2\int_{]0,t]} X_{s-}\mathrm{d}X_s = P\text{-}\lim_n \left(X_0^2 + \sum_{i=0}^n (X_{t_{i+1}} - X_{t_i})^2 \right).$$

It should be emphasised that the partition (t_i) does not depend on $\omega\in\Omega$. The 'true' second variation on X could still be infinite (see [33].)

(2) The integration by parts formula can be proved independently of the Itô formula and the latter deduced via approximation of $\mathscr{C}^2$-functions by polynomials – see [19], [17] for a detailed description of this approach to stochastic calculus.

(3) The continuity property (*) stated in (1) above actually *characterises* semimartingales in the following sense: let (H^n) be as in (*) and define a linear operator $I_X:\mathscr{E}\to L^0$ by $I_X(H^n)=H_0^n X_{t_1}+H_1^n(X_{t_2}-X_{t_1})+\ldots+H_{n-1}^n(X_{t_n}-X_{t_{n-1}})$ for some adapted cadlag process X with $X_0=0$. Then the sequence $(I_X(H^n))_n$ converges to 0 in probability *iff* X is a semimartingale. This was proved by Dellacherie, Mokobodski, Metivier and Pellaumail (see [17]) and independently by Bichteler [2]. See also [19].

This identification of semimartingales as the most general integrators for which the integral map I_X has a reasonable bounded continuity property (thus allowing extension from $\mathscr{E}$ to bounded predictable processes) allows one to develop stochastic integration theory *de novo* directly for semimartingales, without using much of the martingale theory developed in Chapter 3. Nonetheless, it involves careful consideration of different topologies on spaces of processes and perhaps lacks some of the 'comfortable' familiarity associated with $\mathscr{M}^2$-integrals. On the other hand, it is now immediate that the notion of semimartingales is *invariant* under passage to an equivalent probability measure, and also if we replace the filtration $(\mathscr{F}_t)$ by a smaller one $(\mathscr{G}_t)$ with respect to which X is adapted. The interested reader is urged to begin a study of these ideas with Dellacherie's beautiful survey [17].

4.6. Representation of martingales as stochastic integrals

Recall (Proposition 0.2.6) that standard Brownian motion $BM_0(\mathbf{R})$ is a continuous process with $B_0=0$ and such that for $0<s<t$ (B_t-B_s) is normally distributed with mean 0 and variance $(t-s)$, independent of $\{B_u: u\leq s\}$. Thus to show that a given adapted process B is $BM_0(\mathbf{R})$ it will suffice to prove that $\mathbf{E}(e^{iu(B_t-B_s)}|\mathscr{F}_s)=e^{-u^2(t-s)/2}$ for $0<s<t$, by the determining properties of the characteristic function and the corresponding characterisation of normal distributions (see [84]). The Itô formula allows us to deduce the following martingale characterisation of $BM_0(\mathbf{R})$, due to Paul Lévy:

4.6.1. ***Theorem:*** Let B be a continuous $(\mathscr{F}_t)$-martingale such that B_t^2-t also defines a martingale. Then B is a Brownian motion.

Proof: Fix $X \in L^2(\mathcal{F})$ and consider the martingale $X_t = \mathbf{E}(X|\mathcal{F}_t)$. We shall show that if XB is a martingale, then for $u \in \mathbf{R}$, $s \leqslant t$ in $\mathbf{R}^+$

$$\mathbf{E}(e^{iu(B_t - B_s)}X_t|\mathcal{F}_s) = X_s e^{-(t-s)\cdot u^2/2} \tag{4.16}$$

There is no loss of generality in assuming that $s=0$ and $B_0=0$, since we can use $\mathcal{F}'_v = \mathcal{F}_{s+v}$, $X'_v = X_{s+v}$, $t' = t-s$ instead, if needed. Now apply the Itô formula with $f(x) = e^{iux}$ on $[0,t]$:

$$e^{iuB_t} = e^{iu\cdot 0} + iu\int_0^t e^{iuB_s}\,\mathrm{d}B_s - \frac{u^2}{2}\int_0^t e^{iuB_s}\mathrm{d}\langle B\rangle_s. \tag{4.17}$$

By hypothesis $B_t^2 - t$ is a martingale. Since $(t,\omega)\mapsto t$ is increasing and continuous, hence predictable, $\langle B\rangle_t = t$ by the uniqueness of $\langle B\rangle$. For fixed $A \in \mathcal{F}_0$ define $g(s) = \int_A e^{iuB_s}X_s\,\mathrm{d}P$. Multiplying both sides of (4.17) by X_t and integrating over A, we obtain

$$g(t) = \int_A X_t\,\mathrm{d}P + iu\int_A X_t\left(\int_0^t e^{iuB_s}\,\mathrm{d}B_s\right)\mathrm{d}P - \frac{u^2}{2}\int_A X_t\left(\int_0^t e^{iuB_s}\,\mathrm{d}s\right)\mathrm{d}P.$$

Now $\int_A X_t\,\mathrm{d}P = \int_A X_0\,\mathrm{d}P$ since X is a martingale, while the assumption that XB is a martingale means that $X \perp\!\!\!\perp B$ and so (by Remark 4.2.14(2)) X is orthogonal to the stable subspace generated by B, hence also to $\int_0^t e^{iuB_s}\,\mathrm{d}B_s$. Finally we can exchange the order of integration in the third term by Fubini's theorem. We obtain $g(t) = \int_A X_0 \mathrm{d}P - (u^2/2)\int_0^t g(s)\mathrm{d}s$, so that g satisfies $g'(t) = -(u^2/2)g(t)$, or $g(t) = ce^{-tu^2/2}$, where the constant c is given by the initial condition $g(0) = \int_A X_0\mathrm{d}P = c$. As $A \in \mathcal{F}_0$ was arbitrary we have proved $\mathbf{E}(e^{iuB_t}X_t|\mathcal{F}_0) = X_0 e^{-tu^2/2}$, which is (4.16) with $s = 0 = B_0$. So (4.16) is valid in general, and, in particular, with $X \equiv 1$, we obtain the desired result.

4.6.2: Moreover, the same idea can be used to show that martingales w.r.t. the natural filtration of $BM_0(\mathbf{R})$ can be represented as stochastic integrals of $BM_0(\mathbf{R})$, in the following sense: let $\mathcal{G}_t$ be the completion of $\sigma(B_s : s \leqslant t)$ for each $t \in \mathbf{T}$ and suppose that $X \in L^2(\mathcal{G}_\infty)$. Then X can be represented in the form $X = \mathbf{E}(X|\mathcal{G}_0) + \int_0^\infty H_s \mathrm{d}B_s$, where $H = (H_t)$ is predictable with respect to $(\mathcal{G}_t)$ and $\mathbf{E}(\int_0^\infty H_s^2 \mathrm{d}s) < \infty$.

To prove this, restrict attention to $(\mathcal{G}_t)$ and fix $X_0 = \mathbf{E}(X|\mathcal{G}_0) = 0$. Let Y^n be the orthogonal projection of the martingale $X = (X_t)$, where $X_t = \mathbf{E}(X|\mathcal{G}_t)$, onto the subspace I_{B^n} generated in $\mathcal{M}^2$ by the square-integrable martingale $B_t^n = B_{t\wedge n}$ (see Remark 4.2.14(3)). By Remark 4.2.14(4), $Y^n = H^n \cdot B^n$, where H^n is a predictable density (of $\mathrm{d}\langle B^n, X\rangle$ w.r.t. $\mathrm{d}\langle B^n\rangle$). The processes H^n and H^{n+1} coincide on $[0,n]$, so there is a predictable $H \in \mathcal{L}^2(B)$ defining $Y = H \cdot B$.

Using $(X-Y)$ instead of X, we can thus suppose that $X \perp\!\!\!\perp B$. So we must now show that $X=X_\infty=0$. To do this, use (4.16) repeatedly: with $0=t_0<t_i<\ldots<t_n<\infty$, $u_1,\ldots,u_n\in\mathbf{R}$ and $X_{t_n}=\mathbf{E}(X_\infty|\mathscr{G}_{t_n})$ we obtain, as B is $\mathscr{G}$-adapted,

$$\begin{aligned}
&\mathbf{E}(\exp(iu_1B_{t_1})\exp(iu_2(B_{t_2}-B_{t_1}))\ldots\exp(iu_n(B_{t_n}-B_{t_{n-1}}))X_\infty|\mathscr{G}_0)\\
&=\mathbf{E}(\exp(iu_1B_{t_1})\exp(iu_2(B_{t_2}-B_{t_1}))\ldots\exp(iu_{n-1}(B_{t_{n-1}}-B_{t_{n-2}}))\times\\
&\quad\mathbf{E}(\exp(iu_n(B_{t_n}-B_{t_{n-1}}))X_\infty|\mathscr{G}_{t_{n-1}})|\mathscr{G}_0)\\
&=\mathbf{E}(\exp(iu_1B_{t_1})\exp(iu_2(B_{t_2}-B_{t_1}))\times\\
&\quad\ldots\exp(iu_{n-1}(B_{t_{n-1}}-B_{t_{n-2}}))X_{t_{n-1}}\exp(-(t_n-t_{n-1})u_n^2/2)|\mathscr{G}_0)\\
&=\ldots=\mathbf{E}(X_0\exp(-t_1u_1^2)/2\cdot\exp(-(t_2-t_1)u_2^2/2)\times\\
&\quad\ldots\cdot\exp(-(t_n-t_{n-1})u_n^2/2))=0
\end{aligned}$$

since $X_0=0$. Integrating, we have $\mathbf{E}(\exp(iu_1B_{t_1})\ldots\exp(iu_n(B_{t_n}-B_{t_{n-1}}))X)=0$, so that relative to the measure μ with $\mathrm{d}\mu/\mathrm{d}P=X$ all joint distributions of B have trivial Fourier transforms. Hence $\mu\circ(B_{t_1},B_{t_2}-B_{t_1},\ldots,B_{t_n}-B_{t_{n-1}})^{-1}$ is 0 for all random vectors $(B_{t_1},B_{t_2}-B_{t_1},\ldots,B_{t_n}-B_{t_{n-1}})$. Since the increments are independent, the Fourier transforms are additive (see [84]) and hence $\mu\circ(B_{t_1},\ldots,B_{t_n})=0$. By the Daniell–Komogorov theorem (0.1.7) we conclude that $\mu(f)=0$ for all $\mathscr{G}_\infty$-measurable f, hence $\mu=0=X$.

The restriction to $(\mathscr{G}_t)$ is only artificial: in fact we show that $\mathbf{E}(X_\infty|\mathscr{G}_t)=\mathbf{E}(X_\infty|\mathscr{F}_t)$ for all $t\in\mathbf{T}$. For let Z be a right-continuous modification of $(\mathbf{E}(X_\infty|\mathscr{G}_t))$. We have just proved that $Z_t=\mathbf{E}(Z_\infty|\mathscr{G}_0)+\int_0^t H_s\mathrm{d}B_s$ for some $\mathscr{G}$-predictable $H\in\mathscr{L}^2(B)$. But $(B_t,\mathscr{F}_t)$ is a martingale by hypothesis, hence so is $(Z_t,\mathscr{F}_t)$ by Theorem 4.1.9, and since $H\in\mathscr{L}^2(B)$, Z is also bounded in L^2. So $Z_\infty\in L^2$ and $Z_t=\mathbf{E}(Z_\infty|\mathscr{F}_t)$. Now X_∞ is $\mathscr{G}_\infty$-measurable and $Z_t=\mathbf{E}(X_\infty|\mathscr{F}_t)$, hence $Z_\infty=X_\infty$ a.s. This proves the result. Thus we can write $X_t=X_0+\int_0^t H_s\mathrm{d}B_s$ for all $t\in\mathbf{T}$. To summarise:

4.6.3. ***Theorem:*** If $(X_t,\mathscr{F}_t)$ is a square-integrable martingale with $X_\infty\in L^2(\mathscr{G}_\infty)$, where $\mathscr{G}_t$ is the completion of $\sigma(B_s: s\leqslant t)$ for each $t\in\mathbf{T}$ and B is $BM_0(\mathbf{R})$, then there exists $H\in\mathscr{L}^2(B)$, predictable relative to $(\mathscr{G}_t)$, such that $X_t=X_0+\int_0^t H_s\mathrm{d}B_s$ for all $t\in\mathbf{T}$.

4.6.4. ***Remark:*** Results analogous to Theorems 4.6.1 and 4.6.3 hold for Poisson processes, and these have proved to be fundamental in the analysis of jump processes (see [30], [9]). Of course, the integrals involved in that case are simply Stieltjes integrals. Nonetheless, the 'martingale formulation' has produced interesting results.

4.7. An application to financial decision-making

Our representation of martingales on the natural filtration of $BM_0(\mathbf{R})$ as stochastic integrals of $BM_0(\mathbf{R})$ barely touches the surface of a general theory of martingale representation theorems which has been extensively described in [42]. This theory, though abstract in its conception, has considerable scope for application. We describe briefly one of the simplest of these, following the recent formulation by Harrison and Pliska [39] of a model for continuous trading of shares in an idealised stock market.

4.7.1: The idea is to produce a mathematical framework for *options pricing*: an investor may wish to evaluate at time $t < T$ the option of buying at time T a share of a given stock at a fixed price c. The option would obviously not be taken up if the price S'_T at time T is less than c. So he must evaluate what to pay now for the option of receiving a payment of $X = (S'_T - c)^+$ at time T. Moreover, the value of this future payment must be discounted against the option of investing instead in a riskless bond S^0, which accumulates at a fixed rate, i.e. $S^0_t = S^0_0 e^{rt}$ for some $t > 0$. Assuming that the stock S' behaves like a diffusion, Black and Scholes [6] showed that there was indeed a *unique* rational valuation for X.

This result can be formulated as a martingale representation problem: the problem is to find a *trading strategy* $\phi = (\phi_t)$ requiring an initial investment or *price* $\tau = V_0(\phi)$ which yields value $X = V_T(\phi)$ for the portfolio at time T. Here we assume that the investor is able to hold a portfolio of shares $S^0, S^1, \ldots, S^k$ (stocks and their prices are identified) and $V_t(\phi) = \sum_{k=0}^{K} \phi^k_t S^k_t$ denotes the *value* of the *portfolio* $\phi_t = (\phi^0_t, \ldots, \phi^k_t)$ held at time t. We *assume* that the component processes (ϕ^k_t) are *predictable* and *locally bounded* (relative to some fixed filtration $(\mathscr{F}_t)$ with $\mathscr{F}_0$ trivial and $\mathscr{F}_T = \mathscr{F}$ satisfying the usual conditions). We also *assume* (for now) that $S = (S^k)$ is a (vector) *semimartingale*. The *gains process* $G_t(\phi) = \sum_{k=0}^{K} \int_0^t \phi^k_t dS^k_t$ then takes account of all variations in the value of the investment over time. We demand that

$$V_t(\phi) = V_0(\phi) + G_t(\phi) \tag{4.18}$$

for all $t < T$, so that the strategy must be *self-financing* (no new investments or withdrawals are allowed in $]0, T[$).

4.7.2: Assume now that S^0 is given by a riskless bond with $S_0^0=1$, so the value and prices of stocks are discounted by $\beta_t=1/S_t^0$. The *discounted price processes* $Z_t^k=\beta_t S_t^k, k=1,\ldots,K$, the *discounted value process* $V^*(\phi)=\beta V(\phi)=\phi^0+\sum_{k=1}^K\phi^k Z^k$ are thus introduced, together with the *discounted gains process* $G^*(\phi)=\sum_{k=1}^K\int\phi^k \mathrm{d}Z^k$.

The first application of the stochastic calculus is to show that (4.18) implies (and is in fact equivalent to) the identity

$$V^*(\phi)=V_0^*(\phi)+G^*(\phi)$$

for each trading strategy ϕ: the crucial assumption made is that β is a *continuous* process of *bounded variation*, so that by Definition 4.3.13 and an extension of Remark 4.3.3, $[\beta,V(\phi)]=\sum_s\Delta\beta_s\Delta V(\phi)_s=0$, and the integration by parts formula (4.15) applied to $\beta V(\phi)$ thus reduces to the usual one. Using the continuity of β again we obtain (in differential form)

$$\mathrm{d}V^*(\phi)=\mathrm{d}(\beta V(\phi))=\beta_-\mathrm{d}V(\phi)+V_-(\phi)\mathrm{d}\beta+[\beta,V(\phi)]=\beta\mathrm{d}V(\phi)+V_-(\phi)\mathrm{d}\beta.$$

On the other hand, if $V_t(\phi)=V_0(\phi)+G_t(\phi)$, then $\Delta V(\phi)=\phi\Delta S$ and so $V_-(\phi)=\phi S_-$. Applying this to the above identity, $\mathrm{d}V^*(\phi)=\beta\phi\mathrm{d}S+\phi S_-\mathrm{d}\beta=\phi(\beta\mathrm{d}S+S_-\mathrm{d}\beta)$. Now $\mathrm{d}Z=\mathrm{d}(\beta S)=\beta\mathrm{d}S+S_-\beta+[\beta,S]$ and $[\beta,S]=0$ as above. So $\mathrm{d}V^*(\phi)=\phi\mathrm{d}Z$, which is the differential form of the required identity. The converse is also clear, so that ϕ is self-financing iff $V^*(\phi)=V_0(\phi)+G^*(\phi)$.

In particular, since $G^*(\phi)$ is a stochastic integral, martingale properties of the discounted price function will be preserved and hence carried over to the value function $V^*(\phi)$ whenever ϕ is self-financing. Harrison and Pliska now *assume the existence of a measure Q equivalent to P under which Z is a martingale.* (We shall indicate below what this 'means' in a simpler, finite model.)

With this assumption Z is certainly a P-semimartingale (if we work entirely with discounted processes the above assumption can therefore replace our original assumption that S is a semimartingale, which was made merely to introduce the gains process $G(\phi)$). In particular, $V^*(\phi)$ is now a positive local martingale under Q whenever $V(\phi)\geqslant 0$, and hence is also a supermartingale under Q. (*Exercise*: show that every positive local martingale is a supermartingale.)

4.7.3: Thus the model constructed so far will not allow the creation of 'something from nothing': if $V_0^*(\phi)=0$ we have, for all $t>0$, $0\leqslant\int_\Omega V_t^*(\phi)\mathrm{d}Q\leqslant\int_\Omega V_0^*(\phi)\mathrm{d}Q=0$, so we cannot reach $V_T^*(\phi)>0$ on any set of positive Q-measure (or P-measure). On the other hand, investors *can* commit 'suicide': specialise to a diffusion model with $r=0$, $T=K=S_0^1=S^0=1$, fix $b>0$ and set

$$\phi_t^k=\begin{cases}1+b & \text{if } k=0, \quad 0\leqslant t\leqslant \tau(b)\\ -b & \text{if } k=1, \quad 0\leqslant t\leqslant \tau(b)\\ 0 & \text{otherwise,}\end{cases}$$

where $\tau_b=\inf\{t: S_t^1=1+1/b\}=\inf\{t: V_t(\phi)=0\}$ is the 'time of ruin'. (In other words, start with £1, sell short (owe!) £b shares and buy £$(1+b)$ bonds, which you hold until τ_b or $t=1$, whichever is first. Clearly you are insolvent when $S_t^1\geqslant 1+1/b$.) The probability of ruin in $[0,1]$ is $p=P(\tau_b<1)$, which increases to 1 as $b\to\infty$. An adjustment of this strategy, starting with $b=1$ on $[0,\frac{1}{2}]$ and consecutively rearranging the portfolio on $[1-(\frac{1}{2})^n,1-(\frac{1}{2})^{n+1}]$ to produce constant ruin probabilities eliminates the probability of survival. So the model allows for (undesirable) suicide strategies.

4.7.4: To eliminate these, Harrison and Pliska assume the existence of a fixed reference measure P^* and consider as *admissible* only self-financing strategies ϕ with $V^*(\phi)\geqslant 0$ such that $V^*(\phi)$ is a *P^*-martingale.* (Of course the latter is equivalent to $\int_\Omega V_T^*(\phi)\mathrm{d}P^*=V_0^*(\phi)$.) At the same time ϕ is no longer required to be locally bounded: we can allow $\phi^k\in\mathscr{L}^2_{\text{loc}}(Z^k)$ for $k=1,\ldots,K$. Write Φ^* for the set of admissible strategies ϕ.

A positive random variable (or *claim*) $X\in L_+^0$ is then said to be *attainable* by some $\phi\in\Phi^*$ at price π if $\pi=V_0^*(\phi)$, $\beta_T X=V_T^*(\phi)$. Then the assumption imposed on the model at once yields the *uniqueness* of π: if $X=V_T^*(\phi)$ for some $\phi\in\Phi^*$, $\int_\Omega V^*(\phi)\mathrm{d}P^*=V_0^*(\phi)=\pi$ is determined. We can rephrase this as: if a claim X is attainable, the unique price associated with X is $\pi=\int_\Omega\beta_T X\mathrm{d}P^*$. (Note that we think of X as 'given in today's currency' as $\beta_T X$.) The determination of a reference measure P^* is a non-trivial problem in applications but again the theory of semimartingales provides a useful technique for locating P^* via the Girsanov theorem (see [19],[30],[42]). A discussion of these results would take us too far afield. We note simply that attainable claims are simply stochastic integrals relative to Z, if we assume X *integrable*, i.e. $\int_\Omega\beta_T X\,\mathrm{d}P^*<\infty$:

4.7.5. ***Theorem:*** If X is an integrable claim and V^* is the cadlag modification of the martingale $(\mathbf{E}^*(\beta_T X|\mathscr{F}_t))_t$ – where the conditional expectation is taken with respect to P^* – then X is attainable iff $V^*=V_0^*+\int H\,\mathrm{d}Z$ for some $G\in\mathscr{L}^2_{\text{loc}}(Z)$. In that case $V^*(\phi)=V^*$ whenever X is attainable by $\phi\in\Phi^*$.

Proof: If ϕ attains X, set $H^k = \phi^k$ for $k \geqslant 1$, then $G^*(\phi) = \int H\,dZ$ and $V_t^* = \mathbf{E}^*(\beta_T X | \mathscr{F}_t) = \mathbf{E}(V_T^*(\phi) | \mathscr{F}_t) = V_t^*(\phi)$ since V^* is a P^*-martingale. But $V_t^*(\phi) = V_0^* + G_t^*(\phi) = V_0^* + \int_0^t H\,dZ$.

Conversely, if V^* is defined as in the statement for a given $X \in L_+^0$, then $V^* = V_0^* + \int H\,dZ$ allows us to set $\phi^k = H^k$ for $k \geqslant 1$ and $\phi_t^0 = V_0^* - \sum_{k=1}^K \int_0^t \phi_s^k dZ_s^k - \sum_{k=1}^K \phi_t^k Z_t^k$. It is easy to see that $V^*(\phi) = V^*$ and since V^* is a positive martingale by its definition, we see that ϕ attains X and belongs to Φ^*.

Consequently in the Harrison–Pliska model, every integrable claim X in L_+^0 is attainable iff every P^*-martingale is a stochastic integral with respect to Z. We say the model is *complete* if this holds. In particular, if the price process $S = (S^k)_{k \geqslant 1}$ is given as a vector Brownian motion with drift vector, it is not hard to show, using the representation theorem (Theorem 4.6.3), that the corresponding model is complete. Similar results obtain for examples of Poisson type – we refer to [39] for details.

4.7.6. ***Remarks:***

(1) The assumption that a reference measure exists may seem difficult to relate to 'economic reality'. Its role is clearer if one considers a discrete-time model with *finite* sample space Ω: dealing only with measures P with $P(\{\omega\}) > 0$ for all $\omega \in \Omega$ means that our investors agree which states of the market are possible, but not necessarily on how likely each is to occur. The filtration $(\mathscr{F}_t)_{t=0,1,\ldots,T}$ again represents evolving knowledge of prices and the *portfolio* $(\phi_t^k)_{k=0,\ldots,K}$ is again predictable, i.e. fixed with knowledge of prices S_{t-1}^k at time $(t-1)$. Self-financing strategies thus satisfy

$$\phi_t S_t = \phi_{t+1} S_t$$

This simple identity leads easily to the result that the set $\mathbf{P}$ of measures equivalent to P for which βS is a martingale is in one-to-one correspondence with the set of consistent price systems Π (maps $\pi: L_+^0 \to [0,\infty[$ which are positive linear, have $\pi^{-1}(\{0\}) = \{0\}$, and satisfy $\pi(V_T(\phi)) = V_0(\phi)$ for all self-financing ϕ with positive value process). It then turns out that the possibility of turning 'nothing' ($V_0(\phi) = 0$) into 'something' ($\mathbf{E}(V_T(\phi)) > 0$) is ruled out iff $\mathbf{P} \neq \varnothing$ (equivalently $\Pi \neq \varnothing$). One way is trivial: if $\mathbf{P} \neq \varnothing$, hence $\Pi \neq \varnothing$, take $\pi \in \Pi$ and ϕ with $0 = V_0(\phi) = \pi(V_T(\phi))$, then $V_T(\phi) = 0$. So no arbitrage opportunities can exist if $\mathbf{P} \neq \varnothing$. The converse is an easy application of (finite-dimensional) separating hyperplanes. (See [39; p. 229].)

(2) The above model may not in itself contribute to martingale theory, but the analogies between the continuous-time and discrete-time models

deserve further investigation. In the discrete case, Harrison and Pliska derive a characterisation of complete models involving the fine structure (partition cells) of the filtration $(\mathscr{F}_t)$. No such refinement of martingale representation theorems is yet available in the general case. Perhaps a more detailed study of the relations between complete models and martingale representation theory will yield deeper results in the latter theory. The proper characterisation of 'convergence' of more 'realistic' models to complete continuous-time models is another mathematical problem of considerable interest generated by the above applications of martingale theory.

4.8. Stochastic integration and vector measures

In this section we outline some of the ideas in an alternative approach to stochastic integrals, as developed independently by Metivier and Pellaumail ([71],[60],[61]) and Kussmaul ([49],[50]). The recent book [62] by Metivier and Pellaumail describes many extensions of this approach, as well as applications to solutions of stochastic differential equations for Banach-space valued random variables. Our remarks are drawn principally from [50].

4.8.1: The theory begins with *Doléans function* $i_X(]\!]s_F,t_F]\!])=\mathbf{E}(1_F(X_t-X_s))$ of a process $X=(X_t)$, where $F\in\mathscr{F}_s$, $s<t$ in $\mathbf{T}$ (note that if A is an increasing process, $i_A=\mu_A$ in the notation of Definition 3.6.7) and examines the associated vector-valued set function $I_X:]\!]s_F,t_F]\!]\rightarrow 1_F(X_t-X_s)$. If the process X allows the extension of i_X and I_X to *measures* on the predictable σ-field Σ_p, we have the basis for a theory of integration. The Doob–Meyer decomposition theorem is then a result on the structure of these measures. Further, the extension of the theory to E-valued random variables S (where E is a Banach space) is more natural than a direct analysis of E-valued semimartingales. We therefore formulate the first result for E-valued processes:

4.8.2. ***Definitions:*** Let $(\Omega,\mathscr{F},P,(\mathscr{F}_t),\mathbf{T})$ be a stochastic base. A map $X:\mathbf{T}\times\Omega\rightarrow E$ is

(i) a *stochastic process* if $\omega\rightarrow X_t(\omega)$ is weakly measurable for all $t\in\mathbf{T}$ (that is, $x'\circ X_t$ is $\mathscr{F}$-measurable for all $x'\in E'$);

(ii) *measurable* if X is weakly $\mathscr{B}(\mathbf{T})\times\mathscr{F}$-measurable;

(iii) *adapted* if X_t is also weakly $\mathscr{F}_t$-measurable for all $t\in\mathbf{T}$;

(iv) *predictable* if X is weakly measurable with respect to Σ_p;

(v) *optional* if X is weakly measurable with respect to Σ_0;

(vi) *right-continuous* if $t\rightarrow X_t$ is right-continuous as a map from $\mathbf{T}$ to

$L^1_E(P)$. (It turns out that such a map has a unique *modification* $X(t,\cdot)$ with a.s. right-continuous paths, see [50].)

We consider the following two properties of an adapted process X in L^1_E:

(a) there exists $c \geqslant 0$ such that $\sum_i \|X_{t_{i+1}} - X_{t_i}\|_1 \leqslant c$ for all finite partitions (t_i) of $\mathbf{T}$;

(b) there exists $c \geqslant 0$ such that $\sum_i \|\mathbf{E}(X_{t_{i+1}} - X_{t_i} | \mathscr{F}_{t_i})\|_1 \leqslant c$ for all finite partitions (t_i) of $\mathbf{T}$.

If X satisfies (vi) and (*a*) we say X has *integrable variation.* If X satisfies (vi) and (*b*) we call X a *quasimartingale.*

When $c=0$, it is clear that X is a martingale in L^1_E. The set functions i_X, I_X can be extended by additivity to additive set functions on the ring $\mathscr{R}_p$ generated by the stochastic intervals $[\![0_F]\!]$, $]\!]S,T]\!]$, where S,T are *simple* stopping times. The role of quasimartingales is evident from

4.8.3. ***Theorem:*** Let X be an adapted right-continuous L^1_E-bounded process. Then X is a quasimartingale iff the set function i_X has bounded variation on $\mathscr{R}_p$.

Proof: Recall that the *variation* of an E-valued measure μ on a ring $\mathscr{R}$ is defined as $\mathrm{var}(\mu,A) = \sup(\sum_i \|\mu A_i\|)$, where the supremum is taken over all finite partitions (A_i) of $A \in \mathscr{R}$. This again defines an additive set function on $\mathscr{R}$. If $u \in L^1_E(P)$ is given, the E-valued measure $\mu(A) = \mathbf{E}(u1_A)$ has total variation $\mathrm{var}(\mu,\Omega) = \|u\|_1$. (See [27].) Hence if $u = \mathbf{E}((X_t - X_s)|\mathscr{F}_s)$ for fixed $s \leqslant t$ in $\mathbf{T}$, we have

$$\|\mathbf{E}((X_t - X_s)|\mathscr{F}_s)\|_1 = \mathrm{var}(\mu,\Omega) = \sup\left(\sum_i \|\mu F_i\| : F_i \in \mathscr{F}_s, \text{ disjoint}, \bigcup_i F_i = \Omega\right)$$

$$= \sup\left(\sum_i \|\mathbf{E}(1_{F_i}(X_t - X_s))\| : F_i \in \mathscr{F}_s, \text{ disjoint}, \bigcup_i F_i = \Omega\right) \tag{4.19}$$

as now $u \in L^1_E(\mathscr{F}_s)$. In particular, since for each partition (F_i)

$$\sum_i \|i_X(]\!]s_{F_i}, t_{F_i}]\!])\| = \sum_i \|\mathbf{E}(1_{F_i}(X_t - X_s))\|, \tag{4.20}$$

$\mathrm{var}(i_X,]\!]s,t]\!]) \geqslant \|\mathbf{E}((X_t - X_s)|\mathscr{F}_s)\|_1$.

Now suppose that i_X has bounded variation, so that $\mathrm{var}(i_X, \mathbf{T} \times \Omega) = c < \infty$. Since variation is an additive set function, we have for any finite partition (t_i) of $\mathbf{T}$ that $c \geqslant \sum_i \mathrm{var}(i_X,]\!]t_i, t_{i+1}]\!])$ $\geqslant \sum_i \|\mathbf{E}(X_{t_{i+1}} - X_{t_i} | \mathscr{F}_{t_i})\|_1$. Hence X is a quasimartingale.

Conversely, let X be a quasimartingale. We may assume that $X_0 = 0$. Fix $A \in \mathscr{R}_p$ and let (A_n) be a finite partition of A. Choose $t_1 \leqslant t_2 \leqslant \ldots \leqslant t_m$ in $\mathbf{T}$ and $F_{i,1}, \ldots, F_{i,k_i}$ in $\mathscr{F}_{t_i}$ $(i \leqslant m)$ such that $\{]t_i, t_{i+1}] \times F_{i,j} : j \leqslant k_i, i \leqslant m\}$ *refines* the partition (A_n). Then we have

$$\sum_n \|i_X(A_n)\| \leqslant \sum_{i,j} \|i_X(]t_i,t_{i+1}] \times F_{i,j})\| = \sum_{i,j} \|\mathbf{E}(1_{F_{i,j}}(X_{t_{i+1}} - X_{t_i}))\|$$
$$\leqslant \sum_i \|\mathbf{E}(X_{t_{i+1}} - X_{t_i}|\mathscr{F}_{t_i})\|_1,$$

where we have applied (4.20) to each i separately and then used (4.19). Since X is a quasimartingale, the final quantity is bounded by some $c \geqslant 0$ for all (t_i), hence $\sum_n \|i_X(A_n)\| < c$ for all choices of partition (A_n). Hence i_X is of bounded variation on $\mathscr{R}_p$.

4.8.4. ***Corollary:*** X is a martingale if i_X is carried by $[\![0]\!]$.

Proof: We have $i_X(]\!]s_F,t_F]\!]) = \mathbf{E}(1_F(X_t - X_s)) = 0$ for all $F \in \mathscr{F}_s$, $s \leqslant t$ in $\mathbf{T}$, iff X is a martingale. Hence $i_X = 0$ outside sets of the form $[\![0_F]\!]$, $F \in \mathscr{F}_0$.

This result explains why martingales will be 'trivial' for stochastic integration, allowing i_X to have 'mass' only at 0. In the real case Pellanmail uses this to deduce structure theorems such as the Riesz and Doob–Meyer decompositions [71]. (On the other hand, it should be noted that quasimartingales are semimartingales. In fact, it can be shown (see [42]) that any *local* quasimartingale is a *special* semimartingale (see Remark 4.4.12) and conversely.)

4.8.5. ***Examples:*** (with $E = \mathbf{R}$):

(1) If $M, N \in \mathscr{M}^2$, their product MN is a real quasimartingale: for if $\sup_t \|M_t\|_2 = K_1$, $\sup_t \|N_t\| = K_2$, then for $t_1 < t_2 < \ldots < t_{n+1}$

$$\sum_{i=1}^n \|\mathbf{E}(M_{t_{i+1}} N_{t_{i+1}} | \mathscr{F}_{t_i}) - M_{t_i} N_{t_i}\|_1$$
$$= \sum_{i=1}^n \|\mathbf{E}((M_{t_{i+1}} - M_{t_i})(N_{t_{i+1}} - N_{t_i}) | \mathscr{F}_{t_i})\|_1$$
$$\leqslant \sum_{i=1}^n \|(M_{t_{i+1}} - M_{t_i})(N_{t_{i+1}} - N_{t_i})\|_1$$
$$\leqslant \left(\sum_{i=1}^n \|M_{t_{i+1}} - M_{t_i}\|_2^2\right)^{1/2} \left(\sum_{i=1}^n \|N_{t_{i+1}} - N_{t_i}\|_2^2\right)^{1/2},$$

where the last inequality follows by first applying the Hölder inequality to each term in the sum and then using the Schwarz inequality on finite sums. Now since M and N are martingales, $\|M_{t_{i+1}} - M_{t_i}\|_2^2 = \mathbf{E}((M_{t_{i+1}} - M_{t_i})^2) = \mathbf{E}(M_{t_{i+1}}^2 - M_{t_i}^2)$ for $i \leqslant n$, and similarly for N. Hence the final sums in the above inequalities reduce to $||M_{t_{n+1}}-M_{t_1}||_2||N_{t_{n+1}}-N_{t_1}||_2 \leqslant 4K_1K_2$. This shows that MN is a quasimartingale, belonging to class (D), since $M^*,N^* \in L^2$ implies $M^*N^* \in L^1$.

(2) If X is a right-continuous process with independent increments and natural filtration $(\mathscr{F}_t)$, we have $\mathbf{E}(X_t|\mathscr{F}_s)=\mathbf{E}(X_t-X_s)+X_s$ for $s \leqslant t$ in $\mathbf{T}$. (*Exercise*!) Hence X is a quasimartingale iff $t \rightarrow \mathbf{E}(X_t)$ is of bounded variation. In that case, set $A_t=\mathbf{E}(X_t)$. So $A=(A_t)$ is a deterministic, hence certainly predictable, bounded variation process and $X-A$ is a martingale: $\mathbf{E}(X_t-A_t|\mathscr{F}_s)=\mathbf{E}(X_t-X_s)+X_s-\mathbf{E}(X_t)=X_s-A_s$. So this quasimartingale can be decomposed into the sum of a martingale and a predictable process of bounded variation. Such a decomposition is valid for class (D) quasimartingales in the real case and under mild path regularity assumptions in the E-valued case, if E has the Radon–Nikodym property. See [50].

4.8.6: The standard recipe for integration can now be used to extend the integral operator I_X from simple $\mathscr{R}_p$-measurable processes, i.e. processes Z of the form $Z=\sum_i \beta_i 1_{]\!]S_i,T_i]\!]}+\beta_0 1_F 1_{[\![0]\!]}$, where $F \in \mathscr{F}_0$, $S_i \leqslant T_i$ are *simple* stopping times and $\beta_i \in \mathbf{R}$, to more general processes. For Z as above, $I_X(Z)=\sum_i \beta_i(X_{T_i}-X_{S_i})+\beta_0 1_F X_0$. The class $\mathscr{S}$ of these simple processes is closed under multiplication by indicator processes $1_{[\![0,t]\!]}$, so we can also define the *process* Y by $Y_t:=I_X(1_{[\![0,t]\!]}Z)$ directly.

Now consider a subspace $\mathscr{U}$ of the predictable processes endowed with a topology such that

(i) $\mathscr{S}$ is dense in $\mathscr{U}$ and $Z \in \mathscr{U}$ implies $1_{[\![0,t]\!]}Z \in \mathscr{U}$;

(ii) I_X is a continuous map on $\mathscr{S}$;

(iii) I_X, extended to $\mathscr{U}$, allows a unique modification for each map $t \rightarrow I_X(1_{[\![0,t]\!]}Z)$, for $Z \in \mathscr{U}$. (This condition is necessary for the definition of the process Y when Z no longer belongs to $\mathscr{S}$. The usual rule is to restrict attention to cadlag processes.)

If we want to ensure that the sup-norm closure of $\mathscr{S}$ is contained in our set $\mathscr{U}$ of permissible integrands, we must require that the set function I_X has bounded *semivariation*: the semivariation svar(μ,A) of an E-valued bounded set function μ defined on a ring of sets $\mathscr{R}$, with $A \in \mathscr{R}$, is defined by svar$(\mu,A)=\sup||\sum_i \alpha_i \mu A_i||$, where (A_i) ranges over finite partitions of A into sets of $\mathscr{R}$ and $|\alpha_i| \leqslant 1$ for all i. (In the case $E=\mathbf{R}$ we can use $\alpha_i=\pm 1$ to show that var$(\mu,A)=$svar(μ,A), but in general $0 \leqslant$svar$(\mu,A) \leqslant$var(μ,A).) The set function svar(μ,A) need not be additive but is always bounded by $2\sup_{C \in \mathscr{A}}||\mu C||$. (Why?) Moreover, if $\bar{\mathscr{S}}$ denotes the sup-norm closure of $\mathscr{S}$, then the integral operator I_X, extended to $\bar{\mathscr{S}}$, has norm equal to svar(I_X,$\mathbf{T}\times\Omega$) (see [49],[27]). If $E=\mathbf{R}$, svar(I_X,$\mathbf{T}\times\Omega$) will be bounded in $L^1(P)$ iff svar(i_X,$\mathbf{T}\times\Omega$)$=$var(i_X,$\mathbf{T}\times\Omega$) is bounded in $\mathbf{R}$, since $i_X=\mathbf{E}\circ I_X$. But this implies that X is a quasimartingale, by Theorem 4.8.3. This shows that a sensible integration theory cannot be built using I_X unless X is a quasimartingale.

4.8.7: This raises the question: which quasimartingales yield a set function I_X with bounded semivariation? A general answer is given in [62] for Hilbert-space valued processes. We shall restrict attention to the case $E=\mathbf{R}$, where there is a neat solution to this problem. (This result is taken from [50] and is an improvement of [49; Th. 12.13].)

4.8.8. ***Theorem:*** Suppose X is an adapted, real, cadlag process. The following are equivalent:

(i) I_X has bounded semivariation (call X *summable* in this case);
(ii) I_X has a unique extension to a σ-additive measure on Σ_p;
(iii) X is a quasimartingale with $X^*=\sup_t|X_t|\in L^1(P)$;
(iv) XF is a quasimartingale whenever $F=(F_t)$ is a cadlag modification of the martingale $(\mathbf{E}(f|\mathscr{F}_t))_t$ for some $f\in L^\infty(P)$.

Sketch proof: Because $L^1(P)$ is weakly sequentially complete any weakly σ-additive set function $\mu:\mathscr{R}_p\to L^1(P)$ with bounded range has a unique extension to a (norm-)σ-additive measure on Σ_p. (See [49; Prop. 10.10].) Moreover, it is clear that svar(μ,Ω) is bounded iff μ has bounded range, by the remarks preceding 4.8.7. So to prove that (ii)$\Rightarrow$(i) we need only note that a σ-additive set function defined on a σ-algebra has relative weakly compact, hence norm-bounded, range. The more suprising result is (i)$\Rightarrow$(ii), which is proved via (i)$\Rightarrow$(iii)$\Rightarrow$(ii). Before indicating the ideas of this proof, which takes up much of Chapters 9–12 of [49], we note that (i)$\Leftrightarrow$(iv): regarding $f\in L^\infty$ as a linear functional on L^1, and taking simple stopping times $S\leqslant T$, we have

$$(f\circ I_X)(]\!]S,T]\!])=f(X_T-X_S)=\mathbf{E}(f(X_T-X_S))$$

$$=\mathbf{E}(\mathbf{E}(f|\mathscr{F}_T)X_T-\mathbf{E}(f|\mathscr{F}_S)X_S)=i_{XF}(]\!]S,T]\!]),$$

where F is defined as in (iv). Further, for $A\in\mathscr{F}_0$,

$$(f\circ I_X)([\![0_A]\!])=\mathbf{E}(1_A(fX_0))=\mathbf{E}(1_A(XF)_0)=i_{XF}([\![0_A]\!]).$$

Hence the measures $f\circ I_X$ and i_{XF} coincide on $\mathscr{R}_p$.

Now I_X has bounded semivariation iff I_X is (weakly) bounded iff $f\circ I_X$ is bounded for all $f \in L^\infty$ iff i_{XF} is bounded for all $f \in L^\infty$. By Theorem 4.8.3 the final statement holds iff (XF) is a quasimartingale.

For the implications (i)⇒(iii)⇒(ii) we refer primarily to [49]. The implication (iii)⇒(ii) is immediate if we note that (iii) ensures that X belongs to class (D), which is in turn equivalent to the σ-additivity of i_X, hence implies the weak σ-additivity and boundedness of I_X. To prove (i)⇒(iii) we decompose X, which (using (iv) with $f \equiv 1$) is a quasimartingale, into $X = M + V$, where M is a local martingale and V is a process of integrable variation. Then I_M is bounded, since I_V is clearly bounded and I_X is bounded by hypothesis. Instead of X^* we can consider $M^* = \sup_t |X_t| \in L^1$, since $V^* \in L^1$ trivially. M has a cadlag modification so X_S, X_T are well-defined for arbitrary bounded stopping times, and we extend I_M to the ring $\bar{\mathscr{R}}_p$ so generated: $I_M([\![0_F]\!]) = 1_F X_0$, $I_M(]\!]S,T]\!]) = X_T - X_S$. It can be checked that the extension has bounded semivariation. Find an increasing sequence of stopping times (S_n) such that the stopped processes M^{S_n} have $(M^{S_n})^* \in L^1$ for all n. Then $\|(M^{S_n})^*\|_1 \leqslant 40 \mathrm{svar}(I_M, [\![0, S_n]\!])$ by [49; 12.12] so that $\sup_t |M_t| = \sup_n (M^{S_n})^* \in L^1$ by monotone convergence.

The inequality $\|(M^{S_n})^*\|_1 \leqslant 40 \mathrm{svar}(I_M, [\![0, S_n]\!])$ is by no means trivial. It implies the equivalence of the semivariation norm on the space of summable martingales and the space $\mathscr{H}^1$-norm on the space of martingales with $M^* \in L^1(P)$ – see [49; 12.12 Corollary]. Various equivalent proofs of this result exist, see [66], [19].

4.8.9: The *stochastic integral operator* I_X is the extension of I_X to Σ_p. Define $L^1(I_X)$ as the space of I_X-integrable predictable processes. (This means that $L^1(I_X)$ is the completion in the semivariation norm of simple predictable processes, see [49], [50]. Fix $Z \in L^1(I_X)$ and let $m_Z(A) := I_X(1_A Z)$ for $A \in \Sigma_p$. We see that $\|Z\| = \mathrm{svar}(m_Z, \mathbf{T} \times \Omega)$.) Since m_Z is a bounded measure, $t \to I_X(1_{[\![0,t]\!]} Z)$ defines a quasimartingale for each $Z \in L^1(I_X)$. This process has a unique cadlag modification $Y = (Y_t)$ with bounded associated measure I_Y. We write $Y_t = \int_0^t Z_s \mathrm{d} X_s$ for this *stochastic integral.*

Thus Y is the unique cadlag quasimartingale for which $f \circ I_Y$ and $A \to \int_A Z \mathrm{d}(f \circ I_X)$ agree on Σ_p for each $f \in L^\infty(P)$. (The proof of this is left to the reader.) Now if $f \in L^\infty(P)$ and (F_t) is a cadlag modification of $(\mathbf{E}(f|\mathscr{F}_t))$, then $f \circ I_X$ extends i_{XF} uniquely to Σ_p (by Theorem 4.8.8) and i_{XF} has extension $f \circ I_Y : A \to \int_A Z \mathrm{d}(f \circ I_X)$. In particular, if X is a martingale with $X_0 = 0$, then $i_X \equiv 0$, so $f \circ I_X \equiv 0$ for $f \equiv 1$. Hence $i_Y : A \to \int_A Z \mathrm{d}(1 \circ I_X)$ is the zero measure so that Y is a martingale. We have now shown that the stochastic integral of a martingale is again a martingale.

4.8.10: To see how the above theory relates to that developed in sections 4.1–4.4 we must quote the extension of the Doob–Meyer decomposition to quasimartingales: that each class (D) quasimartingale can be written as $Q = M + V$, where M is a uniformly integrable martingale with $M_0 = 0$ and V a cadlag predictable process of integrable variation whose associated measure μ_V is, on Σ_p, the extension of i_Q. (For a proof, see [49]. This result is no surprise if we note that each quasimartingale is the difference of two supermartingales, as may be seen by considering the positive and negative parts of i_Q.) Hence if a product XY of two cadlag processes is a quasimartingale of class (D), we can write $V = \langle X, Y \rangle$. In particular, if $X = Y \in \mathcal{M}^2$, $V = \langle X \rangle$ in the notation of section 4.2.

4.8.11. ***Theorem:*** If X is a summable process and F is a (cadlag) bounded martingale, then XF is a quasimartingale in class (D) and for all $Z \in L^1(I_X)$ the stochastic integral $Y = \int Z\,dX$ is characterised as a quasimartingale by the following equivalent conditions:

(i) $\langle Y, F \rangle_t = \int_0^t Z_s d\langle X, F \rangle_s$ for all $t \in \mathbf{T}$ and bounded cadlag martingales F;

(ii) $\mathbf{E}(Y_t F_t) = \mathbf{E}(\int_0^t Z_s d\langle X, F \rangle_s$ for all $t \in \mathbf{T}$ and bounded cadlag martingales F.

This result shows that the stochastic integral defined through the operator I_X coincides with that defined earlier, using the quadratic variation processes. We omit the simple proof. It should now be clear how the vector measure approach yields the same Itô-calculus as before. The first step is to extend the theory from quasimartingales to semimartingales by localisation. The principal lemma here is to show that processes of *bounded* variation can be approximated by *integrable* variation processes: the paths of a cadlag adapted process X have bounded variation in each $[0,t]$ iff there exist stopping times (T_n) increasing a.s. to $+\infty$ such that $1_{[\![0,T_n[\![}X$ has integrable variation for each n. (This is valid for E-valued processes.) Then a predictable process Z is said to be *locally X-integrable* if for some $(T_n)\uparrow\infty$, $X^{(n)} = 1_{[\![0,T_n[\![}X$ is summable and $Z \in L^1(I_X(n))$ for all n. Again one can show that all locally bounded predictable processes are locally X-integrable (cf. Section 4.4) for any real semimartingale X. In [50], Kussmaul extends many of the above results to E-valued semimartingales when E has the Radon–Nikodym property, and proves an Itô formula for Hilbert-space valued semimartingales. The interested reader is also referred to [60], [62], where many further references may be found on applications of the above ideas.

Appendix

A non-commutative extension of stochastic integration

by C. Barnett and I. F. Wilde

List of sections

A.0. Preliminary remarks
A.1. Introduction
A.2. The commutative case
A.3. The non-commutative case
A.4. Some results for non-commutative processes
A.5. Non-commutative stochastic integration on $[0,\alpha]$
A.6. A special case
A.7. The Itô-Clifford integral
A.8. Concluding remarks
A.9. References

A.0. Preliminary remarks

In this appendix we shall present a non-commutative extension of stochastic integration. Unfortunately the non-commutative context involves the technical difficulties associated with unbounded operators on Hilbert space and so a treatment that is at once full and brief is not possible. Moreover, the ideas involved in non-commutative processes are not made that much clearer by a rigorous treatment of unbounded operators. Hence we present a brief, but in parts incomplete, exposition of some of the elementary theory. We shall assume that the reader is familiar with a little of the theory of von Neumann algebras, and Chapter 2 and sections 0.2, 4.1, 4.2. We shall divide what follows into sections and give references and comments for all sections in the final remarks.

A.1. Introduction

We can think of a stochastic process as a family, (X_α), $\alpha \in \mathbf{R}^+$, of random variables on a probability space $(\Omega,\mathscr{F},P)$. Often it is assumed that

$(X_\alpha)\subseteq L^1(\Omega,\mathscr{F},P)$, e.g. (X_α) is an L^1-martingale. We might say that (X_α) is a (stochastic) process in the *context* of $L^1(\Omega,\mathscr{F},P)$. There is a way of viewing $L^1(\Omega,\mathscr{F},P)$ that leads to the idea of a process, in a different context. For reasons that will become clear, we refer to processes in the context of $L^1(\Omega,\mathscr{F},P)$ as 'commutative', and to this context as 'the commutative case'. The alternative context, hinted at above, is referred to as 'the non-commutative case' and processes in this context are called 'non-commutative'.

A.2. The commutative case

Let $(\Omega,\mathscr{F},P)$ be a probability space. For $1\leqslant p\leqslant\infty$ let $L^p(\Omega,\mathscr{F},P)$ denote the Lebesgue spaces of (equivalence classes of) **C**-valued functions on Ω. If $x\in L^\infty(\Omega,\mathscr{F},P)$ then the map $T_x: L^2(\Omega,\mathscr{F},P)\to L^2(\Omega,\mathscr{F},P)$ given by $y\to xy$, $y\in L^2(\Omega,\mathscr{F},P)$ (defined by representatives) is a bounded linear operator. It is clear that $L^\infty(\Omega,\mathscr{F},P)$ forms an algebra with the obvious operations and it is not difficult to verify that $x\to T_x$ is an isometric homomorphism of $L^\infty(\Omega,\mathscr{F},P)$ into $\mathscr{B}(L^2(\Omega,\mathscr{F},P))$, the bounded operators on $L^2(\Omega,\mathscr{F},P)$. Now $L^\infty(\Omega,\mathscr{F},P)$ is a *commutative* finite von Neumann algebra. It has a unique predual. Since $L^\infty(\Omega,\mathscr{F},P)$ is the dual of $L^1(\Omega,\mathscr{F},P)$ we can identify $L^1(\Omega,\mathscr{F},P)$ and the predual of $L^\infty(\Omega,\mathscr{F},P)$. As $L^\infty(\Omega,\mathscr{F},P)$ is a finite von Neumann algebra its predual can be identified concretely as a linear space of (possibly unbounded) densely defined linear operators affiliated to $L^2(\Omega,\mathscr{F},P)$ (see $[S_1]$). This gives us a new way of viewing a process, namely, as a family of operators each affiliated to a von Neumann algebra of operators acting on some Hilbert space. The non-commutative case is arrived at by extracting the essential features of the viewpoint outlined above. It is worth noting at this stage that the idea of path of a process and measurability with respect to various σ-fields of $\mathbf{R}^+\times\Omega$ are lost in the transition from the commutative to the non-commutative case. The new context relies on the Banach space structure that is present. Despite this many results can make the transition to the new context.

A.3. The non-commutative case

Let a be a (finite) von Neumann algebra of operators acting on a Hilbert space $\mathscr{H}$. Let ϕ be a faithful normal finite trace on a with $\phi(I)=1$. For $1\leqslant p\leqslant\infty$ let $L^p(a,\phi)\equiv L^p(a)$, denote the non-commutative analogues of the Lebesgue spaces associated with (a,ϕ). These are called non-commutative L^p spaces (see $[Y]$). We have $L^\infty(a)=a$ and for $1\leqslant p<\infty$ that $L^p(a)$ is a Banach space (with the appropriate operations) of operators on $\mathscr{H}$ under the norm given by $x\in L^p(a)$, $\|x\|_p=\phi(|x|^p)^{1/p}$, where $|x|=(x^*x)^{1/2}$. We note that $L^p(a) \subseteq L^1(a)$, $1 \leqslant p \leqslant \infty$ and that $L^p(a)$ may be regarded as the completion of a in $\|\cdot\|_p$.

By analogy with the idea of a stochastic base we consider the following system. Let (a_α), $\alpha \in \mathbf{R}^+$, denote a family of von Neumann subalgebras of a satisfying

A.3.1:

(i) a_s is a von Neumann subalgebra of a_t whenever $s \leqslant t$;
(ii) $\bigcup_\alpha a_\alpha$ is ultraweakly dense in a (i.e. the a_α's generate a);
(iii) $\bigcup_{s<t} a_s$ is ultraweakly dense in a_t (left continuity);
(iv) $\bigcap_{s>t} a_s = a_t$ (right continuity).

We call this a *filtration* or nest of algebras; it is a (restricted) analogue of the family of σ-fields encountered in the commutative case. Condition (iii) is not used in the commutative case in the development of the main text. It is not essential for all of what follows, either: the integral of section A.5 can in fact be constructed without this assumption. The conditions (i)–(iv) imply, for $1 \leqslant p < \infty$:

A.3.2:

(i)′ $L^p(a_s)$ is a closed subspace of $L^p(a_t)$ for $s \leqslant t$;
(ii)′ $L^p(a) = \overline{\bigcup_\alpha L^p(a_\alpha)}^{\|\ \|_p}$;
(iii)′ $L^p(a_t) = \overline{\bigcup_{s<t} L^p(a_s)}^{\|\ \|_p}$;
(iv)′ $L^p(a_t) = \bigcap_{s>t} L^p(a_s)$.

The Radon–Nikodym theorem has its counterpart in the non-commutative context. For $x \in L^1(a)$, $\alpha \geqslant 0$ and $B \in a_\alpha$ there is a unique $M_\alpha(x) \in L^1(a_\alpha)$ such that $\phi(xB) = \phi(M_\alpha(x)B)$. The map $M_\alpha(\cdot)$ is called the *conditional expectation* of $L(a)$ onto $L(a_\alpha)$. It shares all the features of the conditional expectation defined in section 2.1, i.e. for $1 \leqslant p \leqslant \infty$ it is a positivity preserving $\|\ \|_p$ contractive linear idempotent map with $M_\alpha(I) = I$. It follows that $M_s \circ M_t = M_s$ if $s \leqslant t$. We can now say what we mean by a non-commutative process. A family $(X_\alpha) \subseteq L(a)$ is a *process* if $X_\alpha \in L^1(a_\alpha)$, $\alpha \in \mathbf{R}^+$ (i.e. the family is *adapted* to the filtration (a_α)). If (X_α) is a process and $(X_\alpha) \subseteq L^p(a)$ then (X_α) is called an L^p-process. We can now define martingales and related processes.

A.3.3. *Definition:* Let $(X_\alpha) \subseteq L^1(a)$ be a process. Then (X_α) is

(i) a *martingale* if $M_s(X_t) = X_s$ for $s \leqslant t$;
(ii) a *supermartingale* if $X_\alpha = X_\alpha^*$, $\alpha \in \mathbf{R}^+$ and $M_s(X_t) \leqslant X_s$ for $s \leqslant t$. The inequality is meant in the operator sense;
(iii) a *submartingale* if $X_\alpha = X_\alpha^*$, $\alpha \in \mathbf{R}^+$ and $M_s(X_t) \geqslant X_s$ for $s \leqslant t$;

(iv) a *potential* if $X_\alpha > 0$, $\alpha \in \mathbf{R}^+$ and (X_α) is a supermartingale with $\phi(X_\alpha) \to 0$ as $x \to \infty$;

(v) a *positive increasing process* if $X_0 = 0 \leqslant X_s \leqslant X_t$ for $0 \leqslant s \leqslant t$.

The commutative and non-commutative contexts meet when the algebra a is commutative. In this case a is isometrically isomorphic to $L^\infty(\Omega,\mathscr{F},P)$ for appropriate choice of $\Omega,\mathscr{F}$ and P. Under this identification $L^p(a) \simeq L^p(\Omega,\mathscr{F},P)$ and we recover the commutative case. Given our remarks in section A.2 we can see now that the terms commutative and non-commutative can serve to distinguish two viewpoints; the first sees a process as a collection of paths, the second as a curve in the appropriate Banach space. Consequently our construction of the stochastic integral in section A.5 is similar to that described in section 2.9 for martingale transforms, which does not involve a concept of 'paths'.

A.4. Some results for non-commutative processes

We offer this section to give some idea of what kind of results carry over to the non-commutative case. The notation is that of Section A.3. In place of proofs we just indicate how the results are arrived at.

A.4.1. ***Theorem*** (martingale convergence): Let (X_α) be an L^p-martingale.

(i) If $1 < p < \infty$ then there is $X \in L^p(a)$ such that $X_\alpha = M_\alpha(X)$ and $\|\cdot\|_p - \lim_{\alpha\to\infty} X_\alpha = X$ if and only if (X_α) is a weakly relatively compact subset of $L^1(a)$.

(ii) If $p = \infty$ and $X_\alpha = M_\alpha(X)$ for $X \in L^\infty(a)$ then $M_\alpha(X) \to X$ ultrastrongly as $\alpha \to \infty$.

Proof: (i) One can use the proof of Theorem 2.6.13 with only minor changes.

(ii) Since $L^\infty(a) \subseteq L^2(a)$, (X_α) is L^2-bounded and hence weakly relatively compact (for $L^2(a)$ is reflexive). One need only show that $\|\ \|_2$ convergence gives strong operator convergence.

The weak relative compactness is not a new feature. In the commutative case it appears as uniform integrability, see Theorem 1.2.11. As in the commutative case there are similar convergence theorems for reverse martingales.

A.4.2. ***Theorem*** (Riesz decomposition): Let (X_α) be an L^1-submartingale. Suppose that $\sup_{\alpha\in\mathbf{R}^+} \|X_\alpha\|_1 < \infty$. Then (X_α) may be decomposed, $X_\alpha = U_\alpha - V_\alpha$, $\alpha \in \mathbf{R}^+$, where (U_α) is an L^1-bounded martingale and (V_α) is potential.

Proof: Just as in the commutative case. (See Definition 2.9.3, Theorem 3.7.2 and Remark 3.7.3.)

A.4.3. ***Corollary:*** If (X_α) is weakly relatively compact in $L^1(a)$ then there is $X \in L^1(a)$ such that $X = \|\cdot\|_1 - \lim_{\alpha \to \infty} X_\alpha$.

Proof: $X_\alpha = U_\alpha - V_\alpha$ and $\|V_\alpha\| = \phi(V_\alpha) \to 0$, as $\alpha \to \infty$, thus (V_α) is a weakly relatively compact subset of $L^1(a)$. It follows that $\{X_\alpha + V_\alpha : \alpha \in \mathbf{R}\}$ is weakly relatively compact and hence (U_α) converges, thus (X_α) converges too.

A.4.4. ***Theorem:*** With the assumptions (i), (ii), (iii), (iv) of section A.3 any L^p-martingale $(1 \leqslant p < \infty)$ is continuous as a map $\mathbf{R}^+ \to L^p(a)$.

Proof: This follows from the ascending and descending martingale convergence theorems.

The Doob–Meyer decomposition is an important feature of the commutative theory. One attempt to define a suitable notion of class (D) for non-commutative processes proposed the following (see [CB_2]):

A.4.5. ***Definition:*** Let $(X_\alpha) \subseteq L^1(a)$ be a process. We say that (X_α) is of class (D) if the set

$$S(X_\alpha) = \left\{ \sum_{i=1}^{n} M_{t_{i-1}}(X_{t_i} - M_{t_{i-1}}) : n \in \mathbf{N},\ 0 \leqslant t_0 < t_1 < \ldots < t_n < \infty \right\}$$

is weakly relatively compact in $L^1(a)$. (Note the similarity with the definition of a quasimartingale. Definition 4.8.2.)

This idea of class (D) is derived directly from the commutative case (see section 3.6) and we can establish the existence of a decomposition for a class (D) potential by recasting the proof of Theorem 3.7.1 (due to K. M. Rao).

A.4.6. ***Lemma:*** Let (X_α) be a class (D) potential. Then the set

$$S_\infty(X_\alpha) = \left\{ \sum_{i=1}^{\infty} M_{(i-1)/2^n}(X_{i/2^n} - X_{(i-1)/2^n}) : n \in \mathbf{N} \right\}$$

is weakly relatively compact.

Proof: The norm closure of $S(X_\alpha)$ contains the set $S_\infty(X_\alpha)$, and weak closures contain norm closures.

A.4.7. ***Theorem:*** Let (X) be a class (D) potential which is also right-continuous as a map from $\mathbf{R}^+$ into $L^1(a)$. Then (X_α) may be decomposed as $X_\alpha = U_\alpha - A_\alpha$, where (U_α) is a martingale and (A_α) a positive increasing process.

Proof: $S_\infty(X_\alpha)$ is weakly relatively compact. By the Eberlein–Smulian theorem there is a subsequence converging weakly to some $A \in L^1(a)$. Since $S_\infty(X_\alpha)$ consists of positive operators, A is positive too. Let $U_\alpha = M_\alpha(A)$, and $A_\alpha = U_\alpha - X_\alpha$. (See also Theorem 3.7.1.)

A.5. Non-commutative stochastic integration on $[0,\alpha[$

A.5.1. ***Definition:***

(i) Let g be a process. We say that g is *elementary* if $g(s) = h1_{[r,t[}(s)$, where $h \in L^1(a_r)$ and (for our purposes) $0 \leqslant r \leqslant t \leqslant \alpha$. A process is *simple* if on $[0,\alpha[$ it is a finite linear combination of elementary process which, for simplicity, we assume to have disjoint supports.

(ii) Let $X = (X_\alpha)$ be an L^2-martingale and g an elementary *a-valued* process. We define the right stochastic integral of g with respect to X to the

$$\int_0^\alpha g(s)\mathrm{d}X_s = h(X_t - X_r),$$

where $g(s) = h1_{[r,t[}(s)$, and the left stochastic integral of g with respect to X as

$$\int_0^\alpha \mathrm{d}X_s g(s) = (X_t - X_r)h.$$

The right and left integrals of a-valued simple processes are defined by linearity using the above. The integrals are independent of the representation of g.

A.5.2. ***Theorem*** (Contraction property): Let $g(s)$ be a simple a-valued process on $[0,\alpha[$. Let (X_α) be an L^2-martingale.

(i) $\|\int_0^\alpha g(s)\mathrm{d}X_s\|_2^2 \leqslant \int_0^\alpha \|g(s)\|_\infty^2 \mathrm{d}\langle X\rangle_s$, where $\mathrm{d}\langle X\rangle_s$ denotes the Stieltjes measure obtained from the increasing function $s \mapsto \|X_s\|_2^2$.

(ii) $\|\int_0^\alpha \mathrm{d}X_s g(s)\|_2^2 \leqslant \int_0^\alpha \|g(s)\|_\infty^2 \mathrm{d}\langle X\rangle_s$.

Proof: (i) We can suppose $g(s) = \sum_{i=1}^n h_{t_{i-1}} 1_{[t_{i-1},t_i[}(s)$, where $h_{t_{i-1}} \in L^\infty(a_{t_{i-1}})$ and $0 = t_0 < t_1 < \ldots < t_n = \alpha$. By definition $\int_0^\alpha g(s)\mathrm{d}X_s = \sum_{i=1}^n h_{t_{i-1}}(X_{t_i} - X_{t_{i-1}})$ and so

$$\left\|\int_0^\alpha g(s)\mathrm{d}X_s\right\|_2^2 = \phi\left(\left(\sum_{i=1}^n h_{t_{i-1}}\Delta X_{t_i}\right)^* \left(\sum_{j=1}^n h_{t_{j-1}}\Delta X_{t_j}\right)\right) = \sum_i (\Delta X_{t_i}^* |h_{t_{i-1}}|^2 \Delta X_{t_i}),$$

where $\Delta X_{t_i} = X_{t_i} - X_{t_{i=1}}$, by taking expectations, which reduces $i \neq j$ terms to zero, as in 2.9.5. Now $0 \leqslant |h_{t_{i-1}}|^2 \leqslant \| |h_{t_{i-1}}|^2 \|_\infty I = \|h_{t_{i-1}}\|_\infty^2 I$ since $\|\cdot\|_\infty$ is a C^*-norm. In addition ϕ is positive and so

$$\left\| \int_0^\alpha g(s) \mathrm{d}X_s \right\|_2^2 \leqslant \sum_{i=1}^n \|h_{t_{i-1}}\|_\infty^2 \phi(\Delta X_{t_i}^* \Delta X_{t_i})$$

$$= \sum_{i=1}^n \|h_{t_{i-1}}\|_\infty^2 \phi(X_{t_i}^* X_{t_i} - X_{t_{i-1}}^* X_{t_{i-1}})$$

using the martingale property.

Now $s \mapsto \phi(X_s^* X_s)$ is a continuous increasing function on $[0,\alpha[$ and therefore defines a Lebesgue–Stieltjes measure on the complete σ-field generated by finite unions of intervals. We call this measure $\mathrm{d}\langle X \rangle$. It is now clear that $\|\int_0^\alpha g(s) \mathrm{d}X_s\|_2^2 \leqslant \int_0^\alpha \|g(s)\|_\infty^2 \mathrm{d}\langle X \rangle_s$.

We see that (ii) follows from (i) because

$$\left\| \int_0^\alpha \mathrm{d}X_s g(x) \right\|_2^2 = \left\| \left(\int_0^\alpha \mathrm{d}X_s g(s) \right)^* \right\|_2^2 = \left\| \int_0^\alpha g(s)^* \mathrm{d}X_s^* \right\|_2^2$$

$$\leqslant \int_0^\alpha \|g(s)^*\|_\infty^2 \mathrm{d}\langle X \rangle_s = \int_0^\alpha \|g(s)\|_\infty^2 \mathrm{d}\langle X \rangle_s$$

because ϕ is a trace and $\| \ \|_\infty$ is a C^*-norm.

A.5.3. ***Theorem:*** For a simple process on $\mathbf{R}^+$ and an L^2-martingale (X_α), the right and left integrals, considered as a function of α (the upper limit), are martingales with zero trace.

Proof:

$$M_{\alpha_1} \int_0^{\alpha_2} g(s) \mathrm{d}X_s = M_{\alpha_1} \left(\sum_{i=1}^n h_{t_{i-1}} \Delta X_{t_i} \right)$$

so if $\alpha_1 \leqslant \alpha_2$ and $t_{j-1} = \alpha_1 < t_j$

$$= \sum_{i=1}^{j-2} h_{t_{i-1}} \Delta X_{t_i} + M_{\alpha_i} \left(\sum_{i=j-1}^n h_{t_{i-1}} \Delta X_{t_i} \right)$$

$$= \sum_{i=1}^{j-2} h_{t_{i-1}} \Delta X_{t_i} + \sum_{i=j-1}^n M_{\alpha_1} \circ M_{t_{i-1}} (h_{t_{i-1}} \Delta X_{t_i})$$

(because $M_{\alpha_1} \circ M_{t_{i-1}} = M_{\alpha_1}$)

$$= \sum_{i=1}^{j-2} h_{t_{i-1}} \Delta X_{t_i} + \sum_{i=j-1}^{n} M_{\alpha_1}(h_{t_{i-1}} M_{t_{i-1}}(X_{t_i} - X_{t_{i-1}}))$$

$$= \sum_{i=1}^{j-2} h_{t_{i-1}} \Delta X_{t_i} = \int_0^{\alpha_1} g(s) \mathrm{d}X_s.$$

If $t_{j-1} < \alpha_1 < t_j$ then write $h_{t_{j-1}} 1_{[t_{j-1}, t_j[}(s)$ as $h_{t_{j-1}} 1_{[t_{j-1}, \alpha_1[}(s) + h_{t_{j-1}} 1_{[\alpha_1, t_j[}(s)$ and proceed as above with g expressed in the new form. The result for the left integral follows by taking the adjoint of $\int_0^{\alpha} g(s)^* \mathrm{d}X_s^*$. Finally the integral has zero trace because

$$\phi(h_{t_{i-1}} \Delta X_{t_i}) = \phi(M_{t_{i-1}}(h_{t_{i-1}} \Delta X_{t_i})) = \phi(h_{t_{i-1}} M_{t_{i-1}}(\Delta X_{t_i})) = 0.$$

A.5.4. ***Definition:*** Let $\mathscr{P}\langle X \rangle$ denote the class of processes g, where g is the $\mathrm{d}\langle X \rangle$-a.s. limit in $\| \; \|_\infty$ of a uniformly bounded sequence of a-valued simple processes.

(i) We note that $\operatorname{ess\,sup}_{s<\alpha} \|g(s)\|_\infty < \infty$ for $\alpha > 0$.

(ii) We shall say (g_n) is a sequence defining g.

A.5.5. ***Theorem:*** Let $X = (X_\alpha)$ be an L^2-martingale and $g \in \mathscr{P}\langle X \rangle$ be defined by (g_n). Then $(\int_0^{\alpha} g_n(s) \mathrm{d}X_s)$ is a Cauchy sequence in $L(a)$. The limit, which we denote by $\int_0^{\alpha} g(s) \mathrm{d}X_s$, is independent of the sequence (g_n) defining g. We shall call $\int_0^{\alpha} g(s) \mathrm{d}X_s$ the right stochastic integral of g with respect to X.

Proof: It is not difficult to see that the simple processes form a linear space and that the right integral is linear in the integrand. If $m, n \in \mathbb{N}$

$$\left\| \int_0^{\alpha} g_n(s) \mathrm{d}X_s - \int_0^{\alpha} g_m(s) \mathrm{d}X_s \right\|_2^2 = \left\| \int_0^{\alpha} (g_n - g_m)(s) \mathrm{d}X_s \right\|_2^2$$

$$\leqslant \int_0^{\alpha} \left\| g_n(s) - g_m(s) \right\|_\infty^2 \mathrm{d}\langle X \rangle_s.$$

Now $g_n \to g$ $\mathrm{d}\langle X \rangle$-a.s. hence $g_n - g_m \to 0$ $\mathrm{d}\langle X \rangle$-a.s. as $m, n \to \infty$. It follows that $\|g_n(s) - g_m(s)\|_\infty^2 \to 0$ for $\mathrm{d}\langle X \rangle$ almost every $s \in [0, \alpha[$ as $m, n \to 0$. By bounded convergence the right-hand integral converges to zero as $m, n \to \infty$. So $(\int_0^{\alpha} g_n(s) \mathrm{d}X_s)$ is Cauchy in $L^2(a)$. One shows that this limit is independent of the choice of (g_n) by supposing that (f_n) and (g_n) define g and then form

$$h_n = \begin{cases} f_n & \text{if } n=2r, \quad r\in\mathbb{N} \\ g_n & \text{if } n=2r+1, \quad r\in\mathbb{N} \end{cases}$$

and apply the first part to (h_n).

A.5.6. ***Corollary:*** We can define $\int_0^\alpha \mathrm{d}X_s g(s)$ in the same way.

A.5.7. ***Theorem:*** Let (X_α) be an L^2-martingale and $g\in\mathscr{P}\langle X\rangle$, then $(\int_0^\alpha g(s)\mathrm{d}X_s)$ is a martingale and $\phi(\int_0^\alpha g(s)\mathrm{d}X_s)=0$.

Proof: For $0\leqslant\alpha_1<\alpha_2$, we can choose (g_n) defining g such that

$$M_{\alpha_1}\left(\int_0^{\alpha_2} g_n(s)\mathrm{d}X_s\right)=\int_0^{\alpha_1} g_n(s)\mathrm{d}X_s.$$

However M_{α_1} is $\|\cdot\|_2$ continuous and

$$M_{\alpha_1}\left(\int_0^{\alpha_2} g(s)\mathrm{d}X_s\right)=M_{\alpha_1}\left(\lim_{n\to\infty}\int_0^{\alpha_2} g_n(s)\mathrm{d}X_s\right)=\lim_{n\to\infty} M_{\alpha_1}\int_0^{\alpha_2} g_n(s)\mathrm{d}X_s$$

$$=\lim_{n\to\infty}\int_0^{\alpha_1} g_n(s)\mathrm{d}X_s=\int_0^{\alpha_1} g(s)\mathrm{d}X_s.$$

A similar argument shows that $\phi(\int_0^\alpha g(s)\mathrm{d}X_s)=0$.

A.5.8. ***Theorem:*** Let (X_α) be an L^2-martingale, $f,g\in\mathscr{P}\langle X\rangle$ and let $Y_\alpha=\int_0^\alpha g(s)\mathrm{d}X_s$. Then $f\in\mathscr{P}\langle Y\rangle$ and

$$\int_0^\alpha f(s)\mathrm{d}Y_s=\int_0^\alpha f(s)g(s)\mathrm{d}X_s.$$

Proof: First note that off the union of two $\mathrm{d}\langle X\rangle$-null sets we have $f_n(s)g_n(s)\to f(s)g(s)$, where (f_n) and (g_n) define f and g respectively. So $f\cdot g\in\langle X\rangle$. We can form (Y_α) according to Theorem A.5.5 and we note that for an interval $[s,t]$ say, and a *simple process* g,

$$\mathrm{d}\langle Y\rangle([s,t])=\phi(Y_t^*Y_t-Y_s^*Y_s)=\phi((Y_t-Y_s)^*(Y_t-Y_s))$$

$$=\phi\left(\left|\int_s^t g(s)\mathrm{d}X_s\right|^2\right)\leqslant\int_s^t \|g(s)\|_\infty^2\mathrm{d}\langle X\rangle_s\leqslant M^2\mathrm{d}\langle X\rangle([s,t])$$

where M is a bound for $||g(s)||_\infty$ on $[s,t]$. This inequality carries over to $g\in\mathscr{P}\langle X\rangle$ by approximation (see below). It follows that $f\in\mathscr{P}\langle Y\rangle$. If f is elementary, say $f(s)=f_r\chi_{[r,s)}(s)$, then

$$\int_0^\alpha f(s)\mathrm{d}Y_s=f_r(Y_s-Y_r)=f_r\int_r^s g(s)\mathrm{d}X_s=f_r\lim_n\int_r^s g_n(s)\mathrm{d}X_s$$

$$=\lim_n\int_r^s f_r g_n(s)\mathrm{d}X_s$$

because multiplication by elements of a is $||\ ||_2$-continuous and $\int_r^s g_n(s)\mathrm{d}X_s$ is just a finite sum. Hence

$$\int_0^\alpha f(s)\mathrm{d}Y_s=\lim_n\int_r^s f_r g_n(s)\mathrm{d}X_s=\lim_n\int_r^s f(s)g_n(s)\mathrm{d}X_s=\int_0^\alpha f(s)g(s)\mathrm{d}X_s.$$

The result for simple processes follows by linearity. One way of completing the proof is to note that the contraction property, Theorem A.5.2, carries over to elements of $\mathscr{P}\langle X\rangle$ because

$$\left\|\int_0^\alpha g_n(s)\mathrm{d}X_s\right\|_2^2\xrightarrow[n]{}\left\|\int_0^\alpha g(s)\mathrm{d}X_s\right\|_2^2$$

while

$$\left\|\int_0^\alpha g_n(s)\mathrm{d}X_s\right\|_2^2\leqslant\int_0^\alpha ||g_n(s)||_\infty^2\mathrm{d}\langle X\rangle_s$$

and $||g_n(s)||_\infty^2\to||g(s)||_\infty^2$ for $\mathrm{d}\langle X\rangle$ almost every $s\in[0,\alpha]$. One uses bounded convergence to get

$$\left\|\int_0^\alpha g(s)\mathrm{d}X_s\right\|_2^2\leqslant\int_0^\alpha ||g(s)||_\infty^2\mathrm{d}\langle X\rangle_s.$$

Make the observation that $f\cdot g$ is the $\mathrm{d}\langle X\rangle$-a.s. limit of $f_n\cdot g$ where (f_n) defines f. Using bounded convergence again we deduce that

$$\left\|\int_0^\alpha (f(s)g(s)-f_n(s)g(s))\mathrm{d}X_s\right\|_2^2\leqslant\int_0^\alpha ||f(s)g(s)-f_n(s)g(s)||_\infty^2\mathrm{d}\langle X\rangle_s\to 0$$

as $n \to \infty$. We have shown that

$$\int_0^\alpha f(s)\mathrm{d}Y_s = \lim_n \int_0^\alpha f_n(s)\mathrm{d}Y_s = \lim_n \int_0^\alpha f_n(s)g(s)\mathrm{d}X_s = \int_0^\alpha f(s)g(s)\mathrm{d}X_s.$$

A.6. A special case

Results somewhat sharper than those of the preceding discussion can be obtained if the submartingale $(X_t^* X_t)$ has a Doob–Meyer type decomposition into the sum of an L^1-martingale and an increasing process which is a multiple of the identity operator. Indeed, suppose that $X_t^* X_t = Z_t + \sigma_t \mathbf{1}$, where (Z_t) is an L^1-martingale and $t \mapsto \sigma_t \in \mathbf{R}$ is increasing. Then

$$\sigma_t - \sigma_s = \phi(X_t^* X_t) - \phi(X_s^* X_s) = \|X_t\|_2^2 - \|X_s\|_2^2$$

and so we see that the Stieltjes measure associated with σ_t is just $\mathrm{d}\langle X \rangle$. Theorem A.5.2 becomes an isometry property of the stochastic integral.

A.6.1. ***Theorem*** (isometry property): Let (X_t) be an L^2-martingale such that $X_t^* X_t$ has the decomposition $X_t^* X_t = Z_t + \sigma_t \mathbf{1}$ as above. Then for any simple L^2-valued process g on $[0,\alpha[$ we have

$$\left\| \int_0^\alpha g(s)\mathrm{d}X_s \right\|_2^2 = \int_0^\alpha \|g(s)\|_2^2 \mathrm{d}\langle X \rangle_s.$$

Proof: With the notation of Theorem A.5.2, we obtain

$$\left\| \int_0^\alpha g(s)\mathrm{d}X_s \right\|_2^2 = \sum_i \phi(\Delta X_{t_i}^* |h_{t_{i-1}}|^2 \Delta X_{t_i})$$

$$= \sum_i \phi(|h_{t_{i-1}}|^2 \Delta X_{t_i}^* \Delta X_{t_i})$$

$$= \sum_i \phi(|h_{t_{i-1}}|^2 (X_{t_i}^* X_{t_i} - X_{t_{i-1}}^* X_{t_{i-1}}))$$

$$= \sum_i \phi(|h_{t_{i-1}}|^2)(\sigma_{t_i} - \sigma_{t_{i-1}})$$

$$= \int_0^\alpha \|g(s)\|_2^2 \mathrm{d}\langle X \rangle_s \qquad \text{QED.}$$

By considering suitable Cauchy sequences as before, it is clear that the stochastic integral can be defined for any f in the completion of the set of simple L^2-valued processes with respect to the seminorm $\|\cdot\|$ given by

$$\|g\|^2 = \int_0^\alpha \|g(s)\|_2^2 \, d\langle X\rangle_s \tag{A.1}$$

For such f the isometry property remains valid in the form

$$\left\| \int_0^\alpha f \, dX_s \right\|_2 = \|f\|.$$

Denote this completion by $\mathscr{H}([0,\alpha],d\langle X\rangle)$. The appearance of the L^2-norm in the integrand on the r.h.s. of (A.1) allows us to identify $\mathscr{H}([0,\alpha],d\langle X\rangle)$.

A.6.2. ***Theorem:*** $\mathscr{H}([0,\alpha],d\langle X\rangle)$ is canonically identified with the set of processes in $L^2([0,\alpha],d\langle X\rangle;L^2(a))$, the Hilbert space of (classes of) L^2-valued maps on $[0,\alpha]$, square-integrable over $[0,\alpha]$ with respect to $d\langle X\rangle$.

Note that the elements of $L^2([0,\alpha],d\langle X\rangle;L^2(a))$ are equivalence classes of maps equal $d\langle X\rangle$ – almost everywhere. An element of $L^2([0,\alpha],d\langle X\rangle;L^2(a))$ is a process if it contains a representative which is a process. To minimise the tedium we will identify representatives with their classes and, for example, consider continuous L^2-valued maps on $[0,\alpha]$ as being elements of $L^2([0,\alpha],d\langle X\rangle;L^2(a))$.

The proof of the theorem is by a series of lemmas.

A.6.3. ***Lemma:*** For given $f\in L^2(a)$, the map $t\to M_t f$ is continuous: $\mathbf{R}^+\to L^2(a)$.

Proof: This is a consequence of the martingale convergence theorem A.4.4.

Let $\mathscr{C}([0,\alpha])$ denote the collection of continuous complex-valued functions on $[0,\alpha]$.

A.6.4. ***Lemma:*** Linear combinations of elements of the form $\gamma(\cdot)g$ with $\gamma\in\mathscr{C}([0,\alpha])$ and $g\in L^2(a)$ are dense in $L^2([0,\alpha],d\langle X\rangle;L^2(a))$.

Proof: This is a standard result (see, for example, [T]).

A.6.5. ***Lemma:*** Let f be a process in $L^2([0,\alpha],d\langle X\rangle;L^2(a))$ and let $\varepsilon>0$ be given. Then there is a continuous process $g:[0,\alpha]\to L^2(a)$ such that

$$\left\{\int_0^\alpha \|f(s)-g(s)\|_2^2 \mathrm{d}\langle X\rangle\right\}^{\frac{1}{2}}<\varepsilon.$$

Proof: By Lemma A.6.4, there is $n\in\mathrm{N}$ and functions $\gamma_1,\ldots,\gamma_n\in\mathscr{C}([0,\alpha])$ and elements $g_1,\ldots,g_n\in L^2(a)$ such that

$$\int_0^\alpha \left\|f(s)-\sum_{i=1}^n \gamma_i(s)g_i\right\|_2^2 \mathrm{d}\langle X\rangle<\varepsilon^2.$$

Now, by Lemma A.6.3, $s\to g(s)=\sum_{i=1}^n \gamma_i(s)M_s g_i$ is a continuous process, $[0,\alpha]\to L^2(a)$, and we have

$$\int_0^\alpha \|f(s)-g(s)\|_2^2 \mathrm{d}\langle X\rangle=\int_0^\alpha \left\|M_s(f(s)-\sum_{i=1}^n \gamma_i(s)g_i)\right\|_2^2 \mathrm{d}\langle X\rangle$$

since f is adapted

$$\leqslant\int_0^\alpha \left\|f(s)-\sum_{i=1}^n \gamma_i(s)g_i\right\|_2^2 \mathrm{d}\langle X\rangle<\varepsilon^2$$

since $M_s: L^2(a)\to L^2(a)$ is a contraction.

A.6.6. ***Lemma:*** Let g be a continuous process in $L^2([0,\alpha],\mathrm{d}\langle X\rangle; L^2(a))$ and let $\varepsilon>0$ be given. Then there is a simple a-valued process h on $[0,\alpha]$ such that

$$\left\{\int_0^\alpha \|g(s)-h(s)\|_2^2 \mathrm{d}\langle X\rangle\right\}^{\frac{1}{2}}<\varepsilon\left\{\int_0^\alpha \mathrm{d}\langle X\rangle\right\}^{\frac{1}{2}}.$$

Proof: If g is continuous on $[0,\alpha]$ it is uniformly continuous on $[0,\alpha]$. Hence, given $\varepsilon>0$, there is a partition $0=t_0<t_1<\ldots<t_n=\alpha$ of $[0,\alpha]$ such that

$$\|g(s)-g(t_i)\|_2<\varepsilon/2$$

whenever $t_{i-1}\leqslant s<t_i$, $1\leqslant i\leqslant n$.

Since a_{t_i} is dense in $L^2(a_{t_i})$ and g is adapted, there is $h_i\in a_{t_i}$ such that

$$\|g(t_i)-h_i\|<\varepsilon/2,\quad 0\leqslant i\leqslant n.$$

Set

$$h(s)=\sum_{i=1}^{n} h_{i-1}1_{[t_{i-1},t_i)}(s), \qquad h(\alpha)=h_n.$$

Then h is a simple a-valued process on $[0,\alpha]$ and

$$\int_0^{\alpha} ||g(s)-h(s)||_2^2\,\mathrm{d}\langle X\rangle<\varepsilon^2\int_0^{\alpha}\mathrm{d}\langle X\rangle.$$

This proves the lemma.

Theorem A.6.2 now follows immediately from Lemmas A.6.5 and A.6.6.

Denote by $\mathscr{H}_{\mathrm{loc}}(\mathbf{R}^+,\mathrm{d}\langle X\rangle)$ the set of processes f such that the restriction of f to $[0,\alpha]$ belongs to $\mathscr{H}([0,\alpha],\mathrm{d}\langle X\rangle)$ for all $\alpha>0$. Theorem A.5.7 has the analogue:

A.6.7. ***Theorem:*** For each $f\in\mathscr{H}_{\mathrm{loc}}(\mathbf{R}^+,\mathrm{d}\langle X\rangle)$, $Y_\alpha=\int_0^\alpha f(s)\mathrm{d}X_s$ is a centred L^2-martingale.

A.7. The Itô-Clifford integral

A.7.1: As an application of the preceding analysis, we shall describe the Itô-Clifford stochastic integral, which is a non-commutative stochastic integral with the free Fermi quantum field as integrator. To do this, we shall first construct the Fermi field. (See [BSW$_1$] for proofs.)

The fermion Fock space over $L^2(\mathbf{R}^+,\mathrm{d}x)$ is the Hilbert space direct sum $\mathfrak{F}=\oplus_{n=0}^{\infty}\mathfrak{F}_n$, where $\mathfrak{F}_0=C$, and for $n\geqslant 1$, $\mathfrak{F}_n$ is the subspace of $L^2(\mathbf{R}^{+n},\mathrm{d}^n x)$ consisting of totally antisymmetric functions. $\mathfrak{F}_n$ is called the n-particle space. If $\phi=(\phi_0,\phi_1,\ldots)\in\mathfrak{F}$ with $||\phi||=1$, then $||\phi_n||^2$ is interpreted as the probability that there are exactly n particles in the state ϕ, and $|\phi_n(x_1,\ldots,x_n)|^2$ is the probability density for their positions (in $\mathbf{R}^+$). The vector $\Omega=(1,0,0,\ldots)$ is called the no-particle or Fock vacuum state.

For $f\in L^2(\mathbf{R}^+,\mathrm{d}x)$, define the operator $b(f):\mathfrak{F}_{n+1}\to\mathfrak{F}_n$ by $b(f):\mathfrak{F}_0\to\{0\}$, and, for $n>1$,

$$(b(f)\phi_{n+1})(x_1,\ldots,x_n)=\sqrt{n}\int_0^{\infty}\overline{f(s)}\phi_{n+1}(s,x_1,\ldots,x_n)\mathrm{d}s.$$

One verifies that the adjoint of $b(f)$ is given by the operator $b^*(f):\mathfrak{F}_n\to\mathfrak{F}_{n+1}$,

$$(b^*(f)\phi_n)(x,\ldots,x_{n+1})=\frac{1}{\sqrt{n+1}}\sum_{j=1}^{n+1}(-1)^{j-1}f(x_j)\phi_n(x,\ldots,\hat{x}_j,\ldots,x_{n+1}),$$

where $\hat{x}_j$ means omit the variable x_j.

For obvious reasons, $b(f)$ is called an annihilation operator and $b^*(f)$ a creation operator. We note that $b^*(f)^2=0$. This realizes the Pauli exclusion principle – there cannot be more than one fermion in any particular state.

It is not difficult to check from their definitions that

$$b(f)b^*(g)+b^*(g)b(f)=\int_0^\infty \bar{f}(s)g(s)\mathrm{d}s\mathbf{1}$$

and

$$b(f)b(g)+b(g)b(f)=b^*(f)b^*(g)+b^*(g)b^*(f)=0$$

for $f,g\in L^2(\mathbf{R}^+,\mathrm{d}x)$.

These relations are known as the canonical anticommutation relations and play a fundamental role throughout quantum mechanics.

Setting $f=g$, we obtain

$$b(f)b^*(f)+b^*(f)b(f)=\|f\|^2\mathbf{1}.$$

Since both operators on the l.h.s. are non-negative, we see that $\|b^*(f)\|\leqslant\|f\|$. But $b^*(f)\Omega=f$ implies that $\|b^*(f)\|=\|f\|$, and so $\|b(f)\|=\|f\|$ also.

We define the free fermion field $\psi(f)$, for $f\in L^2(\mathbf{R}^+,\mathrm{d}x)$, by $\mu(f)=b(f)+b^*(f)$. Thus $\psi(f)$, $f\in L^2_{\mathbf{R}}(\mathbf{R}^+,\mathrm{d}x)$, is a bounded self-adjoint operator on $\mathfrak{F}$. Evidently, $\psi(\alpha f+g)=\alpha\psi(f)+\psi(g)$ and $\psi(f)\psi(g)+\psi(g)\psi(f)=2\int_0^\infty f(x)g(x)\,\mathrm{d}x$, for $\alpha\in\mathbf{R}$, $f,g\in L^2_{\mathbf{R}}(\mathbf{R}^+,\mathrm{d}x)$. In particular $\psi(f)^2=\int_0^\infty f(x)^2\mathrm{d}x$. (It is conventional to define the field as $(b+b*)/\sqrt{2}$ for normalisation reasons which we will not go into. However, it is more convenient for our purposes not to include the $\sqrt{2}$ factor.)

We shall use the field ψ to construct a filtration of von Neumann algebras. For each $t\geqslant 0$, let $\mathscr{C}_t$ denote the von Neumann algebra generated by the fields $\psi(f)$ with $f\in L^2_{\mathbf{R}}(\mathbf{R}^+,\mathrm{d}x)$ such that $f(x)=0$ for almost all $x>t$; i.e. f is supported in $[0,t]$. Let $\mathscr{C}$ be the von Neumann algebra generated by $\psi(f)$, $f\in L^2_{\mathbf{R}}(\mathbf{R}^+,\mathrm{d}x)$. Clearly, we have $\mathscr{C}_s\subseteq\mathscr{C}_t\subset\mathscr{C}$ for $0\leqslant s\leqslant t$, and, using the norm continuity of the map $f\rightarrow\psi(f)$, one can show that $\mathscr{C}_t$ is generated by the $\mathscr{C}_s$ for $0\leqslant s\leqslant t$ and that $\mathscr{C}$ is generated by $\bigcup_{t\geqslant 0}\mathscr{C}_t$.

Let m denote the state on $\mathscr{C}$ given by $m(a)=(a\Omega,\Omega)$, $a\in\mathscr{C}$.

A.7.2. ***Theorem:*** m is a faithful normal trace state on $\mathscr{C}$.

By virtue of this theorem, we can set up a non-commutative integration theory based on $(\mathscr{C},m)$ as described earlier.

For each $t \geqslant 0$, let $M_t : L^p(\mathscr{C}) \to L^p(\mathscr{C}_t)$ denote the conditional expectation. One can show that, on the fields, M_t is given simply as

$$M_t\psi(f) = \psi(1_{[0,t]}f), \quad f \in L^2_{\mathbf{R}}(\mathbf{R}^+, \mathrm{d}x).$$

(For a proof, see, for example, [BSW$_1$]).

If we write $\psi_t = \psi(1_{[0,t]})$ for $t \geqslant 0$, then it is clear that $M_t\psi_s = \psi_t$ for $0 \leqslant t \leqslant s$. In other words, (ψ_t) is an $L^\infty(\mathscr{C})$-martingale adapted to the filtration $\{\mathscr{C}_t : t \geqslant 0\}$. Furthermore, $\psi_t^*\psi_t = \psi_t^2 = \int_0^\infty 1_{[0,t]}(x)\mathrm{d}x\mathbf{1} = t\mathbf{1}$ gives a Doob–Meyer type decomposition for $\psi_t^*\psi_t$ as the sum of a martingale (identically zero) and an increasing process, namely $(t\mathbf{1})$. We can therefore define (non-commutative) stochastic integration with respect to (ψ_t) – this is called the Itô–Clifford stochastic integral. (This terminology arises from the fact that the canonical anticommutation relations imply that the fields generate a Clifford algebra and the stochastic integral is constructed in exactly the same way as the Itô integral with respect to a Wiener process is constructed.)

In the notation of section A.5, we have that $\mathrm{d}\langle\psi_t\rangle$ is just Lebesgue measure on $[0,\infty)$. Hence, for any $f \in \mathscr{H}_{\mathrm{loc}}(\mathbf{R}^+, \mathrm{d}s; L^2(\mathscr{C}))$, the Itô–Clifford stochastic integral $\int_0^\alpha f(s)\mathrm{d}\psi_s$ is an element of $L^2(\mathscr{C}_\alpha)$, $\alpha \geqslant 0$, and $\|\int_0^\alpha f(s)\mathrm{d}\psi_s\|_2^2 = \int_0^\alpha \|f(s)\|_2^2\mathrm{d}s$.

Furthermore, $X_\alpha = \int_0^\alpha f(s)\mathrm{d}\psi_s$ defines a centred L^2-martingale adapted to the filtration $\{\mathscr{C}_\alpha : \alpha \geqslant 0\}$.

In fact, one can prove a martingale representation theorem (cf. section 4.6):

A.7.3. ***Theorem:*** Let (X_α) be a centred L^2-martingale adapted to $\{\mathscr{C}_\alpha : \alpha \geqslant 0\}$. Then there is a unique element $\tilde{X} \in \mathscr{H}_{\mathrm{loc}}(\mathbf{R}^+, \mathrm{d}s; L^2(\mathscr{C}))$ such that $X_\alpha = \int_0^\alpha \tilde{X}(s)\mathrm{d}\psi_s$.

Using this result, one can show that for any L^2-martingale (X_α), one has a Doob–Meyer type decomposition $X^*X_\alpha = Z_\alpha + A_\alpha$, where (Z_α) is an L^1-martingale and (A_α) is an increasing L^1-process given explicitly in terms of $\tilde{X}$.

It is of interest to note that had we defined the Fock space $\mathfrak{F}$ in terms of symmetric rather than antisymmetric functions we would have the Bose–Fock space. Analogously one defines the boson field ϕ_t, $t \geqslant 0$, which is an unbounded self-adjoint operator. The fields Φ_t, Φ_s commute with each other for all $t \geqslant 0$, $s \geqslant 0$ and can be realised as random variables on a probability space (by the spectral theorem). It turns out that Φ_t is a centred Gaussian random variable. Indeed, (Φ_t) is a Wiener process and the stochastic integral with respect to (Φ_t) constructed as in A.6 above, is precisely the *usual Itô integral*!

A.8. Concluding remarks

Section A.2: There are many texts dealing with von Neumann algebras, the authors have found [S_2] and [T] particularly useful.

Section A.3: There have been a number of different approaches to construction of non-commutative L^p-spaces, notably [S_1], [D], [Y]. There is a vast literature on conditional expectations, for the features relevant to non-commutative processes [U], [D] are worth studying.

Section A.4: For convergence results see [U], [CB_1]. It is difficult to construct a Doob–Meyer decomposition in the non-commutative case but results in §7 of [BSW_1] go some way in this direction.

Section A.5: The construction employed here differs from that in §7 of [BSW_1] because it by-passes the Doob–Meyer decomposition. However the class $\langle X \rangle$ of integrands is the same as those of §7 of [BSW_1] and in fact the integral constructed here agrees with and extends that of §7 of [BSW_1]. A fuller discussion of this integral will appear soon [BSW_2].

Section A.6: The historical order is awry here for the content of section A.7 preceded the realisation that one could proceed as in section A.6. Note that the integrands are *L^2-valued*, an improvement on $\mathscr{P}\langle X \rangle$. In addition, the completion of the simple *a*-valued processes is concretely identified as a Banach space of processes. The results A.6.1 to A.6.7 appear in [BSW_1] in the context of section A.7. The assumptions made here are not all that restrictive: cf. the results of section 4.6.

Section A.7: The construction of the Itô–Clifford integral is the main part of [BSW_1]. Theorem A.7.2 is due to Gross and Segal [G], [S_3]. Theorem A.7.3 is from [BSW_1].

This is not the only attempt to construct a non-commutative analogue of stochastic integration, see [RH] for example. We consider that the integrals constructed in sections A.5, A.6 and A.7 deserve the title 'stochastic' because they integrate processes to give martingales and the construction is very similar to that employed in the commutative case. The authors hope to clarify the relationship between the commutative and the non-commutative integrals in a forthcoming paper.

A.9. References for the Appendix:

[BSW_1] C. Barnett, R. F. Streater, I. Wilde, The Itô–Clifford integral. *J. Funct. Analysis* **48** (2) Sept. 1982, 172–212.

[CB_1] C. Barnett, Supermartingales on semifinite W^*-algebras, *J. London Math. Soc.* (2) **24** (1981) 175–81.

[CB_2] C. Barnett, Thesis, University of Hull, 1980.

[BSW_2] C. Barnett, R. F. Streater, I. Wilde, Decompositions and stochastic integrals in an arbitrary probability gauge space, Preprint, Bedford College, London.

[D] J. Dixmier, Formes lineaires sur un anneau d'operateurs, *Bull. Soc. Math. France* **74** (1953) 9–39.

[K] A. Kussmaul, *Stochastic integration and generalized martingales*, Pitman, London, 1977.

[S_1] I. Segal, A non-commutative extension of abstract integration, *Ann. Math.* **58** (1953) 401–57.

[S_2] S. Sakai, *C^*- and W^*-algebras*, Springer 1971.

[S_3] I. Segal, Tensor algebras over Hilbert spaces I, *Trans. Amer. Math. Soc.* **81** (1956) 106–34.

[G] L. Gross, Existence and uniqueness of physical ground states, *J. Funct. Analysis* **10** (1972) 52–109.

[RH] R. L. Hudson, Quantum mechanical Wiener processes, *J. Multivar. Anal.* **7** (1978) 107–23.

[U] H. Umegaki, Conditional Expectation in an operator algebra II, *Tohoku Math. J.* **8** (1956) 86–100.

[Y] F. J. Yeadon, Non-Commutative L^p-spaces, *Math. Proc. Camb. Phil. Soc.* **77** (1975) 91–102.

[T] M. Takesaki, *Theory of operator algebras, Vol. I*, Springer-Verlag 1979.

References

1 R. M. Anderson, A non-standard representation for Brownian motion and Itô integration, *Israel J. Math.* **25** (1976), 15–46.
2 K. Bichteler, Stochastic integration and L^p-theory of semimartingales, *Ann. Probab.* **9** (1981), 49–89.
3 P. Billingsley, *Ergodic theory and information*, Wiley, New York, 1965.
4 P. Billingsley, *Convergence of probability measures*, Wiley, New York, 1966.
5 J. Bismut and B. Skalli, Temps d'arret optimal, theorie generale des processus et processus de Markov, *Z. Wahrsch. verw. Gebiete* **39** (1977), 301–14.
6 F. Black and M. Scholes, The pricing of options and corporate liabilities, *J. Polit. Econom.* **81** (1973), 637–59.
7 R. M. Blumenthal and R. K. Getoor, *Markov processes and potential theory*, Academic Press, New York, 1968.
8 L. Breiman, *Probability*, Addison-Wesley, Reading, Mass., 1968.
9 P. Bremaud, *A martingale approach to point processes*, Thesis, Elec. Res. Lab., Berkeley, M-345, 1972.
10 G. Choquet, Theory of Capacities, *Ann. Inst. Four. Grenoble* **5** (1953–4), 131–295.
11 G. Choquet, Forme abstraite du theoreme de capacitabilite, *Ann. Inst. Four. Grenoble* **9** (1959), 83–99.
12 Y. S. Chow, H. Robbins and D. Siegmund, *Great expectations: theory of optimal stopping*, Houghton Mifflin, Boston, 1971.
13 K. L. Chung, *A course of probability theory*, Harcourt Brace, New York, 1968.
14 N. J. Cutland, Non-standard measure theory and its applications, *Bull. Lond. Math. Soc.* **15** (1983), 529–89.
15 M. H. A. Davis, *Linear estimation and stochastic control*, Chapman and Hall, London, 1977.
16 C. Dellacherie, *Capacites et processus stochastiques*, Springer, Berlin, 1972.
17 C. Dellacherie, Un survol de la theorie de l'integrale stochastique, *Proc. of the International Congress of Mathematicians, Helsinki* **2** (1978), 733.
18 C. Dellacherie, Mesurabilite des debuts et theoremes de section, Sem. Prob. XV, *Lecture Notes in Mathematics* **850**, Springer, Berlin–New York (1981), 351–70.
19 C. Dellacherie and P. A. Meyer, *Probabilities et potentiel, edition entierement refondue*, Hermann, Paris, Ch. I–IV, 1975, Ch. V–VIII 1980.
20 J. Diestel and J. J. Uhl Jr, Vector measures, *A.M.S. Mathematical Surveys* **15**, 1977.

21 N. Dinculeanu, *Vector measures*, Deutscher Verlag der Wissenschaften, Berlin, 1966.
22 S. Djezzar, *Snell's envelope and optimal stopping*, M.Sc. Thesis, University of Hull, 1982.
23 C. Doleans-Dade and P. A. Meyer, Integrales stochastiques par rapport aux martingales locales, Sem. Prob. IV, *Lecture Notes in Mathematics* **124**, Springer, Berlin, New York, 1970.
24 J. L. Doob, *Stochastic processes*, Wiley, New York, 1953.
25 R. G. Douglas, Contractive projections on an L^1-space, *Pac. J. Math.* **15** (1965), 443–62.
26 L. Dubins and L. Savage, *How to gamble if you must*, McGraw-Hill, New York, 1965.
27 N. Dunford and J. T. Schwartz, *Linear operators, part I*, Wiley-Interscience, New York, 1958.
28 E. B. Dynkin, *Markov processes*, Springer, Berlin–New York, 1965.
29 W. F. Eberlein, Abstract ergodic theorems and weak almost periodic functions, *Trans. Amer. Math Soc.* **67** (1949), 217–40.
30 R. J. Elliott, *Stochastic calculus and applications*, Springer, Berlin–New York, 1982.
31 W. Feller, *Introduction to probability theory and its applications*, 2nd edn., Wiley, New York, 1957.
32 D. L. Fisk, Quasi-martingales, *Trans. Amer. Math. Soc.* **120** (1965), 369–89.
33 D. Freedman, *Brownian motion and diffusion*, Holden-Day, San Francisco, 1971.
34 D. J. H. Garling, The martingale convergence theorem and its applications, edited Lecture Notes, Tripos Part III, University of Cambridge, 1978.
35 A. M. Garsia, *Topics in almost everywhere convergence*, Markham, Chicago, 1970.
36 I. I. Gikhman and A. V. Skorokhod, *The theory of stochastic processes*, 2 vols., Springer, Berlin–New York, 1972.
37 P. R. Halmos, *Measure theory*, Van Nostrand, Princeton, 1954.
38 P. R. Halmos, *A Hilbert space problem book*, Van Nostrand, Princeton, 1967.
39 J. M. Harrison and S. R. Pliska, Martingales and stochastic integrals in the theory of continuous trading. *Stoch. Proc. and Applic.* **11** (1981), 215–60.
40 T. Hida, *Brownian motion*, Springer, Berlin–New York, 1980.
41 R. E. Huff, The Radon–Nikodym property for Banach spaces in measure theory, *Lecture Notes in Mathematics* **541**, Springer, Berlin–New York, 1976, 229–42.
42 J. Jacod, Calcul stochastique et problemes de martingales, *Lecture Notes in Mathematics* **714**, Springer, Berlin–New York, 1979.
43 M. Jerison, Martingale formulation of ergodic theorems, *Proc. Amer. Math. Soc.* **10** (1959), 531–9.
44 G. Johnson and L. L. Helms, Class (*D*) supermartingales, *Bull. Amer. Math. Soc.* **69** (1963), 59–62.
45 G. Kallianpur, *Stochastic filtering theory*, Springer, Berlin–New York, 1980.
46 J. F. C. Kingman and S. J. Taylor, *Introduction to measure and probability*, Cambridge University Press, 1966.
47 P. E. Kopp, A ratio limit theorem for contraction projections and applications, *Glasgow Math. J.* **14** (1973), 80–5.
48 P. E. Kopp, D. Strauss and F. J. Yeadon, Positive Reynolds operators on Lebesgue spaces, *J. Math. Anal. Applic.* **44** (1973), 350–65.
49 A. U. Kussmaul, Stochastic integration and generalized martingales, *Pitman Research Notes in Mathematics* **11**, London, 1977.
50 A. U. Kussmaul, Thesis, University of Tübingen, 1980.

51 H. Kunita and S. Watanabe, On square-integrable martingales, *Nagoya Math. J.* **30** (1967), 209–45.
52 J. Lamperti, *Probability*, Benjamin, New York, 1966.
53 P. Levy, *Processus stochastiques et mouvement Brownien*, Gauthier-Villars, Paris, 1957.
54 J. Lindenstrauss and L. Tzafriri, *Classical Banach spaces*, Springer, Berlin–New York, 1977.
55 R. S. Liptser and A. N. Shiryayev, *Statistics of random processes*, 2 vols., Springer, Berlin–New York, 1977.
56 B. Maurey, Theoremes de factorisation, *Asterisque* **11**, 1974.
57 H. P. McKean, *Stochastic integrals*, Academic Press, New York, 1969.
58 E. J. McShane, *Stochastic calculus and stochastic models*, Academic Press, New York, 1974.
59 M. Metivier, Stochastic integrals and vector-valued measures in *Vector and operator-valued measures and applications* (edited by D. H. Tucker and H. B. Maynard), Academic Press, New York, 1973.
60 M. Metivier, Reelle und vektorwertige Quasi-Martingale und die Theorie der stochastischen Integration, *Lecture Notes in Mathematics* **607**; Springer, Berlin–New York, 1977.
61 M. Metivier and J. Pellaumail, Mesures stochastiques a valeurs dans des espaces L_0, *Z. Wahrsch. verw. Gebiete* **40** (1977), 101–14.
62 M. Metivier and J. Pellaumail, *Stochastic integration*, Academic Press, New York, 1980.
63 P. A. Meyer, *Probability and potentials*, Blaisdell, Waltham, Mass., 1966.
64 P. A. Meyer, Martingales and stochastic integrals I, *Lecture Notes in Mathematics* **284**, Springer, Berlin–New York, 1972.
65 P. A. Meyer, Integrales stochastiques I–IV, Sem. Prob. I, *Lecture Notes in Mathematics* **39**, Springer, Berlin–New York, 1967.
66 P. A. Meyer, Un cours sur les integrales stochastiques, Sem. Prob. X, *Lecture Notes in Mathematics* **511**, Springer, Berlin–New York, 1976.
67 J. Neveu, *Mathematical foundations of the calculus of probability*, Holden-Day, San Francisco, 1965.
68 J. Neveu, *Discrete-parameter martingales*, North-Holland, Amsterdam, 1975.
69 J. Neveu, Relations entre le theorie des martingales et la theorie ergodique, *Ann. Inst. Four. Grenoble*, **15** (1965), 31–42.
70 S. Orey, F-Processes, in *Proc. 5th Berkeley Symposium on Mathematical Statistics and Probability II*, vol. 1, University of California Press, Berkeley, Cal., 1965, 301–14.
71 J. Pellaumail, Sur l'integrale stochastique et la decomposition de Doob–Meyer, *Asterisque* **9**, 1973.
72 S. R. Pliska, Duality theory for some stochastic control models, 2nd Bad Honnef workshop on stochastic differential systems, University of Bonn, 1982.
73 K. M. Rao, On the decomposition theorems of Meyer, *Math. Scand.* **24** (1969), 66–78.
74 K. M. Rao, Quasimartingales, *Math. Scand.* **24** (1969), 79–92.
75 K. M. Rao, Brownian motion and classical potential theory, *Aarhus Lecture Notes Series* **47**, Matematisk Institut, Aarhus, 1977.
76 C. A. Rogers *et al.*, *Analytic sets*, Academic Press, London, 1980.
77 W. Rudin, *Real and complex analysis*, McGraw-Hill, New York, 1966.
78 W. Rudin, *Functional analysis*, McGraw-Hill, New York, 1973.
79 F. Scalora, Abstract martingale convergence theorems, *Pac. J. Math.* **11** (1961), 347–74.

80 C. Stricker, Quasimartingales, martingales locales, semimartingales et filtrations naturelles, *Z. Wahrsch. verw. Gebiete* **39** (1977), 55–64.
81 D. W. Stroock and S. R. S. Varadhan, *Multi-dimensional diffusion processes*, Springer, Berlin–New York, 1979.
82 F. Treves, *Topological vector spaces, distributions and kernels*, Academic Press, New York, 1967.
83 D. Williams, *Diffusions, Markov processes and martingales, vol. 1: Foundations*, Wiley, Chichester, 1979.
84 J. Yeh, *Stochastic processes and the Wiener integral*, Marcel Dekker, New York, 1973.

List of symbols

Symbol	Page
$\mathscr{A}$	*Page* 11, 119
$\mathscr{A}_{\mathrm{loc}}$	119
ΔA	117
$\mathscr{B}(\mathbf{R})$	2
$BM_0(\mathbf{R})$	12
$\mathscr{B}^\infty(\Sigma)$	101
$\mathscr{B}(L^2(\Omega,\mathscr{F},P))$	177
$\mathscr{C},\mathscr{C}_S$	7
$\mathscr{C}[0,\alpha]$	187
$\mathscr{D}_p$	45
D'_B	85
D_A	97
$(E,\mathscr{E})$	124
$\hat{\mathscr{E}}$	126
$\mathbf{E}$	3
$\mathbf{E}(\cdot\|\mathscr{G})$	4
$\mathbf{E}_B(\cdot\|\mathscr{G})$	59
$\mathscr{F}$	1
$\bar{\mathscr{F}}$	1
$\mathscr{F}_t$	14,83
$\mathscr{F}_S$	84
$\mathscr{F}_{S-}$	91
$\mathscr{G}$	1
$H\cdot M$	134
$\mathscr{H}_{\mathrm{loc}}(\mathbf{R}^+,\mathrm{d}\langle X\rangle)$	189
$\mathscr{H}^1$	57
$\mathscr{H}([0,\alpha],\mathrm{d}\langle X\rangle)$	187
i_X,I_X	169
$I(f)$	15
I^f_t	16
L^0,L^p	2
L^p_B	58
$\mathscr{L}^2(B)$	15
$\mathscr{L}^2(M)$	135
$L^1(I_X)$	*Page* 174
$L^2([0,\alpha],\mathrm{d}\langle X\rangle;L^2(a))$	187
$M_\chi(x)$	178
$\mathscr{M}$	32,119
$\mathscr{M}^p$	53,131
$\mathscr{M}^2_c$	132
$\mathscr{M}^2_d,\mathscr{M}^2_0$	132
$\mathscr{M}(T)$	144
$\langle M\rangle$	132
$\langle M,N\rangle$	135
$[M]$	148
$[M,N]$	148
$\mathbf{N}$	natural numbers
Π_o,Π_p	101
Π^*_o,Π^*_p	111
Π_Ω	97
$\mathbf{Q}$	rational numbers
$\mathbf{R}$	real numbers
$\mathbf{R}^{\mathbf{T}}$	7
$\mathscr{R}_n$	76
$]\!]S,T]\!]$	90
$\sigma(E,E')$	22
$\Sigma_0,\Sigma_a,\Sigma_p$	94
Σ_π	94
$\mathbf{T}$	6
T_A	92
U^b_a	48
$V\cdot X$	44
$\mathscr{V}$	142
$\mathscr{W}$	11
X_S,X^S	85
$\mathbf{Z}$	integers
$\perp\!\!\!\perp$	136

Index

accessible process 96
accessible projection 103
accessible σ-field 94
accessible part of a stopping time 93
accessible stopping time 90
adapted process 40, 83
admissible strategy 167
Alaoglu's theorem 24
almost separably valued 58
almost surely 2
annihilation operator 190
announcing sequence for a predictable stopping time 90
attainable claim 167

Baire Category Theorem 31
Banach principle 60
Banach space 22
Banach space-valued martingale 60
Bochner-integrable function 28
Borel–Cantelli lemmas 3
Borel σ-field 2
bounded measurable process 101
Brownian motion 12
  martingale representation theorem for 162
cadlag process 88
canonical representation of a process 8
capacitable set 125
capacity 124
Caratheodory 2, 7
Central Limit Theorem 10
charging a stopping time 103
Chebychev inequality 3
Choquet, G. 124
coin-tossing space 72
class (D) process 105, 180
class (D) potential 114
closable (super-) martingale 51, 61, 89
compensated process 142
compensated jump martingale 145
compensator of a process 142
completion of a measure space 1, 125
conditional expectation 4, 34, 37, 58, 178
conditional independence 14
conditional probability 4
conditional variance 38
continuity property of stochastic integrals 161
continuous martingales 132, 156
continuous part of a semimartingale 151, 155
contraction property of non-commutative integral 181
convex set 25
convex function 38
countably generated σ-field 31
cross-section 97, 98, 124
cylinder sets 7

Daniell–Kolmogorov theorem 8
début of a set 97
  n-début 103
discontinuous martingale 132, 143, 152
discontinuous process 143
discounted processes 166
discrete stochastic integral 70
distribution
  of a process 7
  of a random variable 6
Doléans function 169
dominated convergence theorem 3
Donsker's theorem 11
Doob decomposition 67
Doob L^p-inequality 15, 55, 86
Doob maximal inequality 51, 60
Doob upcrossing inequality 49
Doob–Meyer decomposition 113, 122, 180
downcrossings 49
dual optional projection 111

dual predictable projection 111, 121, 148
dual space 22

Eberlein–Smŭlian theorem 36
elementary process 181
ε-optimal stopping time 81
equivalence of processes 8
essential supremum of a family of functions 76
evanescent process 82
exit time 79
expectation 3

Fatou lemma 3
Fermi field 189
fermion Fock space 189
field of sets 1
filtration 83
  non-commutative 178
finite-dimensional distributions 7
first hitting time 85
fundamental sequence 119

gains process 165
Gaussian family 12
Girsanov theorem 167
graph of a stopping time 90

Haar functions 13
Hahn–Banach theorem 25
Hardy–Littlewood maximal inequality 53
Hölder inequality 23, 56
Hunt, G. A. 66

increasing process 67, 104, 179
independent events 4
independent increments 12
indistinguishable processes 9, 82
inner measure 2
integrable variation, process of 106
integration by parts formula 160
integration of processes 106
Itô formula 15
Itô integral 17
Itô–Clifford integral 189

Jensen inequality 38, 59
joint distribution 6
jump
  of a process at a stopping time 103
  of a process exhausted by a stopping time 103
  process 117
  time, of a Poisson process 18

Kolmogorov inequality 57
Kolmogorov's 0–1 law 66
Kunita–Watanabe inequalities 137
Kussmaul, A. U. 169

Lebesgue, H. 108
localisation of a process 118
locally bounded process 153
locally integrable process 118
local martingale 46, 119, 149
lower semi-continuous function 59
L^p-bounded process 131
Markov process 14
martingale 15, 41, 86, 178
  compensated jump 145
  convergence property 62
  convergence theorems 50, 58, 61, 88, 179
  differences 44
  local 46, 119, 149
  vector-valued 60
maximal lemma 60
measure
  induced by an increasing process 107
  σ-finite 118
  stochastic 107
measurable function 2
  strongly 58
  weakly 169
  process 52
Metivier, M. 169
metrisable space 30
modification of a process 9, 83
monotone convergence theorem 3
mosaic 126

narrow convergence 11
'natural' increasing process 113
non-commutative conditional expectation 178
non-commutative filtration 178
non-commutative increasing process 179
non-commutative martingale 179
non-commutative stochastic integral 176
normalised random walk 10
nowhere dense set 31
nowhere differentiable paths of Brownian Motion 13
null sets 1

optimal stopping rule 75
optional process 45, 96
optional projection 101
optional quadratic variation 148, 152, 153, 155
optional sampling 45, 73
optional stopping 75
optional σ-field 94
options pricing 165
orthogonal projection 36, 132
outer measure 2, 124

paths of process 8, 82
Pauli exclusion principle 190
paved space 124

paving 124
Pellaumail, J. 160
potential 68, 104, 105, 179
predictable process 19, 41, 96
predictable quadratic variation 132, 140, 151
predictable σ-field 94
predictable stopping time 90
probability space 1
process
  adapted 83
  cadlag 88
  compensated 142
  increasing 104
  of integrable variation 106
  jump 117
  localisation of a 118
  locally bounded 153
  measurable 82
  non-commutative 178
  optional 44
  Poisson 17, 33
  predictable 19, 41, 96
  progressive 83
  quadratic variation 132
  right-continuous 83
  stochastic 6
  summable 173
projection of a process
  accessible 103
  optional 101
  predictable 101
projective limit 7
projective system of measures 7

quadratic variation 14, 69, 123
  optional 148, 152, 155
  predictable 132
quasimartingale 170

Radon–Nikodym derivative 63
Radon–Nikodym property 62
Radon–Nikodym theorem 5, 64, 178
random function 9
random variable 2
rate of a Poisson process 18
rational valuation 165
raw increasing process 106
reducing a process by a stopping time 120
relexive Banach space 22
regular supermartingale 118
restriction of a stopping time 92
reverse supermartingale 71
right-stochastic integral 183
Riesz decomposition 68, 116, 123, 179

σ-cylinder 9
σ-field 1
section 98
self-financing strategy 165
semimartingale 149, 154
  special 155
semivariation of a vector measure 172
separable Banach space 24
separating hyperplane 25
Sierpinski game 127
simple process 181
simple random walk 10
skipping 47
Snell envelope of a process 76
square-integrable martingales 132
stable subspace 140
  under stopping and restriction 140
state space 82
stationary increments 17
step process 16
stochastic base 40, 83
stochastic control theory 45
stochastic interval 90
stochastic integral
  as operator 134, 174
  as process 135–8
stochastic measure 107
stochastic process 6, 82
stopping time 43, 84
  accessible 90
  predictable 90
  totally inaccessible 90, 144
strictly prior events 91
strong law of large numbers 72
strongly orthogonal 136, 152
strongly reducing a process 150
submartingale 41, 86
summable process 173
supercapacitable set 128
supermartingale 41, 86

tail σ-field 65
time change of a process 107
tightness of a measure 8
total variation 14
tower property of conditional expectation 37
trace in a von Neumann algebra 177
trading strategy 165
transform of a process 44

uniform absolute continuity 28
uniformly integrable set of functions 27
upcrossings 48
upwards filtering set 57
'usual' conditions 83

variance 3
variation 170
Vitali-Hahn-Saks theorem 32
von Neumann algebra 177

weak topology 22
weak*-topology 23
weakly sequentially compact set 30
Wiener measure 11
winning strategy in a Sierpinski game 128